2

NEUROMETHODS

Series Editor
Wolfgang Walz
University of Saskatchewan
Saskatoon, SK, Canada

For further volumes:
http://www.springer.com/series/7657

Neuromethods publishes cutting-edge methods and protocols in all areas of neuroscience as well as translational neurological and mental research. Each volume in the series offers tested laboratory protocols, step-by-step methods for reproducible lab experiments and addresses methodological controversies and pitfalls in order to aid neuroscientists in experimentation. *Neuromethods* focuses on traditional and emerging topics with wide-ranging implications to brain function, such as electrophysiology, neuroimaging, behavioral analysis, genomics, neurodegeneration, translational research and clinical trials. *Neuromethods* provides investigators and trainees with highly useful compendiums of key strategies and approaches for successful research in animal and human brain function including translational "bench to bedside" approaches to mental and neurological diseases.

Nanomedicines for Brain Drug Delivery

Edited by

Javier O. Morales

Department of Pharmaceutical Science and Technology, University of Chile, Santiago, Chile

Pieter J. Gaillard

2-BBB Medicines BV, Leiden, The Netherlands

 Humana Press

Editors
Javier O. Morales
Department of Pharmaceutical Science
and Technology
University of Chile
Santiago, Chile

Pieter J. Gaillard
2-BBB Medicines BV
Leiden, The Netherlands

ISSN 0893-2336   ISSN 1940-6045   (electronic)
Neuromethods
ISBN 978-1-0716-0840-1   ISBN 978-1-0716-0838-8   (eBook)
https://doi.org/10.1007/978-1-0716-0838-8

Cover illustration: Cover image by Ignacio Morales Soto.

This Humana imprint is published by the registered company Springer Science+Business Media, LLC, part of Springer Nature.
The registered company address is: 1 New York Plaza, New York, NY 10004, U.S.A.

Preface to the Series

Experimental life sciences have two basic foundations: concepts and tools. The *Neuromethods* series focuses on the tools and techniques unique to the investigation of the nervous system and excitable cells. It will not, however, shortchange the concept side of things as care has been taken to integrate these tools within the context of the concepts and questions under investigation. In this way, the series is unique in that it not only collects protocols but also includes theoretical background information and critiques which led to the methods and their development. Thus it gives the reader a better understanding of the origin of the techniques and their potential future development. The *Neuromethods* publishing program strikes a balance between recent and exciting developments like those concerning new animal models of disease, imaging, in vivo methods, and more established techniques, including, for example, immunocytochemistry and electrophysiological technologies. New trainees in neurosciences still need a sound footing in these older methods in order to apply a critical approach to their results.

Under the guidance of its founders, Alan Boulton and Glen Baker, the *Neuromethods* series has been a success since its first volume published through Humana Press in 1985. The series continues to flourish through many changes over the years. It is now published under the umbrella of Springer Protocols. While methods involving brain research have changed a lot since the series started, the publishing environment and technology have changed even more radically. Neuromethods has the distinct layout and style of the Springer Protocols program, designed specifically for readability and ease of reference in a laboratory setting.

The careful application of methods is potentially the most important step in the process of scientific inquiry. In the past, new methodologies led the way in developing new disciplines in the biological and medical sciences. For example, Physiology emerged out of Anatomy in the nineteenth century by harnessing new methods based on the newly discovered phenomenon of electricity. Nowadays, the relationships between disciplines and methods are more complex. Methods are now widely shared between disciplines and research areas. New developments in electronic publishing make it possible for scientists that encounter new methods to quickly find sources of information electronically. The design of individual volumes and chapters in this series takes this new access technology into account. Springer Protocols makes it possible to download single protocols separately. In addition, Springer makes its print-on-demand technology available globally. A print copy can therefore be acquired quickly and for a competitive price anywhere in the world.

Saskatoon, SK, Canada   *Wolfgang Walz*

Preface

Nanomedicines have revolutionized research on drug delivery in multiple diseases, and leading strategies have achieved clinical success. Moreover, a significant number of clinical trials are conducted to continue expanding the reach of nanomedicines to new, more effective and with less side effects therapies. The central nervous system (CNS) has similarly been the focus of extended research in the design and evaluation of novel nanocarriers for brain drug delivery. As a target site, the CNS represents a unique challenge given its anatomy and physiology. The blood brain barrier (BBB) is not only a restrictive limitation for systemically administered drugs but also continues to be a largely restrictive barrier to achieve significant CNS nanocarrier bioavailability. While the BBB represents one of the main limitations for significant CNS biodistribution, overcoming it is not the sole reason for limited bioavailability and targeting effects. Successfully targeting the brain microvasculature, distribution through the CNS (after passage through the BBB), and internalization in target brain cells become important challenges once there is a BBB penetration strategy in place.

As such, this book will be a source for finding the latest research in CNS-targeted nanocarriers, methods for their synthesis and thorough characterization. Moreover, a chapter addressing toxicity aspects to be considered in the design and use of brain-targeted nanocarriers will be of interest to the reader. The first two chapters of the book delve into the most widely investigated nanocarriers as brain-targeted delivery systems, i.e., polymeric nanoparticles and liposomes. With a thorough description of the state of the art as well as key aspects of their characterization, the first two chapters also highlight physiological properties relevant to particle design. Chapter 4 depicts the use of self-assembled peptide-based scaffolds for lesions of the nervous system, while Chapter 5 describes not only the use of peptides as CNS drugs but also as potential carriers to optimize brain-targeted delivery. Chapters 6 and 7 describe inorganic and magnetic nanoparticles used for targeting drugs to the CNS as well as their potential in the design of triggerable and aimed systems. Chapter 8 inspects the long-researched nose-to-brain delivery route, highlighting its potential and how the limitations this route presents could be addressed to harness its clinical potential. Chapter 9 is an excellent compilation of characterization methods to model and assess BBB absorption of drugs and drug delivery systems, and as such, this chapter will be of great use to scientists designing brain-targeted delivery systems to predict brain distribution. Finally, Chapter 10 presents the concerns that the use of nanomaterials raises in the context of brain-targeted systems. As such, the last chapter will be a good source to understand the potential neurotoxic effects and the potential role of nanomaterials in neurodegeneration progress.

The editors are immensely grateful to all the individual contributions and authors for sharing their time, effort, and knowledge to create this book. Their outstanding work in the fields covered in this book we hope will be of great interest to the reader and will help guide and move forward the field of nanomedicines to target the brain and the nervous system.

Santiago, Chile   *Javier O. Morales*
Leiden, The Netherlands   *Pieter J. Gaillard*

Contents

Contributors

HAKTAN ALTINOVA • *Department of Neurosurgery, University Hospital RWTH Aachen, Aachen, Germany*

SOUNAK BAGCHI • *Department of Pharmaceutical Sciences, School of Pharmacy, Texas Tech University Health Sciences Center (TTUHSC), Amarillo, TX, USA*

GARY A. BROOK • *Institute of Neuropathology, University Hospital RWTH Aachen, Aachen, Germany*

MARIA ANTÒNIA BUSQUETS • *Pharmacy and Pharmaceutical Technology and Physical Chemistry Department, Faculty of Pharmacy and Food Sciences, University of Barcelona, Barcelona, Catalonia, Spain; Institute of Nanoscience and Nanotechnology, IN2UB, Barcelona, Catalonia, Spain*

JOHANNA CATALAN-FIGUEROA • *Department of Pharmaceutical Science and Technology, School of Chemical and Pharmaceutical Sciences, University of Chile, Santiago, Chile; Department of Biochemistry, School of Chemical and Pharmaceutical Sciences, University of Chile, Santiago, Chile; Experimental Pharmacology Institute, CONICET, National University of Cordoba, Cordoba, Argentina; Advanced Center for Chronic Diseases (ACCDiS), Santiago, Chile*

TANYA CHHIBBER • *Department of Pharmaceutical Sciences, School of Pharmacy, Texas Tech University Health Sciences Center (TTUHSC), Amarillo, TX, USA*

PAUL D. DALTON • *Department for Functional Materials in Medicine and Dentistry and Bavarian Polymer Institute, University of Würzburg, Würzburg, Germany*

LAURA DE LAPORTE • *DWI—Leibniz-Institute for Interactive Materials, Aachen, Germany; Insitute of Applied Medical Engineering, RWTH Aachen University, Aachen, Germany; Insitute of Technical and Macromolecular Chemistry RWTH Aachen University, Aachen, Germany*

JOAN ESTELRICH • *Pharmacy and Pharmaceutical Technology and Physical Chemistry Department, Faculty of Pharmacy and Food Sciences, University of Barcelona, Barcelona, Catalonia, Spain; Institute of Nanoscience and Nanotechnology, IN2UB, Barcelona, Catalonia, Spain*

ANNA MARIA FADDA • *Department of Life and Environmental Sciences, University of Cagliari, Cagliari, Italy*

REINHARD GABATHULER • *Faculty of Life Sciences and Medicine, Blood-Brain Barrier Group, Kings College London, London, UK*

EDUARDO GALLARDO-TOLEDO • *Departamento de Química Farmacológica y Toxicológica, Facultad de Ciencias Químicas y Farmacéuticas, Universidad de Chile, Santiago, Chile; Advanced Center for Chronic Diseases (ACCDiS), Santiago, Chile*

JOSE L. GERARDO NAVA • *Institute of Neuropathology, University Hospital RWTH Aachen, Aachen, Germany; DWI—Leibniz-Institute for Interactive Materials, Aachen, Germany*

LIUMIN HE • *Guangdong-Hong Kong-Macau Institute of CNS Regeneration, Jinan University, Guangzhou, People's Republic of China*

RAHUL DEV JAYANT • *Department of Pharmaceutical Sciences, School of Pharmacy, Texas Tech University Health Sciences Center (TTUHSC), Amarillo, TX, USA*

ABHIJEET JOSHI • *Centre for Biosciences and Bio-medical Engineering, Indian Institute of Technology Indore (IIT-I), Indore, Madhya Pradesh, India*

KRISTIAN KEMPE • *ARC Centre of Excellence in Convergent Bio-Nano Science & Technology, and Drug Delivery, Disposition and Dynamics, Monash Institute of Pharmaceutical Sciences, Monash University, Parkville, VIC, Australia*

MARCELO JAVIER KOGAN • *Departamento de Química Farmacológica y Toxicológica, Facultad de Ciencias Químicas y Farmacéuticas, Universidad de Chile, Santiago, Chile; Advanced Center for Chronic Diseases (ACCDiS), Santiago, Chile*

BEHNAZ LAHOOTI • *Department of Pharmaceutical Sciences, School of Pharmacy, Texas Tech University Health Sciences Center (TTUHSC), Amarillo, TX, USA*

FRANCESCO LAI • *Department of Life and Environmental Sciences, University of Cagliari, Cagliari, Italy*

YANG LU • *Department of TCM Pharmaceutical, School of Chinese Material Mecica, Beijing University of Chinese Medicine, Beijing, People's Republic of China*

RAHUL MITTAL • *Laboratory of Human Molecular Genetics, Department of Otolaryngology, Miller School of Medicine, University of Miami (UM), Miami, FL, USA*

JAVIER O. MORALES • *Department of Pharmaceutical Science and Technology, School of Chemical and Pharmaceutical Sciences, University of Chile, Santiago, Chile; Advanced Center for Chronic Diseases (ACCDiS), Santiago, Chile; Center of New Drugs for Hypertension (CENDHY), Santiago, Chile; Pharmaceutical and Biomaterial Research Group, Department of Health Sciences, Luleå University of Technology, Luleå, Sweden*

JOSEPH A. NICOLAZZO • *Drug Delivery, Disposition and Dynamics, Monash Institute of Pharmaceutical Sciences, Monash University, Parkville, VIC, Australia*

JONAS C. ROSE • *DWI—Leibniz-Institute for Interactive Materials, Aachen, Germany*

MICHELE SCHLICH • *Department of Life and Environmental Sciences, University of Cagliari, Cagliari, Italy*

CHIARA SINICO • *Department of Life and Environmental Sciences, University of Cagliari, Cagliari, Italy*

HUGH D. C. SMYTH • *Molecular Pharmaceutics and Drug Delivery Division, College of Pharmacy, The University of Texas at Austin, Austin, TX, USA*

MEI MEI TIAN • *Bioasis Technologies Inc., Guilford, CT, USA*

SREE-POOJA VARAHACHALAM • *Department of Pharmaceutical Sciences, School of Pharmacy, Texas Tech University Health Sciences Center (TTUHSC), Amarillo, TX, USA*

CAROLINA VELASCO-AGUIRRE • *Departamento de Química Farmacológica y Toxicológica, Facultad de Ciencias Químicas y Farmacéuticas, Universidad de Chile, Santiago, Chile; Advanced Center for Chronic Diseases (ACCDiS), Santiago, Chile*

ZACHARY WARNKEN • *Molecular Pharmaceutics and Drug Delivery Division, College of Pharmacy, The University of Texas at Austin, Austin, TX, USA*

ROBERT O. WILLIAMS III • *Molecular Pharmaceutics and Drug Delivery Division, College of Pharmacy, The University of Texas at Austin, Austin, TX, USA*

WUTIAN WU • *Guangdong-Hong Kong-Macau Institute of CNS Regeneration, Jinan University, Guangzhou, People's Republic of China; Re-Stem Biotech, Suzhou, People's Republic of China*

NA ZHANG • *Guangdong-Hong Kong-Macau Institute of CNS Regeneration, Jinan University, Guangzhou, People's Republic of China*

Chapter 1

Biodegradable Polymeric Nanoparticles for Brain-Targeted Drug Delivery

Kristian Kempe and Joseph A. Nicolazzo

Abstract

The blood-brain barrier (BBB), formed by endothelial cells lining the cerebral microvessels, remains a formidable challenge for the delivery of many therapeutics into the central nervous system (CNS). In an attempt to enhance the CNS disposition of therapeutics, which either have poor inherent permeability across the BBB or whose brain uptake is limited by the function of efflux transporters, a multitude of polymeric nanocarriers have been exploited. Common natural and synthetic polymers used for the development of these nanocarriers include polysaccharides, poly(alkylcyanoacrylate)s, and polyesters such as poly(lactic-co-glycolic acid). To avoid recognition by circulating macrophages and, therefore, minimize their systemic clearance, these polymeric nanocarriers are often coated with polyethylene glycol or emulsifiers such as polysorbate 80, and to enhance their targeting to the BBB, addition of various targeting entities such as antibodies to brain microvascular receptor-mediated transporters is common. Unlike liposomes however, polymeric nanoparticles are more stable, allow for a detailed control of the carrier properties (e.g., size, shape, charge, surface morphology and chemistry), facilitate the delivery of a range of different cargoes with high capacities, and can be engineered with different drug release mechanisms and modified with various targeting ligands. This chapter will provide an up-to-date account on the various polymeric nanoparticle approaches which have been exploited to target therapeutics to the CNS, with particular focus on biodegradable polymers and practical techniques that can be employed for the preparation of these polymeric nanoparticles.

Key words Polymer, PLGA, Nanoparticle, Blood-brain barrier, Targeting, Surface coating

1 Introduction

Nanotechnology innovations have paved the way to novel therapeutic and diagnostic agents/carriers and tools for pharmaceutical and biomedical research. However, drug delivery to the brain has remained one of the biggest challenges in biomedical research with numerous obstacles which have to be overcome. Drugs for treatment of brain-related diseases have to enter the brain in order to exhibit a therapeutic effect, which can be accomplished by invasive and noninvasive methods. In the latter and more clinically relevant case, the compound has to be able to cross the blood-brain barrier

Javier O. Morales and Pieter J. Gaillard (eds.), *Nanomedicines for Brain Drug Delivery*, Neuromethods, vol. 157,
https://doi.org/10.1007/978-1-0716-0838-8_1, © Springer Science+Business Media, LLC, part of Springer Nature 2021

(BBB), an obstacle which limits the ability of many therapeutics to gain access to their target within the central nervous system (CNS) [1, 2]. The BBB is a selective barrier which controls ionic and fluid movement from the systemic circulation to the neural tissue. It prevents harmful substances from entering the brain and at the same time supplies the brain with essential nutrients. Small gaseous and lipophilic molecules can diffuse through the lipid membrane of brain microvascular endothelial cells forming the BBB, whereas larger compounds such as peptides, macromolecules, and hydrophilic drugs exhibit poor transport across this membrane, unless through the assistance of a membrane transporter. Thus, the BBB represents an obstacle for both efficient therapy and diagnosis of brain-related diseases with most state-of-the-art drugs and diagnostic agents, respectively. Among others, polymeric nanomaterial-mediated brain delivery has been proven to be a suitable strategy to overcome this issue. Colloids, in general, have been widely applied in the field of drug delivery [3, 4]. In recent years, polymeric nanoparticles (PNPs), which are in the size range of 10–300 nm, have emerged as ideal carrier systems as they allow for a high level of modularity and thus can be tailored for individual applications [5]. PNPs can be divided into two major groups, polymer nanocapsules and polymer nanospheres [6]. The latter are composed of a dense polymer matrix, whereas the former represent vesicular-type aggregates which consist of a liquid core surrounded by a polymer layer. PNPs are more stable than other delivery systems and are amenable to a detailed control of the carrier properties, including their size, shape, charge, surface morphology, and chemistry [5]. They facilitate the delivery of a range of different cargo with high capacities, which is advantageous compared to a single or prodrug approach. Nanospheres allow drugs to be uniformly dispersed in the polymer matrix mainly by physical interactions, and nanocapsules can encapsulate drugs in their liquid core. Besides the protection of the cargo from enzymatic or chemical degradation, PNPs also enable the triggered release of the cargo through specific engineering of drug release mechanisms into the carrier [7]. However, challenges associated with nanomedicines include their nonspecific interaction with the human body and their in vivo fate, specificity, and possible toxicity [8]. A range of polymer classes have been developed to overcome these limitations, including the use of coating with poly(ethylene glycol) (PEG) and its alternatives [9]. Carriers composed of, or modified with, these classes enable a prolonged blood circulation time and lower the overall toxicity of the PNPs. Moreover, targeted approaches have increased the therapeutic efficacy of encapsulated drugs through the accumulation of the carrier at the site of action [10]. Altogether, PNPs represent promising drug delivery systems.

This book chapter focuses on PNPs, specifically polymeric nanospheres and their application as brain delivery nanomedicines. It sets out design principles for PNPs, reviews commonly used biodegradable polymer classes, and provides background information on polymerization and formulation techniques employed for the fabrication of polymeric brain delivery vehicles. Moreover, strategies used to improve transport of PNPs across the BBB are highlighted using recent examples of PNPs based on polyesters, poly(alkyl cyanoacrylate)s, and polysaccharides.

2   Design Principles of Commonly Employed Polymeric Nanoparticles for Brain Delivery

Noninvasive methods for crossing the BBB have received significant attention in recent years. However, for this purpose, drugs and carriers have to be designed carefully and need to meet certain criteria. To cross the BBB, the modification of existing drugs and their physicochemical properties and, in particular, the attachment of ligands onto molecules or colloids which target the BBB have proven promising, at least in rodent models. PNPs provide an ideal platform for the delivery of therapeutics to the brain because of their abovementioned characteristics. As highlighted in a recent review article by Saltzmann and coworkers, for drug delivery to the brain, PNPs should possess a number of specific properties; in particular, they should be biocompatible, biodegradable, nontoxic, and non-immunogenic, exhibit a size <100 nm, and allow for surface coatings and modifications [11]. As most of these PNPs will be administered systemically, they must also be stable in the bloodstream, show extended blood circulation times, and avoid opsonization. Tables 1, 2, and 3 provide an overview of the three most investigated types of PNPs. They summarize specifications of selected examples particularly focusing on PNPs properties (i.e., polymer type, surface coatings, preparation technique, size), targeting ligands, cargo, and route of administration. In the following sections, some of these properties will be discussed in more detail in order to establish an understanding of the current state of the art of PNPs for brain delivery.

3   Polymer Classes

Biodegradability has been identified as an important requirement of brain-targeting carriers as it prevents the accumulation of foreign materials in the brain and enables the modulation of drug release kinetics. It is for these reasons that biodegradable polymers have received particular attention as base materials for polymeric

Table 1
Summary of characteristics of selected polyester-based PNPs

Core polymer[a]	Surface material/ coating	Targeting	Cargo	Size (nm)	Route of administration[b]	Reference
PLA (PEG-b-PLA; ~3/ ~50)	PEG	Wheat germ agglutinin	–	85–90[c]	i.n.	[12]
PLA (40)	–	Trans-activating (TAT) peptide	Ritonavir	~300[c]	i.v.	[13]
PLA (10)	Pluronic F127	–	Breviscapine	177, 319[d]	i.v.	[14]
PLA (PEG-b-PLA; ~2 or 5 PEG)	PEG	Cationized bovine serum albumin	Sulpiride	329 ± 44[c]	i.v.	[15]
PLGA (7–17)	Pluronic F127	Similopioid peptide; sialic acid residue	Loperamide	140–200[e]	i.v.	[16]
PLGA (PLGA-PEG-COOH; 50)	PEG	Glutathione	Paclitaxel	~200[e]	i.p.	[17]
PLGA (PLGA-PEG-COOH, 15)	PEG	–	Paclitaxel	156 ± 55[c]	i.v.	[18]
PLGA (PEG-b-PLGA; ~3/ ~30)	PEG	Lactoferrin	Coumarin-6, urocortin	<150[f]	i.v.	[19]
PLGA (PEG-b-PLGA; 3/40)	PEG	Pep TGN (a 12-amino-acid- peptide)	Coumarin-6	104–120[c]	i.v.	[20]
PLGA (35–40)	Tween®80	–	Estradiol	138–172[c]	Oral	[21]
PLGA (PLGA-b-PEG-b- PLGA)	Pluronic F127, Tween®80	–	Loperamide	~150[e]	i.v.	[22]

Polymer		Ligand	Drug	Size	Route	Ref
PLGA (mesoporous silica-magnetic particles)	—	Transferrin	Doxorubicin, paclitaxel	~150+	i.v.	[23]
PLA (PEG-b-PLA; 3/34)	PEG	B6 peptide	NAPVSIPQ (neuroprotective peptide), coumarin-6	103–118^c	i.v.	[24]
PLA (PLA-co-depsipeptide; 12)	—	—	Rivastigmine	142 ± 21^c	–	[25]
PLA (PEG-b-PLA)	PEG	Cationic bovine serum albumin	Coumarin-6	80–90^c	–	[26]
PLA (PEG-b-PLA)	PEG	Cationic bovine serum albumin	Coumarin-6	97–104^f	i.v.	[27]
PLGA (35–40)	Tween®80	—	Donepezil	~90^d	i.v.	[28]
PLGA (PEG-b-PLGA; ~3/30 or 37)	PEG	Lectin	Haloperidol	<135^c	i.n.	[29]
PLGA (~11)	Pluronic F127	Glycopeptide (g7)	–	~170^c	i.p.	[30]
PLGA (PEG-b-PLGA; total: 39 and 42)	PEG	Solanum tuberosum lectin (STL)	Fibroblast growth factor	105, 119^c	i.n.	[31]
PLA (PEG-b-PLA; 3/50 or 70)	PEG	TGN	–	~100^c	i.v.	[32]
PLGA (11)	—	Mutant form of diphtheria toxin (CRM197)	Loperamide	176–200^{c,e}	i.p.	[33]
PLGA (11)	—	Glycopeptide (g7)	FITC-albumin	220–260^f	i.v.	[34]
PLGA	Chitosan, PEG	–	Coumarin-6, paclitaxel	258–355^c	i.v.	[35]
PCL (PEG-b-PCL; 2/10)	PEG	–	Paclitaxel	72.5 ± 2.2^c	i.v.	[36]
PCL (PEG-b-PCL; 2/10)	PEG	Angiopep-2	Paclitaxel	<100^c	i.v.	[37]

(continued)

Table 1
(continued)

Core polymer[a]	Surface material/coating	Targeting	Cargo	Size (nm)	Route of administration[b]	Reference
PCL (PEG-b-PCL; ~3/15)	PEG	TGN peptide	Docetaxel	151–170[c]	i.v.	[38, 39]
PCL (PEG-b-PCL; 2/10)	PEG	Angiopep-2	–	74–93[c]	i.v.	[40]
PCL (65)	–	–	Carboplatin	~312[f]	i.n.	[41]
PCL (PEG-b-PCL; 3/15)	PEG	Angiopep-2	Coumarin-6	126[c]	–	[42]
PTMC (PEG-b-PTMC; ~3/6)	PEG	2-Deoxy-D-glucose	Paclitaxel	71[c]	i.v.	[43]

i.v. intravenous, *i.p.* intraperitoneal, *i.n.* intranasal, *Tween® 80* polysorbate 80, *Pluronic F127* poloxamer 188
[a]Numbers indicate molecular weights (kg/mol) of individual blocks
[b]Where route of administration is not specified, studies were undertaken in vitro only
[c]Prepared by emulsion/solvent evaporation
[d]Prepared by emulsion-diffusion-evaporation
[e]Prepared by nanoprecipitation
[f]Prepared by double emulsion/solvent evaporation

Table 2
Summary of characteristics of selected poly(alkyl cyanoacrylate)s PNPs prepared by emulsion polymerization

Core polymer	Surface material/coating	Targeting	Cargo	Size (nm)	Route of administration[a]	References
PBCA	Tween®80	–	Dalargin	230	i.v.	[44]
PBCA	Tween®80	–	Dalargin	210	i.v.	[45]
PBCA	Tween®80	–	Valproic acid	–	i.v.	[46]
PBCA	Tween®80	–	Dalargin	–	i.v.	[47]
PHDCA, PEG-b-PHDCA	PEG, poloxamine 908, tween®80	–	–	137–164	i.v.	[48]
PBCA	Tween®80	Apolipo-protein AII, B, CII, E, J	Dalargin, loperamide	–	i.v.	[49]
PBCA	Tween®80	–	Dalargin	–	i.v.	[50]
P(HDCA-stat-PEGCA)	PEG	–	–	–	–	[51]
PBCA	Tween®80	–	Doxorubicin	185–252	i.v.	[52]
PBCA	Tween®80	–	Methotrexate	70, 170, 220, 345	i.v.	[53]
PBCA	–	–	Zidovudine, lamivudine	<100	–	[54]
PBCA	Tween®80, Pluronic F127	–	Doxorubicin	202, 246	i.v.	[55]
PBCA	Tween®80	–	Rivastigmine	40.5 ± 6.9	i.v.	[56]
PBCA	Tween®80	–	Tacrine	35.6 ± 4.6	i.v.	[57]
PBCA	Tween®80	–	Gemcitabine	–	i.v.	[58]

(continued)

Table 2
(continued)

Core polymer	Surface material/coating	Targeting	Cargo	Size (nm)	Route of administration[a]	References
PBCA	Tween®80	–	Clioquinol	–	i.v.	[59]
PBCA	DSPE[b]-PEG2k-COOH	Cross-reacting material 197 (CRM197)	Zidovudine	87–195	–	[60]

i.v. intravenous, *Tween® 80* polysorbate 80, *Pluronic F127* poloxamer 188

[a]Where route of administration is not specified, studies were undertaken in vitro only

[b]1,2-Distearoyl-sn-glycero-3-phosphoethanolamine

Table 3
Summary of characteristics of selected chitosan-based PNPs

Core polymer	Surface material/coating	Targeting	Cargo	Size (nm)	Route of administration[d]	Reference
Chitosan (<150 kg/Mol)	PEG	OX26 monoclonal antibody	Z-DEVD-FMK (caspase-3 inhibitor)	149, 590, 637[a]	i.v.	[61]
Chitosan (50 kg/mol)	–	–	Estradiol	269 ± 32[a]	i.n., i.v.	[62]
Chitosan (different MW)	–	–	–	300, 900, 12,000[a,b]	–	[63]
Chitosan (54 kg/mol)	PEG	Transferrin receptor monoclonal antibody	Z-DEVD-FMK (caspase-3 inhibitor)	610–650[a]	i.v.	[64]
Chitosan (low MW)	–		Intermembranous fragments of Aβ (IF-A)	15.23 ± 10.97[c]	i.v.	[65]
Chitosan (110 kg/mol)	–	–	Dopamine	98–148[a]	i.p.	[66]
Chitosan (110 kg/mol), glycol chitosan (68 kg/mol)	Tween®80	–	Methotrexate	125–263[a]	–	[67]
Chitosan (750 kg/mol)	–	–	Rivastigmine	163–3300[a]	i.n.	[68]
Chitosan (low MW)	PEG	Transactivator of transcription (TAT) peptide	siRNA (against the Ataxin-1 gene)	50–100[a]	–	[69]
Chitosan (Magnevist® conjugated; medium MW)	BSA		Cyclophosphamide	239[a]	i.v.	[70]

i.v. intravenous, *i.p.* intraperitoneal, i.n. intranasal, *Tween® 80* polysorbate 80
[a]Prepared by ionic gelation
[b]Prepared by high pressure homogenizer; ULTRA-TURRAX
[c]Prepared by emulsion/chemical cross-linking
[d]Where route of administration is not specified, studies were undertaken in vitro only

BBB-targeted delivery vehicles. In this context, a range of synthetic and naturally occurring polymers have been employed, including poly(lactic acid) (PLA), poly(lactide-co-glycolide) (PLGA) and other polyesters, poly(alkyl cyanoacrylates), and polysaccharides. However, it should be noted that carriers composed of inherently nondegradable polymers such as poly((meth)acrylate)s [71] and poly(ethylene imine)s (PEI) [72] have also been reported for brain delivery. This section will focus on the introduction of the individual biodegradable polymer classes, including a brief discussion of their polymerization techniques and their physicochemical and biomedical properties.

3.1 Polyesters

Polyesters are polymers which possess ester functionalities in their main chain. Generally polyesters are obtained by polycondensation of multifunctional carboxylic acids (or derivatives) and alcohols. Alternatively, ring-opening polymerizations (ROP) of lactones provide access to aliphatic polyesters (Fig. 1). Catalyzed by metal-organic compounds, ROP allow to prepare well-defined polymers with defined molecular weight and molecular weight distributions, as opposed to polycondensations. The most important synthetic biodegradable polyesters are poly(ε-caprolactone) (PCL) and PLA/PLGA which are obtained by ROP of ε-caprolactone and lactide/lactide and glycolide, respectively. In a typical polymerization experiment, the initiator (a hydroxy compound) is placed in a

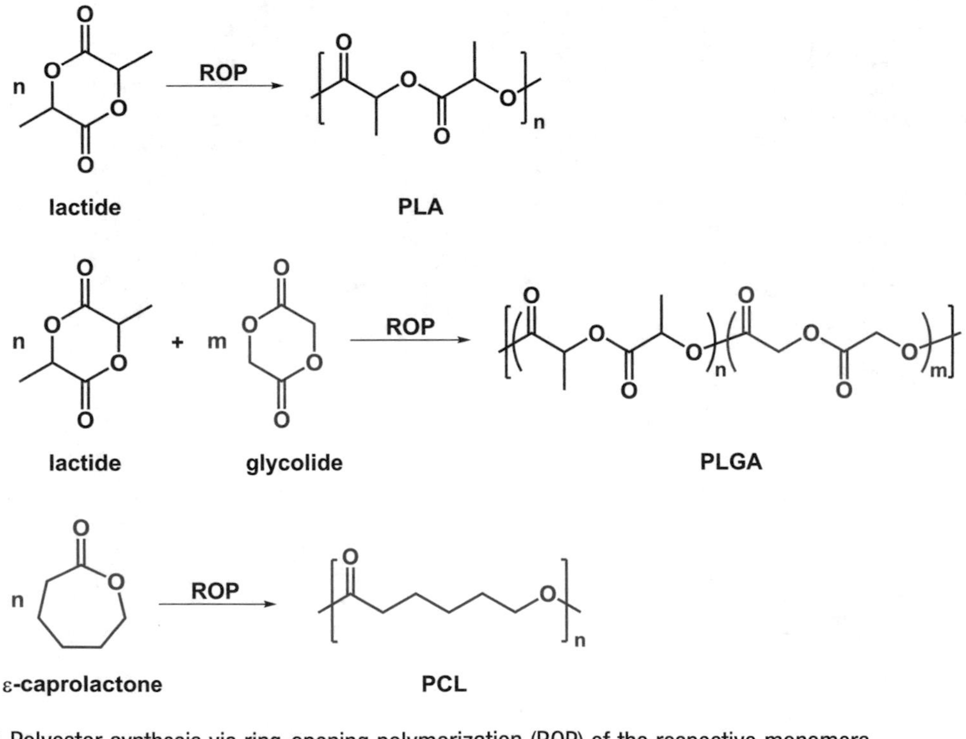

Fig. 1 Polyester synthesis via ring-opening polymerization (ROP) of the respective monomers

three-necked flask and dried under vacuum at elevated temperature for a couple of hours before the two monomers are added in the desired ratio. After the monomers are melted, the catalyst, typically 2-ethylhexanoate, is added, and the mixture is stirred at high temperature for a desired time.

PLA and PLGA have been extensively employed as materials for the fabrication of drug delivery vehicles [73]. They are biocompatible and biodegradable and have been approved by the US Food and Drug Administration for pharmaceutical applications [11, 73]. PLGA degrades in water due to hydrolysis of its ester bonds. Variation of the molar ratios of LA and GA in the copolymer enables the adjustment of the degradation profile with LA-rich PLGA degrading slower. However, small-sized PLGA particles also show enzymatic degradation, most likely by lipases [74]. The pharmacokinetics and biodistribution of PLGA have been reported to follow a dose-dependent and nonlinear behavior [75, 76]. Some PLGA nanoparticle formulations were found to accumulate in the liver, bone marrow, lymph nodes, spleen, and peritoneal macrophages in rodent models. The dose and composition of PLGA also determine the blood clearance and uptake by the mononuclear phagocyte system. However, surface modification of PLGA further allows modulation of their body distribution and therefore alters their performance [77]. This is achieved by using copolymers of PLGA/PLA and PEG as PEG chains of different lengths and terminal hydroxy groups can be exploited for the ROP of lactide and glycolide. In this way, diblock PEG-b-PLGA [78] and triblock PLGA-b-PEG-b-PLGA [79] can be obtained.

3.2 Poly (Alkyl Cyanoacrylate) s (PACAs)

Poly(alkyl cyanoacrylate)s (PACAs) are another class of biodegradable polymers which have attracted significant attention as a base material for BBB-targeted delivery vehicles (Fig. 2). Detailed information about the monomer and polymer synthesis can be found in recent literature [80]. In general, alkyl cyanoacrylate monomers are synthesized from alkyl cyanoacetates and formaldehyde under basic conditions in a Knoevenagel condensation. The PACA oligomers

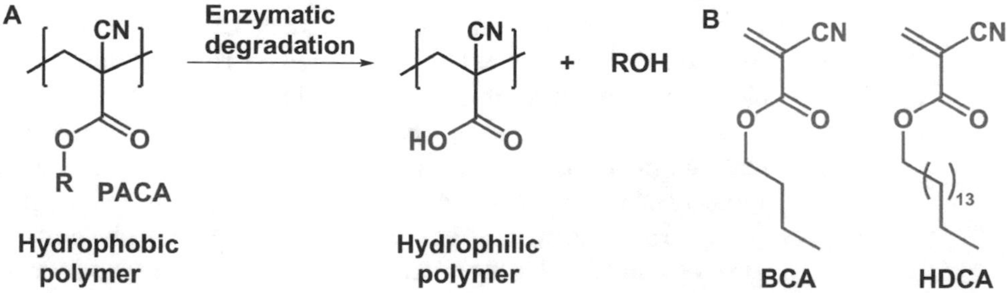

Fig. 2 Poly(alkyl cyanoacrylate)s (PACAs), structure of PACA and degradation product (a), and commonly used monomers for PNPs (b)

obtained are thermally depolymerized using stabilizers such as protonic or Lewis acids to give the individual alkyl cyanoacrylates. To date, a large variety of alkyl cyanoacrylates (ACAs) have been prepared with butyl cyanoacrylate (BCA) and hexadecyl cyanoacrylate (HDCA) being the best studied ones in a BBB carrier context [80]. The electron-withdrawing groups (ester and cyano) located at the α-position of the monomer double bond render the monomer highly reactive toward anions and weak bases. Thus, PACAs are synthesized by a "quasi-instantaneous" anionic polymerization or zwitterionic polymerization. In particular, long alkyl PACAs have been widely used as material for the design of nanosized delivery vehicles. The utilization of PEG-substituted ACAs has given rise to higher bioavailability of the synthesized systems [81, 82]. As opposed to main chain degradable polyesters, PACAs undergo side chain degradation [83], mainly through esterases [84, 85]. Upon cleavage of the side group, the polymers become water-soluble and are susceptive to excretion by the kidneys. Both the resulting PACA and the corresponding alcohols possess low toxicities. In general, the degradation of PACAs is rather fast (within hours) and depends on the length of the alkyl chain. Short alkyl chain PACAs degrade faster but are also more toxic than long alkyl chain PACAs [86, 87]. It has been shown that 24 h after intravenous injection of poly(butyl cyanoacrylate) (PBCA) particles to rodents, 80% of them have been excreted [52, 88], most likely because of the low molecular weight of the PBCA in particulate form [89].

3.3 Polysaccharides

Polysaccharides are natural polymers consisting of monosaccharide repeating units which are connected via glycosidic bonds. They can be divided into two major groups: storage polysaccharides such as glycogen and structural polysaccharides such as chitin. The latter is the precursor system for chitosan, a cationic heteropolymer with broad application potential, including as a mucoadhesive in various formulations [90]. Chitosan is a linear polymer containing randomly distributed d-glucosamine and N-acetyl-D-glucosamine units which are bound by (1,4)-glycosidic linkages. Chitosan is obtained from chitin by (more than 60%) deacetylation of the C2-amino group with typical degrees of deacetylation being between 66% and 95% (Fig. 3a). Due to the free amine groups, chitosan is highly soluble in water and forms complexes with negatively charged molecules. The physicochemical properties of chitosan are mainly dependent on the degree of deacetylation as well its molecular weight [91]. Contrary to low molecular weight chitosan (e.g., 55 kg/mol), high molecular weight chitosan (e.g., 700 kg/mol) is highly viscous and poorly water-soluble under neutral conditions. Thus, mainly low and medium molecular weight chitosans have been used for the fabrication of particulate carriers. In fact, low molecular weight chitosan particles have been

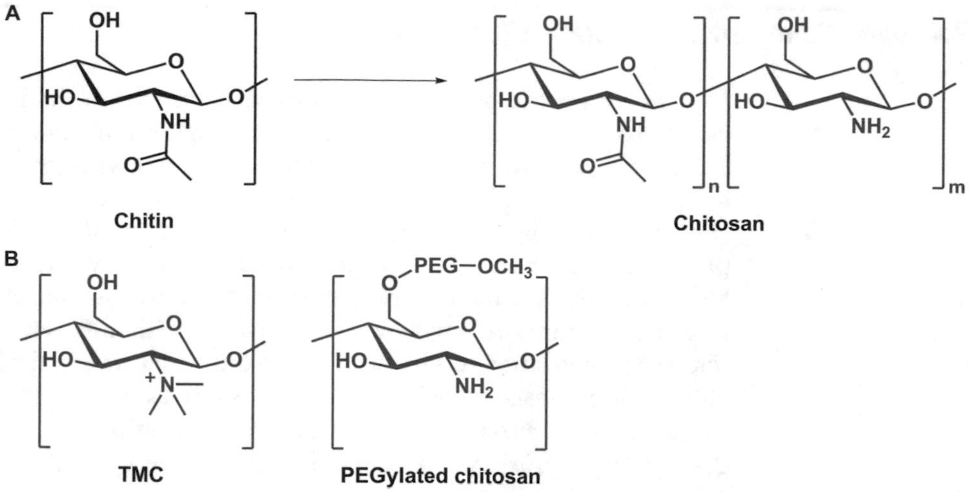

Fig. 3 Preparation of chitosan by deacetylation from chitin (**a**) and structurally simplified examples of modified chitosans, *N,N,N*-trimethyl chitosan and PEGylated chitosan

reported to show better encapsulation efficiencies [92]. Chitosans exist in five different types of helical conformations which greatly determine their properties [93], such as biodegradability and bioactivity. Chitosan can be degraded by a variety of enzymes and esterases, including chitosanase enzymes [94], chitin deacetylase, chitinase, beta-*N*-acetylhexosaminidase [95], and collagenase [96]. Generally, faster degradation is observed for low molecular weight chitosan and for chitosans with a degree of deacetylation lower than 70% [97, 98]. At the same time, low molecular weight chitosan possesses lower cytotoxicity than high molecular weight chitosan [99]. Moreover, the degradation behavior of chitosan NPs is altered depending on the particle formulation technique. Particles stabilized by ionic gelation with tripolyphosphate (TPP) degrade slower than chemically cross-linked chitosan particles [100]. The biodistribution of chitosan after intravenous and intraperitoneal administration and the tissue distribution after oral administration have been summarized in a recent review [100]. The surface chemistry and size of chitosan particles mainly influence their biodistribution upon intravenous administration. The functional groups of chitosan also allow for chemical modification, for example, for the introduction of PEG chains (Fig. 3b) to alter the biodistribution of the corresponding particles [101]. But chitosan, for example, *N,N, N*-trimethyl chitosan (TMC; Fig. 3b), is also used as a coating material for other particles to improve brain targeting due its ability to electrostatically interact with anionic sialic acid residues on the brain microvascular endothelial cells [102].

4 Polymeric Nanoparticle Preparation Techniques

Depending on the polymer type, different techniques have been employed to process and fabricate the polymers introduced in Subheading 3 into particles of different shapes and sizes. There are two main strategies for the fabrication of polymeric nanospheres (Fig. 4): the polymerization of monomers and the dispersion of preformed polymers [103]. Both techniques do allow not only to adjust the physicochemical properties of the particles but also to formulate drugs and engineer the particles with different drug release mechanisms and kinetics. Drugs can be either physically or chemically encapsulated within the carriers or adsorbed or attached to the particle surface. These drugs can be either released by desorption, diffusion, or degradation of the particles. This sub-chapter describes the main techniques used for the fabrication of PLGA (Subheadings 4.1–4.3), PACA (Subheading 4.4), and chitosan (Subheadings 4.5 and 4.6) particles.

4.1 Emulsion/Solvent Evaporation Technique

Hydrophobic polymers such as PLA/PLGA can be formulated into particles using oil-in-water emulsions [104, 105]. This technique is best suited for hydrophobic drug formulations. The polymer is dissolved in a water-immiscible and volatile organic solvent (e.g., dichloromethane is commonly used for PLA/PLGA systems) and emulsified in water using an emulsifier such as polyvinyl alcohol (PVA). Subsequently, the organic solvent is removed by evaporation. This can be accomplished at atmospheric or reduced pressure while controlling the stirring of the emulsion. Further washing steps can be performed to remove residual organic solvent contents.

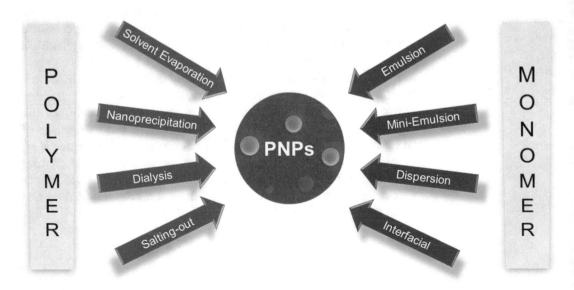

Fig. 4 Overview of the two main strategies and techniques used to fabricate PNPs from pre-synthesized polymers or through polymerization of monomers

4.2 Double Emulsion/Solvent Evaporation Technique

Hydrophilic drugs can be best formulated using water-in-oil-in-water emulsions. The polymer is dissolved in a volatile water-immiscible organic solvent, and the drug dissolved in water is added to the polymer solution to form the primary water-in-oil emulsion [104, 105]. The latter is subsequently added to an aqueous solution containing PVA under continuous stirring yielding the water-in-oil-in-water emulsion. Again, the organic solvent is removed and the particles are washed. The particle size and encapsulation efficiency can be adjusted by the choice of organic solvent and stirring speed.

4.3 Nano-precipitation Technique

Nanoprecipitation is a mild and facile particle fabrication method [106]. To prepare polymer-drug formulations, the polymer and drug are dissolved together in a water-miscible solvent, such as acetone, tetrahydrofuran, or N,N-dimethylacetamide, and either added dropwise to an aqueous solution (in the presence/absence of a surfactant) or water is added to the polymer/drug solution under stirring. Subsequently the organic solvent is removed under atmospheric or reduced pressure or dialysis. Polymer concentration, i.e., the viscosity of the solution, the stirring speed, the order of addition, and the rate of solvent evaporation are factors which can be modified to alter the size of the resulting particles.

4.4 Emulsion Polymerization

Emulsion polymerization is a polymerization technique which is performed in heterogeneous media [107]. It has found widespread application for the fabrication of solid PNPs, specifically nanospheres. It is the method of choice for the preparation of PACA-based nanoparticles [80]. In a typical polymerization, the respective ACA monomer is added dropwise to an acidic aqueous solution ($2 < pH < 3$) containing a surfactant (nonionic, macromolecular, such as dextran). The mixture is stirred for a short number of hours before it is neutralized by the addition of aqueous sodium hydroxide solution, and residual monomer is removed by centrifugation of the PACA particles. Usually nanospheres in the size range of 1–300 nm are obtained. The type and concentration of surfactant and monomer and pH of the polymerization medium have been shown to influence the polymer and particle properties [89, 108–110].

4.5 Ionic Gelation

Ionic gelation is a technique used to physically cross-link polymers through ionic interactions [111, 112]. Particle formation is accomplished by mixing aqueous solutions of a charged polymer, e.g., chitosan, with a polyionic cross-linker under magnetic stirring at room temperature. The pH of the aqueous solutions needs to be carefully adjusted to allow for an optimal cross-linking reaction with the most common cross-linker for the formation of chitosan being tripolyphosphate (TPP).

5   Particle Sizes of PNPs

As stated in the introduction, PNPs can be divided into polymer nanocapsules or polymer nanospheres with the latter being solid particles composed of polymers. Nanosized particles are typically in the size range 10–300 nm, but particles with sizes up to 1000 nm are also often considered nanoparticles. Most review articles suggest that the ideal size for nanoparticles to efficiently traffic the BBB is between 50 nm [113] and 100 nm in diameter [11, 114]. In fact, nanomedicines of a size larger than 100 nm cannot efficiently permeate through the brain extracellular space, once they have permeated the BBB, because of the brain interstitial fluid space width of 38–64 nm [115]. On the other hand, nanoparticles which are smaller than 5.5 nm are prone to be removed from the body through renal clearance [116] and thus will not circulate long enough in the body. Therefore, an optimum range that allows for efficient trafficking across the BBB and through the interstitial fluid of the brain as well as minimal filtration by the kidneys is desired for nanotherapeutics to reach their potential in clinical medicine.

The versatility of the polymer classes and particle preparation methods described in the previous sub-chapters enable the fabrication of PNPs with different sizes. The emulsion/solvent evaporation technique has been established as the method of choice for the preparation of ~100 nm PLGA NPs (Table 1). Nanoprecipitation generally provides slightly larger NPs (100–300 nm) but is a less demanding method. Similarly, the emulsion polymerization of ACAs delivers NPs between 50 and 150 nm (Table 2). Chitosan NPs, synthesized by ionic gelation with oppositely charged compounds, have been used in a wide range of sizes (sub-100 nm to micron-sized) for BBB-targeted delivery (Table 3).

6   Modifications of PNPs Surfaces for Improved Delivery to the Brain

In recent studies, different mechanisms for uptake of nanoparticles into brain microvascular endothelial cells have been discussed. As summarized by Kreuter in reviews, these include processes associated with compromised tight junctions, the prolonged retention of NPs in the brain capillaries, general toxic effects on the brain vasculature, endocytosis and release of cargo within brain endothelial cells followed by transport into the brain interstitial fluid, and transcytosis through the brain microvascular endothelial cells [117–119]. The latter are considered the most likely mechanisms [120]. Modification of the particles and in particular, of their surface properties, has been demonstrated to be a valid approach to exploit the endocytotic/transcytotic pathway.

6.1 Particle Surface Properties

Systemic administration requires carefully designed PNPs with specific surface properties to cross the BBB and eventually deliver their cargo into the brain. Modifications of the particle surface chemistry can be accomplished in two ways, either (1) by introducing functional groups in the monomer/polymer pre-assembly which will be exposed on the surface of the resulting particles or (2) by post-assembly modification of the particle surface. A prominent approach is PEGylation which endows the particle with stealth properties and thus reduces nonspecific interactions with body components, prolongs the blood circulation time, and alters the organ distribution of the particulate system. As shown in Tables 1 and 2, the modification with nonionic surfactants such as polysorbate 80 (Tween®80) and poloxamer 188 is also a common strategy to alter the surface of the particles. In addition, the core polymer itself, the drug (adsorbed onto the surface or incorporated in the core polymer matrix), and the surfactant used for the fabrication of the particles also play crucial roles in the overall performance of the particles [121]. Kreuter and coworkers have studied surfactant-coated PNPs and their potential as brain delivery vehicles in detail. Poly(butylcyanoacrylate) (PBCA) NPs loaded with the drug dalargin (loaded by adsorption on the particle surface) was coated with a set of 12 different surfactants [122]. The particles were injected intravenously into mice, and nociceptive analgesia was measured at different time points after injection. Coating with polysorbate 20, 40, 60, and 80, and not with Cremophor® RH 40, Cremophor® EZ, Brij® 35, poloxamer 184, 388, and 407, and poloxamine 908, resulted in antinociceptive effects, suggesting the latter coating did not assist in BBB trafficking. Post-assembly coating is the dominant approach for the surface modification of PACA NPs (Table 2), most likely due to the more synthetically complicated access to PEG-containing ACA polymers. In contrast, the ROP of LA and GA from PEG initiators enables the straightforward synthesis of PEG-b-PLGA block copolymers of different compositions (Table 1). However, coating with surfactants does not seem to be trivial and requires in-depth investigations for the individual core polymers. Polysorbate 80 has been found to be an ideal coating material for PBCA particles, whereas poloxamer 188 outperformed polysorbate as coating materials of PLGA particles in the design of BBB-targeted carriers [121, 123]. Calvo et al. further studied the distribution of radiolabeled poly(hexadecyl cyanoacrylate) (PHDCA) particles in the brains of mice and rats following systemic administration. PEGylated PHDCA NPs were found to appear in the brain at much higher concentrations than polysorbate 80 or poloxamine-coated PHDCA NPs, highlighting the importance of long-circulating NPs. The authors also observed an interesting species-dependent surfactant effect. Higher concentrations of PHDCA particles were found in mice if the particles were coated with polysorbate 80 compared to poloxamine 908, with the opposite phenomenon observed in rats [48]. This raises an important question as to which coating would be optimum for administration to humans, a question which may be answered through the use of in vitro human BBB models. Another factor which can affect the performance of the PNPs is the drugs which are adsorbed to their surface. Although bare PEG-PHDCA NPs were shown to exhibit stealth properties and thus a prolonged blood circulation time with increased brain accumulation, the loading of doxorubicin (DOx) to their surface diminished the brain targeting significantly [124]. It was suggested that this might be the result of the positive surface charge of the particles which caused an increased association with plasma proteins. In contrast, coating of DOx-loaded PBCA [125] and PLGA [126] NPs with surfactants has shown to shield the DOx charge effectively and thus led to a prolonged survival of rats induced with 9L tumors.

6.2 Attachment of Targeting Ligands

One popular approach to increase the therapeutic index of a drug is to deliver the drug within a polymeric nanocarrier. Nanoparticles are generally designed to improve the pharmacokinetics and biodistribution of drugs; however, on their own, nanoparticles have poor targeting to particular organs. To deliver the cargo to its intended site of action, two different strategies are exploited: passive targeting, i.e., by exploitation of the enhanced permeability and retention (EPR) effect, or active targeting, i.e., by the modification of the particle surface with ligands targeting specific receptors at the site of action. The brain is one of the most challenging targets as it requires the carrier to cross the BBB when systemically administered. There are five main routes to cross the BBB: (a) the paracellular pathway, (b) the transcellular pathway, (c) transport proteins which are substrate specific, and (d) adsorptive-and (e) receptor-mediated transcytosis [11, 127]. Despite the multitude of delivery routes, many drugs cannot be delivered to the brain in sufficient therapeutic amounts as they often lack the ideal physicochemical properties for optimum passive diffusion across the brain microvascular endothelial cells. In particular, the delivery of large-molecule pharmaceuticals is heavily restricted. PNPs represent a formidable solution to these issues as they can be designed to cross the BBB by adsorptive- or receptor-mediated transcytosis. There are a range of receptors and transporters present at the cerebral endothelium which have been exploited for this purpose [128]. To this end, the surface of the PNPs can be modified covalently or non-covalently with ligands which specifically interact with receptors/transporters that are highly expressed at the BBB. To date, different targeted PNPs for the transport across the BBB have been reported as can be also seen in Tables 1, 2, and 3.

One commonly used targeting ligand group is transferrin (Tf) or anti-transferrin receptor monoclonal antibodies (e.g., OX26) which target the transferrin receptor [129, 130]. Chang et al. prepared Tf-targeted PLGA NPs and studied their uptake in a THP-1 (human monocytes) cell line, differentiated THP-1 macrophages, F98 glioma cells, and astrocytes [131]. Internalization of the PNPs was observed in all cell types except THP-1 monocytes, and a caveolae- and clathrin-mediated uptake in F98 cells was revealed. In vivo studies further revealed the preferential retention of the Tf-modified PLGA NPs in the brain parenchyma as opposed to BSA-modified PLGA NPs. However, this was observed in healthy rats, while in rats with a developed brain tumor, both NP types were found, likely because of a disturbed paracellular route in the cancer model. Cui et al. prepared doxorubicin (DOX)- and paclitaxel (PTX)-loaded, Tf-modified magnetic silica PLGA NPs and studied their antiproliferative effect on malignant brain glioma [23]. An increased uptake and cytotoxicity in U-87 glioma cells were observed by applying a magnetic field and in the presence of Tf ligands on the surface of the particles. Tf-modified magnetic particles loaded with both drugs showed the highest anti-glioma activity in an intracranial U-87 MG-luc2 xenograft of BALB/c nude mice as opposed to single-loaded PNPs or nontargeted PNPs.

Apolipoproteins have been found to be suitable targeting ligands to improve delivery to the brain [49]. Initially, numerous drugs bound to PBCA NPs coated with polysorbate 80 have been shown to be delivered across the BBB and exert a pharmacological effect after intravenous administration [122, 132–134]. Protein adsorption studies in citrate-stabilized plasma revealed the adsorption of apolipoprotein E (ApoE) to polysorbate-coated NPs, which were able to cross the BBB, whereas non-coated or poloxamer 338- and poloxamer 407-coated particles had no ApoE adsorption, nor brain uptake, indicating the involvement of ApoE in the uptake process into brain microvascular endothelial cells. To further evaluate this assumption, PBCA particles loaded with dalargin were directly modified with a range of different Apos including ApoAII, ApoB, ApoCII, ApoE, and ApoJ with and without an additional polysorbate 80 coating [49]. The antinociceptive threshold of all these particles was determined after intravenous injection in ICR mice and ApoE-deficient mice ApoEtm1Unc. Polysorbate coating and/or surface modification of PBCA particles with ApoE and ApoB resulted in an antinociceptive effect, which was significantly reduced in ApoE-deficient mice, further confirming a crucial role of Apo in the overall brain uptake process. Thus, polysorbate-coated PNPs seem to adsorb Apo from the blood, and Apo modification endows the PNPs with lipoprotein properties which results in a receptor-mediated endocytotic uptake of the PNPs via interaction with members of the low-density lipoprotein (LDL) receptor family expressed at the BBB [135]. This concept has also been translated to different types of nanoparticles such as human serum albumin (HSA) particles [136].

Insulin or anti-insulin receptor monoclonal antibodies are another targeting ligand [137] which can be immobilized on the surface of NPs to increase their uptake through interaction with the insulin receptor at the BBB. HSA NPs covalently modified with insulin or the 29B4 antibody and loaded with loperamide triggered antinociceptive effects in ICR mice after intravenous injection [138]. By co-administering these targeted NPs with free 29B4 antibody, loperamide delivery was completely inhibited, revealing the importance of the insulin receptor for the trafficking across the BBB and antinociceptive activity of the insulin-targeted HSA NPs loaded with loperamide.

Glutathione (GSH), a naturally occurring tripeptide, plays an important role as antioxidant in the brain and facilitates the transport across the BBB and thus can be employed as targeting ligand [139]. This approach has been used for the fabrication of GSH-targeted PEGylated liposomes, and its potential to deliver a range of cargoes into the brain has been demonstrated [140, 141]. Sutariya and coworkers used this technology for the preparation of targeted PLGA NPs [17]. The targeted NPs were prepared by nanoprecipitation of PLGA-PEG-COOH followed by incubation in GSH solution. In vitro uptake studies in RG2 glioma cells showed preferential uptake of GSH functionalized particles compared to non-functionalized particles. Preliminary in vivo results in mice revealed a higher brain uptake of dye-loaded GSH-PLGA NPs compared to the free dye.

Besides the natural occurring ligands, peptides which are engineered to targeted specific receptors have also been employed for BBB-targeted PNPs. Angiopeps are peptides which are designed to resemble the structure of ligands, such the Kunitz protease inhibitor (KPI) fragment, which are known to target the low-density lipoprotein receptor-related protein (LRP) receptor. Xin et al. demonstrated that Angiopep-2 coupled to PEG-b-PCL NPs enhances the uptake of the particles in brain capillary endothelial cells through clathrin- and caveolae-mediated endocytosis and the transport of the Angiopep-2 containing particles across a BBB model compared to bare PEG-b-PCL NPs [40]. Rhodamine B-labeled Angiopep-2-modified particles also accumulated more significantly in brain tissue after intravenous injection into mice relative to nontargeted NPs. Pretreatment with free Angiopep-2-inhibited the uptake, crossing, and accumulation of the PNPs and thus suggested the involvement of the LRP receptor in the transcytosis process. Based on this concept, Huile et al. designed cascade BBB-targeting PEG-b-PCL NPs which contain not only Angiopep-2 but also an EGFP-EGF1 protein to target brain neuroglial cells after uptake of the NPs into the brain [42]. In vitro studies revealed that these PNPs were taken up by bEnd.3 brain microvascular endothelial cells and neuroglial cells to a higher extent than non-modified particles. Only Angiopep-2-modified

NPs showed a higher accumulation in the brain in ex vivo studies, and co-localization studies of the NPs with neuroglial cells showed that the dual modified NPs targeted and entered the neuroglial cells.

CRM197, the nontoxic analogue of diphtheria toxin (DT), is another targeting ligand of interest. It targets the membrane-bound precursor of heparin-binding epidermal growth factor (HB-EGF) receptor, also known as the DT receptor (DTR) which is expressed on brain microvascular endothelial cells, glia, and neurons [142]. PBCA NPs modified with CRM197 were prepared to deliver zidovudine across the BBB [60]. Decreasing the particle size and increasing the grafting quantity were reported to enhance the permeability coefficient of zidovudine. Tosi et al. investigated the performance of CRM197-grafted PLGA NPs in vivo [33]. They demonstrated that these particles can cross the BBB and reach all brain areas without compromising the integrity of the BBB and are also able to deliver drugs such as loperamide. Co-administration experiments with g7-modified PLGA revealed a higher accumulation of CRM197-PLGA NPs in GFAP-positive cells and interneurons 30 min after administration; however, after longer times, both modified particle types showed an increased co-localization indicating similar trafficking pathways after initially different pathways.

The second type of transcytosis is the adsorptive transcytosis which does not exploit receptor-mediated processes but rather charge-dependent internalization events. In this context, in particular cationic compounds, exposed on the surface of the PNPs, which enable an enhanced interaction with negatively charged cell surfaces are employed. Typical examples are coatings with the transduction domain of human immunodeficiency virus type-1 (TAT) peptide [13, 69] and cationic bovine serum albumin (CBSA) [15, 26, 27].

7 Conclusions

Given the flexibility associated with manipulation of PNPs, in terms of their size, shape, and surface characteristics, these carrier systems have been shown to be attractive platforms that can be used to target the BBB specifically and deliver therapeutic agents to the CNS. Despite there being a large number of studies demonstrating their ideal characteristics using in vitro and preclinical in vivo models, the translational gap, as with many BBB-targeting approaches, still exists and whether these agents will show eventual utility in the clinic is the next step which requires intense investigation. The future of CNS drug targeting is an interesting one, which comes with many challenges for NP systems, including their biodegradability, long-term accumulation concerns, and manufacturing and scale-up complexities. However, given their biodegradable nature, polymeric NPs presented in this chapter may play a crucial role in filling some of these translational gaps and are a potential source of exciting preclinical-to-clinical studies that may ultimately overcome the challenge of CNS drug delivery.

References

1. Abbott NJ et al (2006) Astrocyte-endothelial interactions at the blood-brain barrier. Nat Rev Neurosci 7(1):41–53

2. Kabanov AV, Gendelman HE (2007) Nanomedicine in the diagnosis and therapy of neurodegenerative disorders. Prog Polym Sci 32 (8–9):1054–1082

3. Mishra B et al (2010) Colloidal nanocarriers: a review on formulation technology, types and applications toward targeted drug delivery. Nanomed Nanotechnol Biol Med 6(1):9–24

4. Beija M et al (2012) Colloidal systems for drug delivery: from design to therapy. Trends Biotechnol 30(9):485–496

5. Elsabahy M, Wooley KL (2012) Design of polymeric nanoparticles for biomedical delivery applications. Chem Soc Rev 41 (7):2545–2561

6. Letchford K, Burt H (2007) A review of the formation and classification of amphiphilic block copolymer nanoparticulate structures: micelles, nanospheres, nanocapsules and polymersomes. Eur J Pharm Biopharm 65 (3):259–269

7. Mura S et al (2013) Stimuli-responsive nanocarriers for drug delivery. Nat Mater 12 (11):991–1003

8. Shi J et al (2017) Cancer nanomedicine: progress, challenges and opportunities. Nat Rev Cancer 17(1):20–37

9. Knop K et al (2010) Poly(ethylene glycol) in drug delivery: pros and cons as well as potential alternatives. Angew Chem Int Edit 49 (36):6288–6308

10. Cheng CJ et al (2015) A holistic approach to targeting disease with polymeric nanoparticles. Nat Rev Drug Discov 14(4):239–247

11. Patel T et al (2012) Polymeric nanoparticles for drug delivery to the central nervous system. Adv Drug Deliv Rev 64(7):701–705

12. Gao X et al (2006) Lectin-conjugated PEG–PLA nanoparticles: preparation and brain delivery after intranasal administration. Biomaterials 27(18):3482–3490

13. Rao KS et al (2008) TAT-conjugated nanoparticles for the CNS delivery of anti-HIV drugs. Biomaterials 29(33):4429–4438

14. Liu M et al (2008) Pharmacokinetics and biodistribution of surface modification polymeric nanoparticles. Arch Pharm Res 31 (4):547–554

15. Parikh T et al (2010) Efficacy of surface charge in targeting pegylated nanoparticles of sulpiride to the brain. Eur J Pharm Biopharm 74(3):442–450

16. Tosi G et al (2010) Sialic acid and glycopeptides conjugated PLGA nanoparticles for central nervous system targeting: in vivo pharmacological evidence and biodistribution. J Control Release 145(1):49–57

17. Geldenhuys W et al (2011) Brain-targeted delivery of paclitaxel using glutathione-coated nanoparticles for brain cancers. J Drug Target 19(9):837–845

18. Guo J et al (2011) Aptamer-functionalized PEG–PLGA nanoparticles for enhanced antiglioma drug delivery. Biomaterials 32 (31):8010–8020

19. Hu K et al (2011) Lactoferrin conjugated PEG-PLGA nanoparticles for brain delivery: preparation, characterization and efficacy in Parkinson's disease. Int J Pharm 415 (1–2):273–283

20. Li J et al (2011) Targeting the brain with PEG–PLGA nanoparticles modified with phage-displayed peptides. Biomaterials 32 (21):4943–4950

21. Mittal G et al (2011) Development and evaluation of polymer nanoparticles for oral delivery of estradiol to rat brain in a model of Alzheimer's pathology. J Control Release 150(2):220–228

22. Chen Y-C et al (2013) Effects of surface modification of PLGA-PEG-PLGA nanoparticles on loperamide delivery efficiency across the blood–brain barrier. J Biomater Appl 27 (7):909–922

23. Gao J-Q et al (2013) Glioma targeting and blood–brain barrier penetration by dual-targeting doxorubincin liposomes. Biomaterials 34(22):5628–5639

24. Liu Z et al (2013) B6 peptide-modified PEG-PLA nanoparticles for enhanced brain

delivery of neuroprotective peptide. Biocon-jug Chem 24(6):997–1007

25. Pagar K, Vavia P (2013) Rivastigmine-loaded L-lactide-depsipeptide polymeric nanoparti-cles: decisive formulation variable optimiza-tion. Sci Pharm 81:865–885

26. Lu W et al (2005) Cationic albumin conju-gated pegylated nanoparticle with its transcy-tosis ability and little toxicity against blood–brain barrier. Int J Pharm 295 (1–2):247–260

27. Lu W et al (2007) Brain delivery property and accelerated blood clearance of cationic albu-min conjugated pegylated nanoparticle. J Control Release 118(1):38–53

28. Md S et al (2014) Preparation, characteriza-tion, in vivo biodistribution and pharmacoki-netic studies of donepezil-loaded PLGA nanoparticles for brain targeting. Drug Dev Ind Pharm 40(2):278–287

29. Piazza J et al (2014) Haloperidol-loaded intranasally administered lectin functionalized poly(ethylene glycol)–block-poly(d,l)-lactic-co-glycolic acid (PEG–PLGA) nanoparticles for the treatment of schizophrenia. Eur J Pharm Biopharm 87(1):30–39

30. Vilella A et al (2014) Insight on the fate of CNS-targeted nanoparticles. Part I: Rab5-dependent cell-specific uptake and distribu-tion. J Control Release 174:195–201

31. Zhang C et al (2014) Intranasal nanoparticles of basic fibroblast growth factor for brain delivery to treat Alzheimer's disease. Int J Pharm 461(1–2):192–202

32. Zhang C et al (2014) Dual-functional nano-particles targeting amyloid plaques in the brains of Alzheimer's disease mice. Biomater-ials 35(1):456–465

33. Tosi G et al (2015) Exploiting bacterial path-ways for BBB crossing with PLGA nanoparti-cles modified with a mutated form of diphtheria toxin (CRM197): in vivo experi-ments. Mol Pharm 12(10):3672–3684

34. Salvalaio M et al (2016) Targeted polymeric nanoparticles for brain delivery of high molec-ular weight molecules in lysosomal storage disorders. PLoS One 11(5):e0156452

35. Parveen S, Sahoo SK (2011) Long circulating chitosan/PEG blended PLGA nanoparticle for tumor drug delivery. Eur J Pharmacol 670(2–3):372–383

36. Xin H et al (2010) Enhanced anti-glioblastoma efficacy by PTX-loaded PEGy-lated poly(ε-caprolactone) nanoparticles: in vitro and in vivo evaluation. Int J Pharm 402(1–2):238–247

37. Xin H et al (2011) Angiopep-conjugated poly (ethylene glycol)-co-poly(ε-caprolactone) nanoparticles as dual-targeting drug delivery system for brain glioma. Biomaterials 32 (18):4293–4305

38. Gao H et al (2012) Precise glioma targeting of and penetration by aptamer and peptide dual-functioned nanoparticles. Biomaterials 33(20):5115–5123

39. Gao H et al (2014) Study and evaluation of mechanisms of dual targeting drug delivery system with tumor microenvironment assays compared with normal assays. Acta Biomater 10(2):858–867

40. Xin H et al (2012) The brain targeting mech-anism of Angiopep-conjugated poly(ethylene glycol)-co-poly(ε-caprolactone) nanoparti-cles. Biomaterials 33(5):1673–1681

41. Alex AT et al (2016) Development and evalu-ation of carboplatin-loaded PCL nanoparti-cles for intranasal delivery. Drug Deliv 23 (7):2144–2153

42. Huile G et al (2011) A cascade targeting strat-egy for brain neuroglial cells employing nano-particles modified with angiopep-2 peptide and EGFP-EGF1 protein. Biomaterials 32 (33):8669–8675

43. Jiang X et al (2014) Nanoparticles of 2-deoxy-d-glucose functionalized poly(ethylene gly-col)-co-poly(trimethylene carbonate) for dual-targeted drug delivery in glioma treat-ment. Biomaterials 35(1):518–529

44. Kreuter J et al (1995) Passage of peptides through the blood-brain barrier with colloidal polymer particles (nanoparticles). Brain Res 674(1):171–174

45. Olivier J-C et al (1999) Indirect evidence that drug brain targeting using polysorbate 80-coated polybutylcyanoacrylate nanoparti-cles is related to toxicity. Pharm Res 16 (12):1836–1842

46. Darius J et al (2000) Influence of nanoparti-cles on the brain-to-serum distribution and the metabolism of valproic acid in mice. J Pharm Pharmacol 52(9):1043–1047

47. Schroeder U et al (2000) Body distribution of 3H-labelled dalargin bound to poly(butyl cya-noacrylate) nanoparticles after I.V. injections to mice. Life Sci 66(6):495–502

48. Calvo P et al (2001) Long-circulating PEGy-lated polycyanoacrylate nanoparticles as new drug carrier for brain delivery. Pharm Res 18 (8):1157–1166

49. Kreuter J et al (2002) Apolipoprotein-mediated transport of nanoparticle-bound drugs across the blood-brain barrier. J Drug Target 10(4):317–325

50. Kreuter J et al (2003) Direct evidence that polysorbate-80-coated poly(butylcyanoacry-late) nanoparticles deliver drugs to the CNS

via specific mechanisms requiring prior binding of drug to the nanoparticles. Pharm Res 20(3):409–416

51. Garcia-Garcia E et al (2005) A relevant in vitro rat model for the evaluation of blood-brain barrier translocation of nanoparticles. Cell Mol Life Sci 62(12):1400–1408

52. Ambruosi A et al (2006) Biodistribution of polysorbate 80-coated doxorubicin-loaded [14C]-poly(butyl cyanoacrylate) nanoparticles after intravenous administration to glioblastoma-bearing rats. J Drug Target 14(2):97–105

53. Gao K, Jiang X (2006) Influence of particle size on transport of methotrexate across blood brain barrier by polysorbate 80-coated polybutylcyanoacrylate nanoparticles. Int J Pharm 310(1–2):213–219

54. Kuo Y-C, Chen H-H (2006) Effect of nanoparticulate polybutylcyanoacrylate and methylmethacrylate–sulfopropylmethacrylate on the permeability of zidovudine and lamivudine across the in vitro blood–brain barrier. Int J Pharm 327(1–2):160–169

55. Petri B et al (2007) Chemotherapy of brain tumour using doxorubicin bound to surfactant-coated poly(butyl cyanoacrylate) nanoparticles: revisiting the role of surfactants. J Control Release 117(1):51–58

56. Wilson B et al (2008) Poly (n-butylcyanoacrylate) nanoparticles coated with polysorbate 80 for the targeted delivery of rivastigmine into the brain to treat Alzheimer's disease. Brain Res 1200:159–168

57. Wilson B et al (2008) Targeted delivery of tacrine into the brain with polysorbate 80-coated poly(n-butylcyanoacrylate) nanoparticles. Eur J Pharm Biopharm 70(1):75–84

58. Wang C-X et al (2009) Antitumor effects of polysorbate-80 coated gemcitabine polybutylcyanoacrylate nanoparticles in vitro and its pharmacodynamics in vivo on C6 glioma cells of a brain tumor model. Brain Res 1261:91–99

59. Kulkarni PV et al (2010) Quinoline-n-butyl-cyanoacrylate-based nanoparticles for brain targeting for the diagnosis of Alzheimer's disease. Wires Nanomed Nanobiotechnol 2(1):35–47

60. Kuo Y-C, Chung C-Y (2012) Transcytosis of CRM197-grafted polybutylcyanoacrylate nanoparticles for delivering zidovudine across human brain-microvascular endothelial cells. Colloids Surf B Biointerfaces 91:242–249

61. Aktaş Y et al (2005) Development and brain delivery of chitosan–PEG nanoparticles functionalized with the monoclonal antibody OX26. Bioconjug Chem 16(6):1503–1511

62. Wang X et al (2008) Preparation of estradiol chitosan nanoparticles for improving nasal absorption and brain targeting. Eur J Pharm Biopharm 70(3):735–740

63. Hombach J, Bernkop-Schnürch A (2009) Chitosan solutions and particles: evaluation of their permeation enhancing potential on MDCK cells used as blood brain barrier model. Int J Pharm 376(1–2):104–109

64. Karatas H et al (2009) A nanomedicine transports a peptide caspase-3 inhibitor across the blood–brain barrier and provides neuroprotection. J Neurosci 29(44):13761–13769

65. Songjiang Z, Lixiang W (2009) Amyloid-beta associated with chitosan nano-carrier has favorable immunogenicity and permeates the BBB. AAPS PharmSciTech 10(3):900

66. Trapani A et al (2011) Characterization and evaluation of chitosan nanoparticles for dopamine brain delivery. Int J Pharm 419(1–2):296–307

67. Trapani A et al (2011) Methotrexate-loaded chitosan- and glycolchitosan-based nanoparticles: a promising strategy for the administration of the anticancer drug to brain tumors. AAPS PharmSciTech 12(4):1302–1311

68. Fazil M et al (2012) Development and evaluation of rivastigmine loaded chitosan nanoparticles for brain targeting. Eur J Pharm Sci 47(1):6–15

69. Malhotra M et al (2013) Synthesis of TAT peptide-tagged PEGylated chitosan nanoparticles for siRNA delivery targeting neurodegenerative diseases. Biomaterials 34(4):1270–1280

70. Agyare EK et al (2014) Engineering theranostic nanovehicles capable of targeting cerebrovascular amyloid deposits. J Control Release 185:121–129

71. Gregori M et al (2015) Investigation of functionalized poly(N,N-dimethylacrylamide)-block-polystyrene nanoparticles as novel drug delivery system to overcome the blood–brain barrier in vitro. Macromol Biosci 15(12):1687–1697

72. Hwang DW et al (2011) A brain-targeted rabies virus glycoprotein-disulfide linked PEI nanocarrier for delivery of neurogenic microRNA. Biomaterials 32(21):4968–4975

73. Li J, Sabliov C (2013) PLA/PLGA nanoparticles for delivery of drugs across the blood-brain barrier. Nanotechnol Rev 2(3):241–257

74. Landry FB et al (1996) Degradation of poly (d,l-lactic acid) nanoparticles coated with

albumin in model digestive fluids (USP XXII). Biomaterials 17(7):715–723

75. Panagi Z et al (2001) Effect of dose on the biodistribution and pharmacokinetics of PLGA and PLGA–mPEG nanoparticles. Int J Pharm 221(1–2):143–152

76. Yang Y-Y et al (2001) Morphology, drug distribution, and in vitro release profiles of biodegradable polymeric microspheres containing protein fabricated by double-emulsion solvent extraction/evaporation method. Biomaterials 22(3):231–241

77. Esmaeili F et al (2008) PLGA nanoparticles of different surface properties: preparation and evaluation of their body distribution. Int J Pharm 349(1–2):249–255

78. Cheng J et al (2007) Formulation of functionalized PLGA–PEG nanoparticles for in vivo targeted drug delivery. Biomaterials 28(5):869–876

79. Ghahremankhani AA et al (2007) PLGA-PEG-PLGA tri-block copolymers as an in-situ gel forming system for calcitonin delivery. Polym Bull 59(5):637–646

80. Nicolas J, Couvreur P (2009) Synthesis of poly(alkyl cyanoacrylate)-based colloidal nanomedicines. Wires Nanomed Nanobiotechnol 1(1):111–127

81. Brambilla D et al (2010) Design of fluorescently tagged poly(alkyl cyanoacrylate) nanoparticles for human brain endothelial cell imaging. Chem Commun 46(15):2602–2604

82. Brambilla D et al (2012) PEGylated nanoparticles bind to and alter amyloid-beta peptide conformation: toward engineering of functional nanomedicines for Alzheimer's disease. ACS Nano 6(7):5897–5908

83. Lenaerts V et al (1984) Degradation of poly (isobutyl cyanoacrylate) nanoparticles. Biomaterials 5(2):65–68

84. Müller RH et al (1990) In vitro model for the degradation of alkylcyanoacrylate nanoparticles. Biomaterials 11(8):590–595

85. Scherer D et al (1994) Influence of enzymes on the stability of polybutylcyanoacrylate nanoparticles. Int J Pharm 101(1):165–168

86. Leonard F et al (1966) Synthesis and degradation of poly (alkyl α-cyanoacrylates). J Appl Polym Sci 10(2):259–272

87. Lherm C et al (1992) Alkylcyanoacrylate drug carriers: II. Cytotoxicity of cyanoacrylate nanoparticles with different alkyl chain length. Int J Pharm 84(1):13–22

88. Ambruosi A et al (2005) Body distribution of polysorbate-80 and doxorubicin-loaded [14C]poly(butyl cyanoacrylate) nanoparticles

after i.v. administration in rats. J Drug Target 13(10):535–542

89. Douglas SJ et al (1985) Molecular weights of poly(butyl 2-cyanoacrylate) produced during nanoparticle formation. Br Polym J 17(4):339–342

90. Hajji S et al (2014) Structural differences between chitin and chitosan extracted from three different marine sources. Int J Biol Macromol 65:298–306

91. Jayakumar R et al (2008) Preparative methods of phosphorylated chitin and chitosan—an overview. Int J Biol Macromol 43(3):221–225

92. Yang H-C, Hon M-H (2009) The effect of the molecular weight of chitosan nanoparticles and its application on drug delivery. Microchem J 92(1):87–91

93. Franca EF et al (2008) Characterization of chitin and chitosan molecular structure in aqueous solution. J Chem Theory Comput 4(12):2141–2149

94. Pechsrichuang P et al (2013) Production of recombinant Bacillus subtilis chitosanase, suitable for biosynthesis of chitosan-oligosaccharides. Bioresour Technol 127:407–414

95. Sanon A et al (2005) N-Acetyl-β-d-hexosaminidase from Trichomonas vaginalis: substrate specificity and activity of inhibitors. Biomed Pharmacother 59(5):245–248

96. Kulish EI et al (2006) Enzymatic degradation of chitosan films by collagenase. Polym Sci Ser B 48(5):244–246

97. Hsu S-H et al (2004) Chitosan as scaffold materials: effects of molecular weight and degree of deacetylation. J Polym Res 11(2):141–147

98. Taşkın P et al (2014) The effect of degree of deacetylation on the radiation induced degradation of chitosan. Radiat Phys Chem 94:236–239

99. Huang M et al (2004) Uptake and cytotoxicity of chitosan molecules and nanoparticles: effects of molecular weight and degree of deacetylation. Pharm Res 21(2):344–353

100. Kean T, Thanou M (2010) Biodegradation, biodistribution and toxicity of chitosan. Adv Drug Deliv Rev 62(1):3–11

101. Sarvaiya J, Agrawal YK (2015) Chitosan as a suitable nanocarrier material for anti-Alzheimer drug delivery. Int J Biol Macromol 72:454–465

102. Wang ZH et al (2010) Trimethylated chitosan-conjugated PLGA nanoparticles for the delivery of drugs to the brain. Biomaterials 31(5):908–915

103. Rao JP, Geckeler KE (2011) Polymer nanoparticles: preparation techniques and size-control parameters. Prog Polym Sci 36 (7):887–913

104. Arshady R (1991) Preparation of biodegradable microspheres and microcapsules: 2. Polyactides and related polyesters. J Control Release 17(1):1–21

105. Makadia HK, Siegel SJ (2011) Poly lactic-co-glycolic acid (PLGA) as biodegradable controlled drug delivery carrier. Polymers 3 (3):1377

106. Schubert S et al (2011) Nanoprecipitation and nanoformulation of polymers: from history to powerful possibilities beyond poly(lactic acid). Soft Matter 7(5):1581–1588

107. Thickett SC, Gilbert RG (2007) Emulsion polymerization: state of the art in kinetics and mechanisms. Polymer 48 (24):6965–6991

108. Seijo B et al (1990) Design of nanoparticles of less than 50 nm diameter: preparation, characterization and drug loading. Int J Pharm 62 (1):1–7

109. Tuncel A et al (1995) Monosize poly(ethylcyanoacrylate) microspheres: preparation and degradation properties. J Biomed Mater Res 29(6):721–728

110. Douglas SJ et al (1985) Particle size and size distribution of poly(butyl 2-cyanoacrylate) nanoparticles. II Influence of stabilizers. J Colloid Interface Sci 103(1):154–163

111. Calvo P et al (1997) Novel hydrophilic chitosan-polyethylene oxide nanoparticles as protein carriers. J Appl Polym Sci 63 (1):125–132

112. Lapitsky Y (2014) Ionically crosslinked polyelectrolyte nanocarriers: recent advances and open problems. Curr Opin Colloid Interface Sci 19(2):122–130

113. Yokel RA (2016) Physicochemical properties of engineered nanomaterials that influence their nervous system distribution and effects. Nanomed Nanotechnol Biol Med 12 (7):2081–2093

114. Zhang T-T et al (2016) Strategies for transporting nanoparticles across the blood-brain barrier. Biomater Sci 4(2):219–229

115. Thorne RG, Nicholson C (2006) In vivo diffusion analysis with quantum dots and dextrans predicts the width of brain extracellular space. Proc Natl Acad Sci 103 (14):5567–5572

116. Soo Choi H et al (2007) Renal clearance of quantum dots. Nat Biotechnol 25 (10):1165–1170

117. Kreuter J (2001) Nanoparticulate systems for brain delivery of drugs. Adv Drug Deliv Rev 47(1):65–81

118. Kreuter J (2012) Nanoparticulate systems for brain delivery of drugs. Adv Drug Deliv Rev 64:213–222

119. Kreuter J (2014) Drug delivery to the central nervous system by polymeric nanoparticles: what do we know? Adv Drug Deliv Rev 71:2–14

120. Kreuter J (2013) Mechanism of polymeric nanoparticle-based drug transport across the blood-brain barrier (BBB). J Microencapsul 30(1):49–54

121. Gelperina S et al (2010) Drug delivery to the brain using surfactant-coated poly(lactide-co-glycolide) nanoparticles: influence of the formulation parameters. Eur J Pharm Biopharm 74(2):157–163

122. Kreuter J et al (1997) Influence of the type of surfactant on the analgesic effects induced by the peptide dalargin after its delivery across the blood–brain barrier using surfactant-coated nanoparticles. J Control Release 49 (1):81–87

123. Kulkarni SA, Feng S-S (2011) Effects of surface modification on delivery efficiency of biodegradable nanoparticles across the blood–brain barrier. Nanomedicine 6 (2):377–394

124. Brigger I et al (2004) Negative preclinical results with stealth® nanospheres-encapsulated doxorubicin in an orthotopic murine brain tumor model. J Control Release 100(1):29–40

125. Ambruosi A et al (2006) Influence of surfactants, polymer and doxorubicin loading on the anti-tumour effect of poly(butyl cyanoacrylate) nanoparticles in a rat glioma model. J Microencapsul 23(5):582–592

126. Wohlfart S et al (2011) Efficient chemotherapy of rat glioblastoma using doxorubicin-loaded PLGA nanoparticles with different stabilizers. PLoS One 6(5):e19121

127. Begley DJ (1996) The blood-brain barrier: principles for targeting peptides and drugs to the central nervous system. J Pharm Pharmacol 48(2):136–146

128. Jones AR, Shusta EV (2007) Blood–brain barrier transport of therapeutics via receptor-mediation. Pharm Res 24(9):1759–1771

129. Pardridge WM, Boado RJ (2012) Reengineering biopharmaceuticals for targeted delivery across the blood–brain barrier. In: Wittrup KD, Gregory LV (eds) Methods in

enzymology, vol 503. Academic, Cambridge, MA, pp 269–292

130. Ulbrich K et al (2009) Transferrin- and transferrin-receptor-antibody-modified nanoparticles enable drug delivery across the blood–brain barrier (BBB). Eur J Pharm Biopharm 71(2):251–256

131. Chang J et al (2012) Transferrin adsorption onto PLGA nanoparticles governs their interaction with biological systems from blood circulation to brain cancer cells. Pharm Res 29 (6):1495–1505

132. Alyautdin RN et al (1997) Delivery of loperamide across the blood-brain barrier with polysorbate 80-coated polybutylcyanoacrylate nanoparticles. Pharm Res 14(3):325–328

133. Gulyaev AE et al (1999) Significant transport of doxorubicin into the brain with polysorbate 80-coated nanoparticles. Pharm Res 16 (10):1564–1569

134. Friese A et al (2000) Increase of the duration of the anticonvulsive activity of a novel NMDA receptor antagonist using poly(butylcyanoacrylate) nanoparticles as a parenteral controlled release system. Eur J Pharm Biopharm 49(2):103–109

135. Willnow TE et al (1999) Lipoprotein receptors: new roles for ancient proteins. Nat Cell Biol 1(6):E157–E162

136. Zensi A et al (2009) Albumin nanoparticles targeted with Apo E enter the CNS by transcytosis and are delivered to neurones. J Control Release 137(1):78–86

137. Coloma MJ et al (2000) Transport across the primate blood-brain barrier of a genetically engineered chimeric monoclonal antibody to the human insulin receptor. Pharm Res 17 (3):266–274

138. Ulbrich K et al (2011) Targeting the insulin receptor: nanoparticles for drug delivery across the blood–brain barrier (BBB). J Drug Target 19(2):125–132

139. Smeyne M, Smeyne RJ (2013) Glutathione metabolism and Parkinson's disease. Free Radic Biol Med 62:13–25

140. Rotman M et al (2015) Enhanced glutathione PEGylated liposomal brain delivery of an anti-amyloid single domain antibody fragment in a mouse model for Alzheimer's disease. J Control Release 203:40–50

141. Gaillard PJ et al (2014) Pharmacokinetics, brain delivery, and efficacy in brain tumor-bearing mice of glutathione pegylated liposomal doxorubicin (2B3-101). PLoS One 9(1): e82331

142. Gaillard PJ et al (2005) Diphtheria toxin receptor-targeted brain drug delivery. Int Congr Ser 1277:185–198

Chapter 2

Liposomes as Brain Targeted Delivery Systems

Francesco Lai, Michele Schlich, Chiara Sinico, and Anna Maria Fadda

Abstract

Liposomes and lipid vesicles are leading strategies in allowing brain targeted drug release with prominent success. Liposome rational design and functionalization have improved the brain bioavailability of several blood-brain barrier (BBB)-impermeable molecules, demonstrating the thrilling therapeutic potential of these nanocarriers. PEGylation continues to be a strategy for enhanced circulation times, and its associated EPR effect could be exploited to target vascularized brain tumors. Surface modifications with antibodies, transferrin, insulin and targeting glucose transporters, among others, are strategies used in research of new potential therapies for cerebral ischemia, brain tumors, and Alzheimer's and Parkinson's diseases. More recent advancements include technologies based on cationic liposomes (untargeted and targeted to the brain microvasculature) as well as surface modifications with cell-penetrating peptides. These strategies to modify liposomes to improve drug bioavailability in the brain are thoroughly presented and discussed in this chapter.

Key words Liposomes, Cationic liposomes, Cell-penetrating peptides, EPR effect, Brain tumor, Alzheimer, Parkinson, Brain targeting

1 Introduction

Drug delivery to the central nervous system (CNS) is still challenging due to the nature of the blood-brain barrier (BBB), which performs several important functions but is characterized by a high selectivity that prevents transport of many active molecules to the CNS as well as their accumulation in the brain parenchyma [1, 2]. The result is that reaching the CNS is difficult for several drug molecules: more than 98% small drugs and about 100% high molecular weight compounds do not overcome the BBB. Several important drugs such as natural, recombinant, or synthesized peptides and proteins, small-interfering RNA (siRNA), monoclonal antibodies, and gene therapeutics with a well-established activity to CNS receptors do not readily permeate into brain parenchyma due to the presence of the BBB [1, 3, 4].

As a consequence, various brain diseases are still undertreated although several strategies have been developed for delivery

Javier O. Morales and Pieter J. Gaillard (eds.), *Nanomedicines for Brain Drug Delivery*, Neuromethods, vol. 157,
https://doi.org/10.1007/978-1-0716-0838-8_2, © Springer Science+Business Media, LLC, part of Springer Nature 2021

therapeutics to the brain, using both invasive and noninvasive techniques. Invasive methods, which include disruption of the BBB [5–7], convection-enhanced delivery [8], intracerebral and intracerebroventricular infusion [9, 10], and use of implants [11, 12], show many limitations such as high costs for anesthesia and hospitalization, low drug diffusion into the brain parenchyma, enhancement of tumor dissemination after tight junctions disruption, lack of efficacy, and others [13]. Noninvasive techniques and, in particular, the use of nanocarriers have shown superior performances in comparison with the invasive ones in terms of patient compliance, higher efficacy, and safety [6, 14–16].

Nanocarriers include different colloidal systems (i.e., lipid and polymeric nanoparticles, micelles, vesicles) and have the advantage of being functionally modifiable in their physicochemical and surface properties in order to favor their CNS uptaking. Moreover, they are able to protect the entrapped drug from endogenous and exogenous agents and can modify and improve drug pharmacokinetics, thus reducing side effects. In addition, their surface can be properly modified and decorated to obtain prolonged circulation time and to target drugs to the CNS [17, 18].

Among these, vesicular systems, especially liposomes, have received a great attention due to their safety and biocompatibility and because of their structure that allows them to load both hydrophilic and lipophilic molecules as well as small and large molecules [19].

In this chapter, after a brief introduction to liposomes, the most recent applications of these lipid vesicles in the treatment of different CNS diseases will be described.

2 Liposomes

Liposomes are small vesicles constituted by an aqueous core enclosed by mono- or multi-lamellar phospholipid bilayer membranes. They have been identified by Alec D. Bangham and his group at the Babraham Institute of Animal Physiology of University of Cambridge, in the 1960s, studying phospholipid dispersions as a model for a cell membrane [20]. Since the observation of Bangham that phospholipids in aqueous systems can form closed bilayered structures, liposomes have arisen as one of the most promising carriers in medical fields for drug, biomolecules, and gene delivery. Molecules with different solubility can be encapsulated in liposomes, where the hydrophilic molecules can be encapsulated while the hydrophobic ones are incorporated into the lipidic membrane.

However, after in vivo administration, the plain liposomes tend to fuse and/or aggregate with each other resulting in immature release of their payload over time. In addition, liposomes underwent rapid systemic clearance due to their uptake by the cells of the mononuclear phagocyte system.

To overcome these problems, liposomes were firstly modified into long circulating vesicles by decreasing particle size (<100 nm) or by linking biocompatible hydrophilic polymers, usually polyethylene glycol chains, to their surface. Such polymers form a protective layer over the liposome surface, thus avoiding liposome recognition by opsonins and, therefore, the rapid clearance (stealth or pegylated liposomes). Then, they were specifically surface decorated to release the encapsulated drug in the target site [21, 22].

Overall, the positive features of liposomes, such as ease of preparation, biocompatibility, low toxicity, high loading capacity, controllable release kinetics, and versatile surface modification, have led to the development of several clinically approved liposomal formulations as well as of many products which are under different clinical trials.

The first successful pegylated liposome-based product was Doxil® in the market from 1995 for the doxorubicin treatment of patients with ovarian cancer and AIDS-related Kaposi's sarcoma. Later, DaunoXome® was approved for the delivery of daunorubicin in the treatment of advanced HIV-associated Kaposi's sarcoma. Subsequently, a few more products have become available for the management of various cancers. These products include DepoCyt®, Myocet®, Mepact®, and Marqibo®. Recently, a fluorouracil and leucovorin combination therapy-based product, marketed as Onivyde™, was approved for metastatic adenocarcinoma of the pancreas. Although cancer was the most widely researched field for liposomal products, several formulations for other diseases were developed. For instance, liposomal formulations of amphotericin B, Amphotec®, and AmBisome®, were produced for fungal infections in the 1990s. Also, liposomes have become important tools for vaccination when products such as Epaxal® and Inflexal® V were approved for vaccination against hepatitis and influenza, respectively [23].

First research on liposomes as drug carriers to CNS dealt with the use of untargeted liposomes in the treatment of brain diseases characterized by an altered BBB permeation with vessel fenestrations (~10 nm), which allowed small liposomes to overcome the BBB. To this purpose pegylated liposomes were successfully tested [24–27].

However, then, it was clear that to target pegylated liposomes specifically to the brain, they needed to be decorated with specific ligands whose receptors are highly expressed on the BBB, such as transferrin (Tf), insulin (I), etc. Coupling a specific ligand to the pegylated liposome surface allows the vesicles to be actively taken up by the target cells and enter the brain by receptor-mediated transcytosis (RMT) mechanism. The Tf and I receptors are the most studied receptors for the targeting of liposomes by RMT, because of their high expression on the BBB [1].

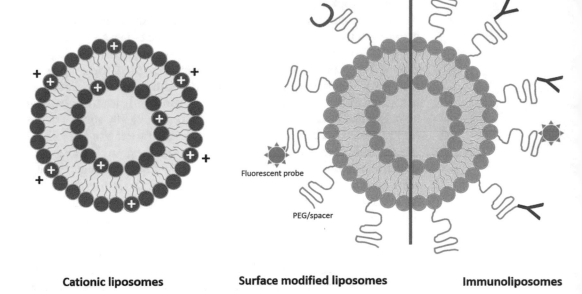

Fig. 1 Schematic representation of different liposomes for drug delivery to the CNS

To achieve an active targeting to the brain, immunoliposomes, i.e., liposomes decorated with an antibody (e.g., OX26 mAb, 83-14 HIR mAb, etc.) on their surface, have also been successfully employed, taking advantage of the high binding of the antibody to the target [28] (Fig. 1).

More recently, cationic liposomes have emerged as a new tool for drug delivery to the brain.

3   Surface-Modified Liposomes

The literature concerning surface-modified liposomes as drug carrier to overcome the BBB encompasses a huge number of research papers. For these reasons, in this chapter we focused only on the most relevant research where the liposomal carriers were designed for a specific brain disease. Therefore, the following discussion was divided based on most common CNS pathologies (for a summary of this research, please see Table 1).

3.1 Cerebral Ischemia

Ischemic stroke is a condition characterized by an insufficient brain blood flow due to occlusion of cerebral vasculature that can cause poor oxygen supply, damaging the neural tissue. During reperfusion, the reoxygenation generates elevated levels of radical oxygen species (ROS) such as superoxide, peroxide, and hydroxyl radicals that can further injure the tissue. Furthermore, cerebral ischemia and reperfusion injury lead to brain vessel fenestrations (about

Table 1
Summary of main results of investigations on liposomes used for brain targeted drug delivery

Liposome composition	Size (nm)	Loaded molecules	Homing devices	Animal model/ species	Activity	References
Cerebral ischemia						
DPPC, DPPS, and cholesterol	49	CDP choline	NI	Rats	CDP choline liposome administration enhanced brain functionality by a rapid recovery of the damaged neuronal cell membranes	[29]
DPPC, DPPS, GMI ganglioside, and cholesterol	50	CDP choline	NI	Rats	CDP choline liposome treatment protected neuronal cells against damage induced by cerebral ischemia	[24]
DPPC, DPPS, GMI ganglioside, and cholesterol	50	CDP choline	NI	Rats	CDP choline liposome treatment reduced cerebral damage after ischemia	[30]
DPPC, DPPS, GMI ganglioside, and cholesterol	50	CDP choline	NI	Rats	CDP choline liposome administration improved the survival rate of rats after ischemia and reperfusion	[25]
DSPC, DSPE-PEG2,000, and cholesterol	100	RhB, Gd, and citicoline	Anti-HSP72 antibody	Rats	Citicoline liposome treatment reduced lesion volumes after ischemia	[31]
DSPC, DSPE-PEG2,000, DSPE-PEG2,000-Mal, and cholesterol	210	Gd and RhB	Anti-ICAM-1 antibody	Mice	ICAM-1-targeted MPIO allowed direct in vivo MRI-based detection after stroke	[32]
PC, DSPE-PEG2,000, DSPE-PEG2,000-Mal, and cholesterol	160	SOD enzyme	Anti-NR1-receptor antibody	Mice	Ligand-targeted liposomes provided protection against ischemia-reperfusion damage and reduced inflammatory markers	[33]

(continued)

Table 1
(continued)

Liposome composition	Size (nm)	Loaded molecules	Homing devices	Animal model/ species	Activity	References
DSPC and DSPE-PEG2,000	NI	DiI or ^{125}I	AEPO	Rats	Pegylated-targeted liposomes provided both brain ischemic zone accumulation and retention for at least 24 h after injection. Liposome administration reduced ischemic cerebral injury and ameliorated motor functions	[34]
DSPC and DSPE-PEG2,000	NI	NI	AEPO	Rats	Pegylated-targeted liposome treatment improved neurological function after ischemia due to neuroprotective action	[35]
POPC, DDAB, and DSPE-PEG2,000	100	VEGF	Tf	Rats	Tf-VEGF liposomes showed neuroprotective activity and vascular regeneration in the chronic stage of cerebral infarction	[36]
Soy lecithin, DSPE-PEG2,000, and cholesterol	74	ZL006	TF peptide	Rats	Ligand-targeted ZL006 liposomes improved neurological function and reduced infarct volume induced by ischemia reperfusion	[37]
PE, cholesterol, and dicetylphosphate	60–90	CDP choline	p-Aminophenyl-α-d-mannoside	Rats	CDP liposomes showed neuroprotection by inhibition of mitochondrial damage in moderate cerebral ischemia reperfusion	[38]
Brain tumors						
HSPC, cholesterol, and DSPE-PEG2000	NI	DOX	NI	Humans	DOX liposome treatment provided long-term stabilization of glioblastoma multiforme	[39]

Composition	Size	Drug	Ligand	Species	Outcome	Ref.
HSPC, cholesterol, and DSPE-PEG2000	NI	DOX and 99mTc-DTPA	NI	Humans	Theranostic liposome treatment provided high concentration of DOX in brain tumors in patients with glioblastoma multiforme and metastatic brain lesions	[40]
DSPC and cholesterol	NI	DNR	NI	Humans	Liposome-encapsulated DNR showed cytotoxicity against human glioma tumors	[41]
DSPC and cholesterol	NI	DNR	NI	Humans	High concentration of DNR was detected in brain tumors after systemic administration of liposomal DNR	[42]
HSPC, cholesterol, and DSPE-PEG2000	80–90 nm	DOX	–	Rats	Long circulating liposomes enhanced brain DOX exposure and improved its therapeutic activity against early small- and large-sized brain tumors	[43]
HSPC, DSPE-PEG2,000, and cholesterol	90	DOX	RGERPPR peptide	Mice	Ligand-targeted DOX liposomes showed an enhanced therapeutic effect on glioblastomas	[44]
L-α-Soya phosphatidyl choline, stearylamine, and cholesterol	990–2100	5FL, 99mTc-DTPA	Tf	Rats	Transferrin-coupled 5FL and 99mTc-DTPA liposomes were selectively uptaken from the brain capillary endothelial cells producing a level of radioactivity 17- and 13-fold higher than the free radioactive agent and non-coupled liposomes, respectively	[45]
DPPC, DSPE-PEG2,000, and cholesterol	100	^{10}B	Tf	Mice	Ligand-targeted liposomes delivered a high concentration of borocaptate sodium into the tumor tissue	[46]

(continued)

Table 1
(continued)

Liposome composition	Size (nm)	Loaded molecules	Homing devices	Animal model/species	Activity	References
HSPC, DSPE-PEG2,000, and cholesterol	100	DOX	CTX	Mice	Ligand-targeted liposomes showed high accumulation into tumors and improved antitumor activity	[47]
SPC, cholesterol, DSPE-PEG2000	<110	Paclitaxel	RGD and R8 cell-penetrating peptide	Mice	R8-RGD tandem enable liposomes to cross the BBB. Moreover, they increased the glioma cellular uptake of liposomes by 2-fold and nearly 30-fold compared to separate R8 and RGD, respectively	[48]
SPC, DSPE-PEG2,000, and cholesterol	130	PTX	TR peptide	Mice	Ligand-targeted PTX liposomes showed targeting ability to cancer stem cells, destroying the vasculogenic mimicry channels	[49]
SPC, DSPE-PEG2,000, and cholesterol	100–120	PTX	Cell-penetrating peptides	Mice	Dual-targeted liposomes exhibited selective targeting and anticancer therapeutic effects	[50]
PC, DSPE-PEG2,000, and cholesterol	123	DNR	MAN and Tf	Rats	Ligand-targeted DNR liposomes improved therapeutic efficacy for gliomas	[51]
EPC, DSPE-PEG2,000, and cholesterol	110	EPI	Tf and TAM	Rats	Evident effect of targeting brain tumor cells in vitro and extended median survival time in brain glioma-bearing rats	[52]
PC, DSPE-PEG2,000, and cholesterol	100–110	Topotecan	TAM and WGA	Rats	Ligand-targeted topotecan liposomes prolonged median survival time of brain tumor-bearing rats	[53]

Composition	Size	Drug	Targeting ligand	Animal	Description	Ref.
DSPC, DSPE-PEG2,000, and cholesterol	180	DOX	Tf and folate	Rats	Ligand-targeted DOX liposomes carried the drug across the BBB and allowed its selected distribution in brain gliomas	[54]
PC, DSPE-PEG2,000, and cholesterol	104	DNR and quinacrine	WGA and TAM	Mice	Ligand-targeted liposomes for the delivery of DNR plus quinacrine exhibited evident capabilities in crossing the BBB, in killing glioma and glioblastoma stem cells, and in diminishing brain gliomas in mice	[55]
DMPC, DOTAP, cholesterol	NI	Dialkyl carbocyanine label DiIC$_{18}$	Intra-arterial injection during transient cerebral hypoperfusion	Rats	Cationic liposomes achieved higher brain concentrations than anionic and neutral ones. CL were observed within glioma, but the retention over time was higher in the peri-tumoral tissue	[56]
HSPC, DOPE, CTAB, DDAB, and cholesterol	187	DOX	Cationic lipids + focused ultrasound	Rats	The combination of cationic liposomes and focused ultrasound increased survival time	[57]
CHETA, phosphatidylcholine	123–129	DOX	Lactoferrin + pro-cationic lipids	Rats	Higher DOX brain uptake as compared to conventional liposomes and prolonged median survival time	[58]
SPC, cholesterol, chol-PEG, chol-PEG-TAT	105	DOX	TAT	Mice	Increased delivery to the brain and survival time as compared to untargeted liposomes and free drug	[59]
Alzheimer's and Parkinson's disease						
POPC, DDAB, DSPE-PEG 2000, and DSPE-PEG 2000-maleimide	85	TH expression plasmid	OX26	Rats	Administration of (TH) expression plasmid OX26 liposomes normalized brain tyrosine hydroxylase enzyme activity. The TH gene expression decayed 50% and 90% at 6 and 9 days after a single i.v. administration	[60]

(continued)

Table 1
(continued)

Liposome composition	Size (nm)	Loaded molecules	Homing devices	Animal model/ species	Activity	References
DPPC, DODAB, and DSPE-PEG2,000	117	GDNF plasmid	OX26	Rats	Sustained therapeutic effects are achieved in experimental PD with the formulation described here	[61]
DSPE-PEG2,000	196	α-Mangostin	Tf	Rats	Targeted liposomes crossed the BBB and delivered α-mangostin into rat brain	[62]
DPPC, cardiolipin [CL]), DSPE-PEG(2000)	90–160	QU	Lf, RMP-7	In vitro model	RMP-7-Lf-QU liposomes were a promising carrier targeting the BBB to prevent Aβ-insulted neurodegeneration. Compared with free QU, RMP-7-Lf-QU liposomes could also significantly inhibit the expression of phosphorylated c-Jun N terminal kinase, phosphorylated p38, and phosphorylated tau protein at serine 202 by SK-N-MC cells	[63]
Cholesterol, DPPC, PE-PEG (2000); DSPE-PEG (2000)	100–155	NGF	Lf	In vitro model	Lf/NGF-liposomes efficiently crossed the BBB and inhibited the degeneration of SK-N-MC cells with Aβ-induced neurotoxicity	[64]
DSPC, cholesterol, DSPE-PEG, DSPE-PEG-Mal	105–121	siRNA	RVG	In vitro model	RVG peptide anionic liposomes were capable of efficiently loading, protecting, and delivering anti-α-syn siRNA to primary cortical and hippocampal cells in vitro	[19]

DOPE, DOTAP, cholesterol, DSPE-PEG-COOH	192	β-galactosidase plasmid	Poly-L-arginine, Tf	Rats	Higher brain accumulation and β-galactosidase expression than transferrin-decorated liposomes after systemic administration	[65]
DOPE, DOTAP, cholesterol, DSPE-PEG-COOH	183–188	DOX	Tf + TAT or Tf + mastoparan or Tf + penetratin	In vitro BBB model, rats	Tf-penetratin liposomes showed the maximum transport to the brain as compared to single targeted and to other combinations	[66]
EPC, cholesterol, DSPE-PEG	139–171	Trientine	Penetratin	Rats	Trientine brain accumulation improved by 16-fold as compared to the free drug	[67]

5FL 5-florouracil, *99mTc* technetium-99m, *AEPO* asialo-erythropoietin, *CL* cardiolipin, 1′,3′-bis[1,2-dimyristoyl-sn-glycero-3-phospho]-sn-glycerol, *CDP* cytidine diphosphate, *CPP* cell-penetrating peptide, *CTAB* cetyltrimethylammonium bromide, *CTX* chlorotoxin, *DCP* dihexadecyl phosphate, *DDAB* dimethyldioctadecylammonium bromide, *DiI* 1,1′-dioctadecyl-3,3,3′,3′-tetramethylindocarbocyanine perchlorate, *DMPC* dimyristoylphosphatidylcholine, *DNR* daunorubicin, *DODAB* dioctadecyldimethylammonium bromide, *DODAP* 1,2-dioleoyl-3-dimethylammonium-propane, *DOPE* 1,2-dioleoyl-sn-glycero-3-phosphoethanolamine, *DOTAP* 1,2-dioleoyl-3-trimethylammonium-propane, *DOX* doxorubicin, *DPPC* dipalmitoylphosphatidylcholine, *DPPS* 1,2-dipalmitoyl-sn-glycero-3-phosphoserine, *DSPC* distearoylphosphatidylcholine, *DSPE* distearoylphosphatidylethanolamine, *DSPE-PEG2000,* 1,2-distearoyl-sn-glycero-3-phosphoethanolamine-N-[carboxy(polyethylene glycol)-2000], *DTPA* diethylenetriaminepentaacetic acid, *EPC* egg phosphatidylcholine, *EPI* epirubicin, *Gd* gadolinium, *HSPC* hydrogenated soy phosphatidylcholine, *iRNA* small interfering RNA, *Lf* lactoferrin, *MAN* P-aminophenyl-α-d-mannopyranoside, *NGF* neuron growth factor, *NI* not informed, *OX26* murine anti-transferrin receptor antibody, *PC* phosphatidylcholine, *PE* phosphatidylethanolamine, *PEG* polyethylene glycol, *POPC* 1-palmitoyl-2-oleoylphosphatidylcholine, *PE-PEG(2000)* 1,2-dipalmitoyl-sn-glycero-3-phosphoethanol-amine-N-[methoxy (polyethylene glycol)-2000], *PTX* paclitaxel, *QU* quercetin, *R8* octa-arginine cell-penetrating peptide, *RhB* rhodamine B, *RMP-7* bradykinin analog, *RVG* rabies virus glycoprotein-derived peptide, *SPC* soybean phospholipids, *TAM* tamoxifen, *TAT* cell-penetrating peptide, *Tf* transferrin, *TH* tyrosine hydroxylase, *VEGF* vascular endothelial growth factor, *WGA* wheat-germ agglutinin, *ZL006* 5-(3, 5-dichloro-2-hydroxybenzylamino)-2-hydroxybenzoic acid

100 nm) caused by a biphasic opening of the BBB [68–71]. Early research on liposomes as drug delivery systems to the brain concerned the use of small liposomes for the delivery of cytidine-5I-diphosphate choline (CDPc) [24, 25, 30]. CDPc is a commercially available therapeutic agent for treatment of stroke. This drug restores the cell membrane phosphatidylcholine, degraded during brain ischemia to free radicals and fatty acids, acting as an intermediate in its biosynthesis from choline. Moreover, CDPc has been shown to restore the activity of mitochondrial ATPase and membrane Na1/K1 ATPase, to inhibit activation of phospholipase A2, and to accelerate reabsorption of cerebral edema in various experimental models [72]. Unfortunately, due to its rapid peripheral hydrolyzation, polar nature, and absence of a specific transport mechanism, CDPc shows a BBB accumulation far lower than desired. It has been calculated that the CDPc brain uptake in rat is 2% when administered i.v. and 0.5% following oral route [30].

CDPc encapsulated in long circulating liposomes improved the survival rate of Wistar rats subjected to ischemia and reperfusion by approximately 66% compared to free CDPc and protected the brain against peroxidative damage caused by post-ischemic reperfusion. To explain these results, the authors suggested a passage of liposomal vesicles through the fenestrations of the BBB caused by the ischemic event [24, 25].

Later, the same authors suggested that the ability of long circulating liposomes to improve the accumulation of non-degraded CDPc into the brain parenchyma is also related to their ability to act as a sustained drug delivery system [29].

Many other research groups have been using long circulating liposomes or liposomes targeted to a specific target upregulated in different brain or cerebral vessels cell damaged after the stroke event, to increase accumulation of different drugs or diagnostic molecules in the ischemic area [31–35]. In all these specific cases, liposomes were not designed to cross the BBB using a specific transport mechanism (RMT, CMT, AMT, etc.), but they accumulated in the brain parenchyma by escaping through the brain vessel fenestrations or by targeting vessel endothelial cells in the ischemic region, thus remaining in the brain vasculature.

On the other hand, some authors have used modified liposomes to allow or to improve the drug transport capacity through the BBB, using a specific transport mechanism.

Zhao et al. used transferrin-coupled liposomes to deliver the vascular endothelial growth factor (VEGF) in a post-ischemic treatment study. They found a decrease in infarct volume and better neurological function in the animal group treated with transferrin-coupled liposomes in comparison with the group treated with saline or non-modified VEGF-loaded liposomes [36].

Wang et al. designed pegylated liposomes for brain delivery of 5-(3, 5-dichloro-2-hydroxybenzylamino)-2-hydroxybenzoic acid (ZL006), a new tested neuroprotectant in ischemic stroke [37]. Because of the ZL006 low permeability across the BBB, liposomes carrying the drug were modified with a peptide (HAIYPRH) able to target the transferrin (Tf) receptor and mediate the transport of the nanocarriers across the BBB by receptor-mediated endocytosis (RMT) [73, 74].

Mannosylated liposomes are able to cross the BBB through an endocytosis mechanism that involves the glucose transporter 1 (GLUT1), which is mainly expressed by both BBB endothelial cells and glioma cells in the brain [75]. Mannosylated liposomes transported CDPc through the BBB and prevented the mitochondrial damage induced by moderate cerebral ischemia-reperfusion to a higher extent than free CDP or non-modified liposomes [38].

3.2 Brain Tumors

Brain cancer accounts for more than 100 different types of tumors, the most common of which are gliomas, representing about the 30% of primary brain and CNS tumors and 80% of all malignant brain tumors.

Brain tumor therapy consists of partial or total surgery followed by radiotherapy and/or chemotherapy. In general, due to the high toxicity of antineoplastic agents, the challenge of tumor chemotherapy is to increase accumulation of the active agent in the tumor cells, thus reducing its distribution in the healthy ones. Anyway, the design of an efficient brain tumor-targeted drug delivery system should consider the distinctive characteristics from peripheral tumors. In particular, the different barriers that prevent the drug from reaching the tumor tissue must be considered. At the early stage of the brain tumor development and around the tumor edge of the infiltrating glioma, the BBB is still intact. However, the tumor growth and infiltration cause BBB impairment and development of new vessels (angiogenesis) characterized by the presence of the blood-brain tumor barrier (BBTB) less permeable than the blood-tumor barrier (BTB) of the peripheral tissue malignant solid tumors. With the deterioration of brain tumor, the abnormality of these new microvessels enhances the permeability of the BBTB and the enhanced permeation and retention (EPR) effect appears, thus allowing the possible liposome accumulation by passive targeting. It is, therefore, clear that according to the different stages of the tumor growth, an unlike strategy should be adopted.

The ability of liposomes to accumulate passively in the brain tumors has been investigated by many research groups in both animal models and clinical studies [39–41, 43]. In general, the authors report both long-term survival and inhibition of tumor growth in treated animals or patients.

A plasma pharmacokinetics study was performed in eight patients with recurrent glioblastoma treated with a non-pegylated liposomal formulation containing daunorubicin. The authors found that the encapsulation into liposomes allowed to achieve potentially cytotoxic drug concentrations in human gliomas associated with a low level of systemic exposure and toxicity [41, 42].

Pegylated liposomes were used to increase the accumulation of doxorubicin (Caelyx®) via EPR effect [39, 40] in two different clinical studies. A 7- to 19-fold higher accumulation of ^{99m}Tc-DTPA-radiolabeled Stealth® liposomal doxorubicin compared with the free drug was found in both glioblastomas and metastatic tumors in 15 patients undergoing radiotherapy [40].

Many studies have been focusing on liposomes for the delivery of antineoplastic or contrast agents in brain cancer therapeutic or diagnostic approaches.

Depending on the type and stage of the brain tumor, the BBB could be still intact. Moreover, because the fenestrations in the vessels of the intracranial tumors are smaller than those of peripheral cancers, the EPR effect of intracranial tumors is relatively weak compared with that of the peripheral tumors [44, 76].

Therefore, modification of the vesicle surface with various ligands, whose receptors are highly expressed on the BBB, is a well-known strategy to target the PEG liposomes specifically to CNS as well as to enhance the accumulation of the carried drug in the brain parenchyma.

Attachment of specific molecules to the liposomal surface-granted PEG chains allows vesicles to be actively taken up by the endothelial target cells, and, therefore, they can enter the brain by an effective transport across the intact BBB.

Because of its high expression on the BBB, transferrin has been used as a homing device to target liposomes to the brain in the glioma chemotherapy.

Soni et al. used Tf-modified liposomes to achieve an enhanced delivery of the anticancer drug 5-fluorouracil to the brain [45]. Results of the in vivo studies suggested a selective uptake of the Tf-modified liposomes from the brain capillary endothelial cells. An average of 17-fold increase in the brain uptake of 5-fluorouracil was observed after administration of the Tf-modified liposomes, while non-modified liposomes caused only a 10-fold increase in the brain uptake compared with the free drug.

Boron neutron capture therapy (BNCT) is a successful treatment of various cancers, which requires the selective delivery of relatively high concentrations of boron-10 (^{10}B) to the tumor tissue. The ability of transferrin-modified PEG liposomes to enhance sodium borocaptate (BSH, a source of ^{10}B) delivery to malignant glial tumor cells was investigated in nude mice transplanted with U87D human glioma cells [46].

Compared with free BSH and PEG liposomes, only Tf-modified PEG liposomes maintained a potent ^{10}B concentration in the tumor, while the ^{10}B concentration in blood and normal brain decreased. The same research group prepared transferrin-modified PEG liposomes encapsulating both sodium borocaptate and iomeprol, an iodine contrast agent, and studied their biodistribution after intratumoral convection-enhanced delivery (CED) in a rat glioma tumor model. The combined use of CED and Tf-PEG liposomes not only enabled the precise and potent boron delivery to the tumor tissue but also allowed to visualize the brain boron distribution by real-time computed tomography [77].

Tf-modified carriers must compete for transferrin receptor (TfR) binding side with the endogenous Tf present in high amount in plasma (2.6 mg/ml), which saturates the BBB transferrin receptor [78].

For this reason, other molecules capable of binding TfR were tested, and monoclonal antibodies (mAb) have been largely studied. Jefferies et al. identified a mouse mAb, OX26, that can react in vitro and in vivo with rat TfR but cannot block Tf binding because it reacts in a binding site different from that of Tf [79, 80].

Huwyler et al. first developed stealth immunoliposomes (PEG immunoliposomes) for brain delivery of radiolabeled daunomycin ([^{3}H] daunomycin) mediated by the binding of OX26 to the rat transferrin receptor [81].

Pharmacokinetics and brain uptake of the [^{3}H] daunomycin-loaded PEG immunoliposomes were compared to those of the drug, conventional liposomes, and PEG liposomes used as controls.

Free [^{3}H] daunomycin showed high BBB permeability, but its brain tissue accumulation was poor (less than 0.01% injected dose/brain gram at 60 min) due to its rapid clearance from circulation. Using conventional liposomes, the [^{3}H] daunomycin AUC increased fourfold compared with free daunomycin, but the permeability greatly decreased resulting in a low brain tissue accumulation similar to that of the free drug. The use of PEG liposomes conferred no advantage in brain drug delivery: indeed, while the plasma AUC was greatly increased, the permeability was reduced to a value of about 0, hence demonstrating that no brain uptake of the PEG liposomes had been achieved.

On the contrary, when OX26 PEG immunoliposomes were used as carriers of [^{3}H]daunomycin, the permeability value increased and AUC moderately decreased, compared to PEG liposomes, leading to the greatest value of brain tissue accumulation (about 0.03% injected dose/brain gram) at 60 min, in comparison to the controls [81, 82].

Daunomycin was also incorporated in OX26 biotinylated immunoliposomes prepared by a non-covalent (biotin-streptavidin) coupling procedure, and its cellular uptake and pharmacological and cytotoxic effects were tested on RBE4 cells and compared to the free drug [83]. Results demonstrated daunomycin uptake increased (two- to threefold) using the biotinylated immunoliposomal carriers [84].

Immunoliposomes conjugated with 83-14 HIR mAb (murine 83-14 mAb to human insulin receptor) have been used in vitro to deliver antisense gene for epidermal growth factor receptor (EGFR) to U87 cells, high-grade human brain glioma cells overexpressing EGFR [85]. By confocal fluorescent microscopy, it was demonstrated that HIR mAb targets the insulin receptor on the U87 plasma membrane, thus enabling the endocytosis of immunoliposomes. Targeting the EGFR antisense gene to U87 glioma cells with the 83-14 PEG immunoliposomes inhibited EGFR expression and cell proliferation, as demonstrated by the reduction in [^{3}H] thymidine cell incorporation. In vivo targeting of U87 was achieved using immunoliposomes conjugated with both mouse mAb 8D3 (able to bind to the BBB TfR) and 83-14 mAb to human insulin receptor. A brain tumor model was realized implanting about 500,000 human U87 glioma cells in the caudate-putamen nucleus of adult female SCID mice. After implantation of glioma cells, the induced tumor was perfused by microvasculature of mouse brain origin expressing mouse transferrin receptors. A 100% increase of tumor-bearing mice lifespan was obtained after intravenous injection of EGFR antisense gene-loaded 8D3/83-14 PEG immunoliposomes [86].

Other research strategies in the glioma therapy provide delivery of anticancer drugs across the BBTB.

Tumor vessels and glioma cells overexpress many kinds of receptors that have been exploited to design liposomes modified with specific ligands.

Chlorotoxin (ClTx) is a scorpion-derived peptide, able to bind to a lipid raft-anchored complex that mainly contained the glioma-specific chloride ion channel and matrix metalloproteinase 2 (MMP-2) endopeptidase.

Xiang et al. evaluated the ability of doxorubicin-loaded ClTx-modified liposomes to target glioma and to increase antitumor activity [47]. Data obtained using a U87 tumor-bearing mouse model demonstrated that ClTx-modified liposomes, in comparison with non-modified liposomes, enhance cellular uptake against murine and human gliomas and brain microvascular endothelial cells. The author indicated that these results are mainly due to the ability of ClTx to bind the specific receptors on the glioma cells or glioma neovascular endothelial cells promoting the cellular uptake via a receptor-mediated endocytosis mechanism.

Tumor-penetrating peptide (RGERPPR) is a specific ligand of neuropilin-1, a transmembrane glycoprotein, that acts as a co-receptor for a number of extracellular ligands and that is overexpressed on glioblastoma and endothelial cells in tumor vessels [87].

RGERPPR peptide-functionalized liposomes have been designed to enhance the antitumor effect of doxorubicin after i.v. administration in a glioblastoma-bearing nude mice model. The results indicated that RGERPPR functionalization enhances the transport of liposomes through the tumor vessels into the glioblastoma tissue. Moreover, functionalized liposomes, compared with unmodified liposomes and free doxorubicin, significantly prolonged the survival time of nude mice bearing intracranial glioblastoma. In particular, this result seems to be related to the ability of RGERPPR peptide to enhance liposome penetration from tumor vessel through the tumor stroma into the deep glioblastoma tissue [44].

Shi et al. designed vesicles modified with a multifunctional peptide TR, a tandem peptide consisting of cRGD and TH peptide. cRGD is a cyclic Arg-Gly-Asp peptide able to selectively target integrin $\alpha v \beta 3$ overexpressed on the endothelial cells as well as on glioma cells [48, 49]. TH is a histidine-rich peptide that can help a nanocarrier to escape efficiently from lysosomes in the endothelial cells and penetrate deeply in the glioma.

In an intracranial glioma-bearing mice model, paclitaxel-loaded TR-modified liposomes showed enhanced therapeutic efficacy compared to free paclitaxel, paclitaxel-loaded PEG liposomes, paclitaxel-loaded TH liposomes, and paclitaxel-loaded cRGD liposomes. The experimental data showed the enhanced ability of TR-modified liposomes to cross BBB and to selectively accumulate in the glioma, even in the deep glioma spheroids [49]. Moreover, mice treated with paclitaxel-loaded TR-modified liposomes showed a median survival time (45 days) longer than those treated with free paclitaxel (25.5 days), paclitaxel-PEG-liposomes (30.5 days), paclitaxel-RGD-liposomes (38.5 days), and paclitaxel-TH-liposomes (27 days).

Aiming at increasing nanocarrier penetration into the BBB and, contemporary, their selective accumulation in the glioma cells, different research groups are involved in the development of the so-called dual-targeting liposomes. This more sophisticated delivery system consists of drug-loaded liposomes modified with two different homing devices targeting to a receptor in the BBB and to a receptor on the brain cancer cells. The challenge is to create a nanocarrier not only able to cross the BBB but also capable of targeting and crossing the tumor cell membranes via a specific mechanism of transport [50–55].

The anthracycline antibiotic daunorubicin was encapsulated in liposomes whose surface was modified with p-aminophenyl-α-D-manno-pyranoside (MAN) and transferrin (Tf) [51]. MAN showed a specific affinity to the glucose transporter GLUT1 that mediates transport of glucose-like substances via a carrier-mediated transcytosis mechanism [88]. Moreover, previously, another study had shown that Tf-modified liposomes are highly internalized in tumor cells overexpressing the TF receptor [89]. In vitro studies demonstrated the ability of MAN-Tf liposomes to cross the BBB and then to target brain glioma cells. Furthermore, dual-targeting liposomes improved therapeutic efficacy in brain glioma-bearing male Sprague Dawley rats, in terms of tumor volume reduction and improvement in median survival time, in comparison with free daunorubicin, daunorubicin liposomes, daunorubicin liposomes modified with MAN, and daunorubicin liposomes modified with Tf.

In the same way, the potential dual-targeting effects of doxorubincin-loaded PEG liposomes modified with folate (F) and transferrin (Tf) were studied using a C6 cells-bearing rat model [54]. The use of folate as homing device is justified because of its overexpression on the tumor cell surface [90]. Doxorubicin-modified F-Tf-modified PEG liposomes demonstrated an improved transport through the BBB as well as glioma uptake. Compared with the free doxorubicin and doxorubicin-loaded liposomes, the dual-targeting liposomes greatly increased the survival of brain tumor-bearing animals and demonstrated less toxicity.

3.3 Alzheimer's Disease

Alzheimer (AD) or dementia is a chronic neurodegenerative disease that affects the functioning of the central nervous system. The main symptoms of AD involve loss of memory, behavioral disturbance, and language problems. Although the exact cause of AD is not yet identified, aging, genetic mutation, and family background are considered important factors responsible for the dementia [91, 92].

The neuropathogenesis of AD involves the accumulation of insoluble protein with formation of neurofibrillary tangles of hyperphosphorylated tau protein and β-amyloid plaques leading to brain atrophy and neurodegeneration [92]. AD is also associated with defective BBB function due to disruption of tight and adherens junctions, an increase in bulk-flow fluid transcytosis, and/or enzymatic degradation of the capillary basement membrane [93].

Surface-modified liposomes have been used to target different drugs used in the therapy of Alzheimer's disease (acetylcholinesterase inhibitors, antioxidant, etc.). The aim of these strategies was to reduce some severe side effects associated with accumulation of these drugs in peripheral tissues and/or to enable the accumulation in brain parenchyma for those active molecules with a low BBB permeability [62–64].

Tf-modified liposomes were used to improve the brain accumulation of polyphenolic xanthone α-mangostin (α-M) [62]. α-M demonstrated a good anticholinesterase effect and a high ability to inhibit Aβ aggregation [94]. However, its accumulation in the CNS is limited because of its poor penetration through the BBB. Pharmacokinetics studies in rats demonstrated the ability of Tf-modified liposomes to enhance the accumulation of α-M in the brain and to reduce its distribution in the heart, compared to α-M solution and unmodified liposomes.

Kuo et al. prepared Tf-modified liposomes to enhance the delivery of quercetin and neuron growth factor (NGF) across the blood-brain barrier (BBB) in vitro [63, 64]. In particular, the authors demonstrated that the presence of Lf on the NGF-loaded liposome surface determines an increased ability to cross the BBB, leading to an enhanced inhibition of the degeneration of SK-N-MC cells after an insult with cytotoxic β-amyloid (Aβ) fibrils [64].

3.4 Parkinson's Disease

Parkinson's disease (PD) is a long-term neurodegenerative disorder characterized by motor symptoms, which affects four million people worldwide.

In particular, PD patients show a neurodegeneration in the nigral-striatal pathway of the brain leading to a reduction of the striatum tyrosine hydroxylase (TH) activity.

OX26 PEG immunoliposomes carrying TH expression plasmid were used to reversibly restore tyrosine hydroxylase (TH) activity in the striatum of adult rats, in 6-hydroxydopamine-induced PD model [60].

TH expression plasmid-loaded OX26 PEG immunoliposomes were injected intravenously, and the TH enzyme activity or immunoreactive TH levels were measured.

As demonstrated by confocal microscopy of fluoresceinated DNA, OX26 mAb first triggers receptor-mediated transcytosis (RMT) of the immunoliposomes through the BBB in vivo. Then, since TfR is widely expressed on neurons throughout the CNS [95], OX26 PEG immunoliposomes can enter into neurons using a second RMT. Finally, the fusogenic lipids of the immunoliposomes allow the plasmid DNA to be released and transported into the nuclear compartment for transgenic expression. The TH enzyme activity is normalized both in the dorsal striatum ipsilateral to the 6-hydroxydopamine injection and in the contralateral striatum when the plasmid is encapsulated in OX26 immunoliposomes. When TH expression plasmid was encapsulated in immunoliposomes coupled to a non-specific TfR antibody (IgG2a), no increase in TH enzyme activity in the ipsilateral striatum was observed, showing that the presence of OX26 murine monoclonal antibody (mAb) to the rat transferrin receptor is crucial to permit TH activity normalization [96].

When administered in the early stages of the PD disease, glial cell-derived neurotrophic factor (GDNF) is able to prevent the progression of PD by inducing survival, growth, and regeneration of substantia nigra dopamine neurons. Xia et al. intravenously administrated iOX26-targeted PEGylated liposomes to deliver GDNF plasmid into the CNS [61].

The expression of GDNF genes, under the influence of a rat tyrosine hydroxylase promoter, was observed in the substantia nigra, adrenal gland, and liver, where the TH gene is highly expressed. Moreover, prolonged therapeutic delivery was achieved at the neurons of the nigrostriatal tract in experimental PD.

More recently, knockdown of alpha-synuclein through non-viral delivery of short interfering RNA (siRNA) has been explored as a potential therapeutic strategy for PD. To this aim, in addition to other nanocarriers (exosomes, peptides, magnetic nanoparticles), rabies virus glycoprotein-targeted anionic liposomes showed high transfection efficacy in primary neurons in vitro and prolonged serum stability, making them promising candidates for in vivo testing [19].

4 Cationic Liposomes

The electrostatic charge on the surface of liposomes plays a crucial role in the fate of these nanosystems after systemic injection. Indeed, it is well-known that cationic nanoparticles can interact with negatively charged entities in the bloodstream, such as plasma proteins, immunoglobulins, or endothelial cells, leading to a broad spectrum of different consequences [97]. For example, the adsorption of plasma proteins on the surface of cationic liposomes will ultimately lead to their recognition by the mononuclear phagocyte system and removal from the bloodstream though phagocytosis [92]. Thus, to allow a sufficient plasmatic concentration of the nanocarriers, high doses would be required, increasing the risk of toxicity. On the other hand, the electrostatic interaction between a positively charged nanoparticle and the anionic glycoproteins and proteoglycans on the surface of endothelial cells would result in the close contact of the two, favoring the clathrin-mediated endocytosis of the nanoparticle [98]. This beneficial interaction could also take place at the luminal side of brain capillary endothelial cells, which show a high density of clathrin-rich regions characterized by a negative charge, thus representing a possible strategy for brain delivery through cationic nanocarriers.

It is well-known that cationic proteins such as protamine, as well as neutral proteins modified with basic amino acids, can cross the BBB and distribute in the brain parenchyma through a mechanism called adsorptive-mediated transcytosis (AMT). Despite being not completely elucidated yet, it has been reported that the AMT process does not involve a specific receptor-ligand recognition, but rather an interaction between opposite electrostatic charges [99]. This pathway has been exploited for brain delivery of positively charged liposomes, which demonstrated to accumulate in the central nervous system to a higher extent than untargeted neutral or anionic vesicles [56]. AMT is also the mechanism hypothesized for BBB penetration of cell-penetrating peptides (CPPs), a heterogeneous group of cationic peptides which share an amphipathic character and the ability to change their secondary structure upon association with membrane lipids [100]. In the following section, recent results from the application of untargeted cationic liposomes as well as liposomes decorated with CPP for brain delivery will be reviewed, devoting special attention to in vivo results. The "cationic strategy" has been explored to a much lower extent as compared to the surface-modified liposomes, discussed in the previous section. Thus, the following discussion will not be divided basing on the treated pathology but rather on the agent (lipid component or CPP) bearing the cationic charge.

4.1 Untargeted Cationic Liposomes

The surface charge of liposomes can be fine-tuned by the inclusion of a lipid component bearing a positive charge, most often induced by an amino or ammonium group located in the hydrophilic domain of the molecule. The other domain of a cationic lipid typically has a hydrophobic character and can be either composed of one or more alkyl chains or of a sterol moiety, linked to the cationic polar head by a short spacer. Cationic lipids can be further classified, basing on the number of ionizable groups, in monovalent and multivalent lipids. Several examples of cationic lipids are commercially available, and many more have been synthesized and mainly tested as transfection reagents for siRNA and DNA due to their ability to complex nucleic acids and enhance their cell uptake [101]. Accordingly, the only structure-activity relationship data available refer to the transfection efficiency in vitro and in vivo on several cell types and different tissues (especially tumors), not taking into account the possible brain targeting skills of the novel cationic lipids, which therefore have an unexplored potential to be unveiled.

The cationic lipids most often employed for the formulation of brain-directed liposomes are dioleoyl-trimethylammonium-propane (DOTAP), didodecyl-dimethylammonium bromide (DDAB), dioctadecenyl-trimethylammonium propane (DOTMA), and stearylamine (SA) as well as cholesterol derivatives modified with an ionizable group.

Joshi et al. evaluated the differential brain retention of anionic, neutral, and cationic liposomes after intra-arterial (IA) administration during transient cerebral hypoperfusion (TCH) [102]. IA injections allow skipping the liver first-pass effects, thus conveying larger amounts of the drug to the target. Moreover, the administration during TCH reduces liposome contact with blood proteins, thereby increasing the interactions with the negatively charged endothelial cells and boosting the brain uptake of approximately three times [103]. As expected, cationic vesicles showed a 3- to 15-fold increase in the brain accumulation parameters (peak concentration, end concentration, AUC) as compared to neutral or anionic liposomes. In addition, the authors describe a better performance of cationic liposomes both in the case of an intact BBB and when the same was opened by focused ultrasound.

In a subsequent study, increasing amounts of DOTAP have been employed to produce cationic liposomes bearing different charge densities, and their brain accumulation after IA administration in rats was evaluated [56]. As expected, a significant difference in the brain accumulation of liposomes containing 5% DOTAP, as compared to vesicles composed of 25% DOTAP, could be evidenced. However, a further increase in cationic lipid (up to 50%) showed a higher peak plasma concentration coupled with the most rapid clearance, thus providing a final concentration not significantly different from that of 25% DOTAP liposomes. This result clearly shows the dual effect of positively charged vesicles when administered in vivo, raising the need of an extensive formulation study to find the optimal balance between brain tissue accumulation and quick plasma clearance.

A combined strategy for brain delivery was also employed by Lin et al. In their work, cationic liposomes (prepared using 9.5% DDAB) loaded with doxorubicin were administered by intravenous injection in rats during transient BBB opening induced by focused ultrasound (FUS) [57]. The association of the physical BBB disruption and the use of cationic vesicles produced an increase in median survival time in glioma-bearing rats, as compared to the liposomes treatment alone, which anyway proved a more positive outcome than the free, unencapsulated drug.

In all these works, the "cationic strategy" was coupled either with an efficient administration route (IA) and the induction of transient cerebral hypoperfusion or with the temporary BBB disruption by focused ultrasound, providing successful pharmacokinetic and clinical outcomes in vivo. However, the increased brain uptake and retention provided by these methods should be carefully balanced against the potential risk of stroke (in the case of TCH) and all the risk connected to the loss of the barrier function, such as local inflammatory reactions and hemorrhages.

In the attempt to attenuate the undesired effects of positively charged vesicles, Chen et al. designed novel pro-cationic liposomes, which possess a neutral charge at physiological pH but turn into cationic entities following an enzymatic reaction occurring on the surface of endothelial cells [58]. This enzyme-responsive formulation could be obtained by the incorporation of a carbamate derivative of cholesterol (CHETA) and the decoration with lactoferrin to promote the close contact with the cell membrane. A CHETA molecule comprises a hydrophobic region (cholesterol residue) and a hydrophilic region composed of a cationic and an anionic site, separated by a disulfide bond. The disulfide bond can be reduced by disulfide isomerase or thioredoxin reductase expressed on the surface of brain endothelial cells, thus removing the anionic site and leaving a net positive charge on CHETA and, as a consequence, on the liposome [104]. Pro-cationic liposomes loaded with doxorubicin enhanced the drug accumulation in the brain as compared to conventional liposomes and significantly prolonged median survival time in glioma-bearing rats. Interestingly enough, both receptor- and adsorptive- mediated transcytosis were found to be involved in the brain uptake of these vesicular systems.

Overall, due to the variety of biological consequences mediated by a positive surface charge, it is hard to imagine a clinical translation of simple cationic liposomes, administered i.v. for brain delivery of actives [105]. On the other hand, in the attempt of limiting the noxious effects of positive charge while preserving its benefits, more sophisticated formulations, the combination with physical methods, and alternative administration routes (e.g., intra-arterial, intranasal, intracranial) have been explored, showing promising results on relevant animal models and in early-stage clinical trials [106, 107]. However, since positive charge is only one of several attributes regulating nanoparticle toxicity, a better clarification of the mechanisms involved in cationic charge-induced damages is urgently required to guide the design of novel surface modifications and formulation strategies [108].

4.2 CPP-Decorated Liposomes

The term cell-penetrating peptides (CPPs) is used to refer to a group of short (typically 5–30 amino acid residues), cationic and/or amphipathic peptides, possessing an intrinsic ability to cross the cell membrane and mediate uptake of a wide range of cargoes [100]. The positive charge of these peptides is induced by sequences consisting of arginine, lysine, and histidine, and it has been identified as an essential feature for the mechanism governing their cell entry [109]. Indeed, no specific receptor has been found to be involved in CPP transport across plasma membranes. However, different pathways such as macropinocytosis, clathrin-dependent or clathrin-independent endocytosis, heparan sulfate proteoglycan (HSPG)- mediated endocytosis, as well as direct penetration have been found to be involved depending on the CPP sequence, the transported cargo, and the cell membrane composition [110].

Similar to cationic vesicles, CPP-decorated liposomes have been employed to transport drugs and macromolecules across the BBB, exploiting the AMT pathway [99]. CPPs can be exposed on the liposome surface after being linked to a phospholipid polar head or to cholesterol, either with or without the interposition of a hydrophilic linker such as PEG.

A number of CPPs have been discovered or rationally synthesized and used for molecular delivery to mammalian cells. Among them, the mostly used for brain targeting are TAT, SynB3, penetratin, poly-L-arginine, transportan, and mastoparan [111]. The potential for BBB crossing mediated by these CPPs is reported by several studies which, however, lack homogeneity since many different cargoes and various in vitro/in vivo assays are employed [111]. Accordingly, high variability in CPP-mediated brain accumulation can be found depending on the transported molecule or nanoparticle. Thus, in the following paragraph, only the literature concerning the brain delivery of CPP-decorated liposomes will be presented.

The brain targeting properties of TAT peptide (AYGRKKRRQRRR)-modified liposomes were reported by Qin et al. In two consecutive papers, the authors describe the ability of TAT-modified liposomes to cross the BBB and distribute in the brain after systemic injection. In the first work, the authors investigate the mechanism underlying the penetration of liposomes through an in vitro model of BBB by using inhibitors of the different pathways, demonstrating the involvement of an energy-dependent mechanism. Moreover, the presence of TAT on the vesicles turned out to increase the delivery to several brain regions after 24 h from the systemic injection but also to induce the accumulation of liposomes in the liver and kidneys [112]. In a subsequent research, the same nanocarrier was exploited to deliver doxorubicin to an orthotopic xenograft mouse model of glioblastoma [59]. The TAT-modified liposomes increased the accumulation of doxorubicin in the brain while decreasing the drug delivered to the heart as compared to untargeted vesicles and the free drug. This is particularly relevant because of the well-known cardiotoxicity of doxorubicin. In addition, the targeted formulation produced an improvement in median survival time of the tumor-bearing mice when compared to controls. Some preliminary structure-activity relationships of TAT peptides for brain delivery of liposomes were defined by the same authors [113]. In more detail, the positively charged residues (lysines and arginines) were found to be essential for the uptake of TAT-modified liposomes by endothelial cells and their distribution in the brain parenchyma, while the randomization of the amino acid sequence had little or no effect on the activity. Moreover, the absence of a specific receptor for the internalization of TATs was confirmed by the observation of a similar uptake of the different peptides in various organs (heart, spleen, liver, and lung).

In the attempt to increase the brain specificity of CPP-decorated liposomes, Sharma et al. used a combination of poly-L-arginine and transferrin (Tf) for liposomal brain gene delivery [65]. The dual-targeted formulation showed a higher brain accumulation and β-galactosidase expression than transferrin-decorated liposomes after systemic administration in rats. Moreover, the authors indicated that the anionic charge of transferrin could attenuate the positive charge on the liposomal surface, thus increasing the stability in serum and reducing the hemolytic potential of the formulation. The research was then expanded to study the combination of other CPPs (TAT, mastoparan, and penetratin) with transferrin for liposome decoration and increased brain delivery [66]. At first, the authors set up a novel in vitro blood-brain tumor barrier model to test the dual-targeted formulations, all of which showed an increased ability to cross the endothelial layer as compared to the single-ligand liposomes. Then, the biodistribution after a single injection in rats was studied, showing some interesting differences among the CPP. Among the dual-targeted formulations, the Tf-penetratin liposomes showed the maximum transport to the brain, while the administration of Tf-mastoparan resulted in lower brain accumulation. The authors suggested that the more hydrophobic character of mastoparan could have a role in the higher hemolytic activity toward red blood cells. Thus, as a consequence of the interaction with erythrocytes, mastoparan liposomes showed a higher accumulation in lungs and liver, thus reducing the fraction available for brain accumulation.

A dual-targeted approach was also exploited by Zong et al., who included a peptidic ligand for the transferrin receptor (T7) and a cell-penetrating peptide (TAT) on their liposomal formulations [114]. The added value of using the short T7 peptide rather than the transferrin is the lack of competition with endogenous transferrin for the receptor binding, as the two have distinct binding sites. T7-TAT liposomes loaded with doxorubicin were successful in increasing the median survival time of C6 glioma-bearing mice as compared to the single-targeted formulation or to the PEGylated liposomes. Moreover, the presence of transferrin receptors on tumor cells allowed the deeper penetration of the dual-targeted formulation within tumor spheroids in vitro.

Enhanced drug delivery to brain cancer was the main task for Liu et al., who conjugated the cyclic RGD peptide (an $\alpha v \beta 3$ integrin ligand for tumor targeting) to an octa-arginine (cell-penetrating peptide) to create a novel entity (R8-RGD) possessing both the ability to cross the BBB and penetrate in the tumor mass [48]. The R8-RGD-modified liposomes showed an increase in cellular uptake of 2-fold and nearly 30-fold compared to separate R8 and RGD, respectively. Moreover, the conjugate peptide induced a redistribution of liposomes from the liver to the brain when compared to the R8 modification, which promotes unspecific interactions due to its highly cationic charge. Finally, paclitaxel-loaded R8-RGD liposomes successfully prolonged the survival of intracranial glioma-bearing mice.

The use of cell penetration peptides for liposome brain delivery is not limited to the treatment of CNS malignancies, as witnessed by a recent paper where penetratin liposome were employed to deliver the copper-chelating trientine (TETA) as a potential treatment for Wilson's disease [67]. The authors produced also a control formulation composed of cationized albumin-decorated liposomes, which is known to improve the interactions of vesicles with the brain endothelium and, consequently, the brain uptake.

Interestingly, despite the higher positive charge, the albumin-modified liposomes were much less efficient than penetratin in inducing the brain accumulation of TETA, which was improved by 15-fold as compared to the free drug in the case of the CPP, producing a therapeutically relevant concentration.

Overall, CPP-decorated liposomes are able to increase the brain delivery of actives but also the accumulation in other organs. To attenuate this effect, it has been shown that the addition of a BBB receptor-specific ligand or antibody on the liposome surface could improve the brain specificity of the system. In the case of anionic proteins such as transferrin, this approach would give the additional benefit of reducing the cationic charge, which was often linked with toxicity and unspecificity. Finally, basing on the discussed literature, some structure-activity relationship can be drawn for CPP:

1. Changing the amino acid order in the CPP sequence has no effect on the brain penetration.

2. Cationic residues play a key role in cell uptake.

3. Increasing the hydrophobic character may lead to hemolysis and toxicity.

References

1. Lai F, Fadda AM, Sinico C (2013) Liposomes for brain delivery. Expert Opin Drug Deliv 10:1003–1022. https://doi.org/10.1517/17425247.2013.766714

2. Abbott NJ, Rönnbäck L, Hansson E (2006) Astrocyte–endothelial interactions at the blood–brain barrier. Nat Rev Neurosci 7:41–53. https://doi.org/10.1038/nrn1824

3. Pardridge WM (2006) Molecular Trojan horses for blood-brain barrier drug delivery. Curr Opin Pharmacol 6:494–500. https://doi.org/10.1016/j.coph.2006.06.001

4. Skaper S (2008) The biology of neurotrophins, signalling pathways, and functional peptide mimetics of neurotrophins and their receptors. CNS Neurol Disord Drug Targets 7:46–62. https://doi.org/10.2174/187152708783885174

5. Deli MA (2009) Potential use of tight junction modulators to reversibly open membranous barriers and improve drug delivery. Biochim Biophys Acta 1788:892–910. https://doi.org/10.1016/j.bbamem.2008.09.016

6. Garcia-Garcia E, Andrieux K, Gil S, Couvreur P (2005) Colloidal carriers and blood-brain barrier (BBB) translocation: a way to deliver drugs to the brain? Int J Pharm 298:274–292. https://doi.org/10.1016/j.ijpharm.2005.03.031

7. Neuwelt EA, Maravilla KR, Frenkel EP et al (1979) Osmotic blood-brain barrier disruption. Computerized tomographic monitoring of chemotherapeutic agent delivery. J Clin Invest 64:684–688. https://doi.org/10.1172/JCI109509

8. Bobo RH, Laske DW, Akbasak A et al (1994) Convection-enhanced delivery of macromolecules in the brain. Proc Natl Acad Sci U S A 91:2076–2080

9. Chauhan NB (2002) Trafficking of intracerebroventricularly injected antisense oligonucleotides in the mouse brain. Antisense Nucleic Acid Drug Dev 12:353–357. https://doi.org/10.1089/108729002761381320

10. Chamberlain MC, Kormanik PA, Barba D (1997) Complications associated with intraventricular chemotherapy in patients with leptomeningeal metastases. J Neurosurg 87:694–699. https://doi.org/10.3171/jns.1997.87.5.0694

11. Guerin C, Olivi A, Weingart JD et al (2004) Recent advances in brain tumor therapy: local intracerebral drug delivery by polymers. Investig New Drugs 22:27–37

12. Wang PP, Frazier J, Brem H (2002) Local drug delivery to the brain. Adv Drug Deliv Rev 54:987–1013

13. Gabathuler R (2010) Approaches to transport therapeutic drugs across the blood-brain barrier to treat brain diseases. Neurobiol Dis 37:48–57. https://doi.org/10.1016/j.nbd.2009.07.028

14. Wong HL, Wu XY, Bendayan R (2012) Nanotechnological advances for the delivery of CNS therapeutics. Adv Drug Deliv Rev 64:686–700. https://doi.org/10.1016/j.addr.2011.10.007

15. Invernici G, Cristini S, Alessandri G et al (2011) Nanotechnology advances in brain tumors: the state of the art. Recent Pat Anticancer Drug Discov 6:58–69

16. Yang H (2010) Nanoparticle-mediated brain-specific drug delivery, imaging, and diagnosis. Pharm Res 27:1759–1771. https://doi.org/10.1007/s11095-010-0141-7

17. Barbu E, Molnàr E, Tsibouklis J, Górecki DC (2009) The potential for nanoparticle-based drug delivery to the brain: overcoming the blood-brain barrier. Expert Opin Drug Deliv 6:553–565. https://doi.org/10.1517/17425240902939143

18. Krishnaiah YSR (2010) Pharmaceutical technologies for enhancing oral bioavailability of poorly soluble drugs. J Bioequivalence Bioavailab 2:28–36. https://doi.org/10.1002/med.20201

19. Schlich M, Longhena F, Faustini G et al (2017) Anionic liposomes for small interfering ribonucleic acid (siRNA) delivery to primary neuronal cells: evaluation of alpha-synuclein knockdown efficacy. Nano Res 10:3496–3508. https://doi.org/10.1007/s12274-017-1561-z

20. Bangham AD, Horne RW (1964) Negative staining of phospholipids and their structural modification by surface-active agents as observed in the electron microscope. J Mol Biol 8:660–IN10. https://doi.org/10.1016/S0022-2836(64)80115-7

21. Béduneau A, Saulnier P, Benoit JP (2007) Active targeting of brain tumors using nanocarriers. Biomaterials 28:4947–4967. https://doi.org/10.1016/j.biomaterials.2007.06.011

22. Laquintana V, Trapani A, Denora N et al (2009) New strategies to deliver anticancer drugs to brain tumors. Expert Opin Drug Deliv 6:1017–1032. https://doi.org/10.1517/17425240903167942

23. Bulbake U, Doppalapudi S, Kommineni N, Khan W (2017) Liposomal formulations in clinical use: an updated review. Pharmaceutics 9:12. https://doi.org/10.3390/pharmaceutics9020012

24. Fresta M, Wehrli E, Puglisi G (1995) Enhanced therapeutic effect of cytidine-5-'-diphosphate choline when associated with GM1 containing small liposomes as demonstrated in a rat ischemia model. Pharm Res 12:1769–1774

25. Fresta M (1996) Biological effects of CDP-choline loaded long circulating liposomes on rat cerebral post-ischemic reperfusion. Int J Pharm 134:89–97. https://doi.org/10.1016/0378-5173(95)04448-5

26. Craparo EF, Bondì ML, Pitarresi G, Cavallaro G (2011) Nanoparticulate systems for drug delivery and targeting to the central nervous system. CNS Neurosci Ther 17:670–677. https://doi.org/10.1111/j.1755-5949.2010.00199.x

27. Schmidt J, Metselaar JM, Wauben MHM et al (2003) Drug targeting by long-circulating liposomal glucocorticosteroids increases therapeutic efficacy in a model of multiple sclerosis. Brain 126:1895–1904. https://doi.org/10.1093/brain/awg176

28. De Luca MA, Lai F, Corrias F et al (2015) Lactoferrin- and antitransferrin-modified liposomes for brain targeting of the NK3 receptor agonist senktide: preparation and in vivo evaluation. Int J Pharm 479:129–137. https://doi.org/10.1016/j.ijpharm.2014.12.057

29. Fresta M, Puglisi G (1999) Reduction of maturation phenomenon in cerebral ischemia with CDP-choline-loaded liposomes. Pharm Res 16:1843–1849

30. Muralikrishna Adibhatla R, Hatcher JF, Tureyen K (2005) CDP-choline liposomes provide significant reduction in infarction over free CDP-choline in stroke. Brain Res 1058:193–197. https://doi.org/10.1016/j.brainres.2005.07.067

31. Agulla J, Brea D, Campos F et al (2014) In vivo theranostics at the peri-infarct region in cerebral ischemia. Theranostics 4:90–105. https://doi.org/10.7150/thno.7088

32. Deddens LH, van Tilborg GAF, van der Toorn A et al (2013) MRI of ICAM-1 upregulation after stroke: the importance of choosing the appropriate target-specific particulate contrast agent. Mol Imaging Biol 15:411–422. https://doi.org/10.1007/s11307-013-0617-z

33. Yun X, Maximov VD, Yu J et al (2013) Nanoparticles for targeted delivery of antioxidant enzymes to the brain after cerebral ischemia

and reperfusion injury. J Cereb Blood Flow Metab 33:583–592. https://doi.org/10.1038/jcbfm.2012.209

34. Ishii T, Asai T, Oyama D et al (2012) Amelioration of cerebral ischemia–reperfusion injury based on liposomal drug delivery system with asialo-erythropoietin. J Control Release 160:81–87. https://doi.org/10.1016/j.jconrel.2012.02.004

35. Ishii T, Asai T, Fukuta T et al (2012) A single injection of liposomal asialo-erythropoietin improves motor function deficit caused by cerebral ischemia/reperfusion. Int J Pharm 439:269–274. https://doi.org/10.1016/j.ijpharm.2012.09.026

36. Zhao H, Bao X, Wang R et al (2011) Post-acute ischemia vascular endothelial growth factor transfer by transferrin-targeted liposomes attenuates ischemic brain injury after experimental stroke in rats. Hum Gene Ther 22:207–215. https://doi.org/10.1089/hum.2010.111

37. Wang Z, Zhao Y, Jiang Y et al (2015) Enhanced anti-ischemic stroke of ZL006 by T7-conjugated PEGylated liposomes drug delivery system. Sci Rep 5:12651. https://doi.org/10.1038/srep12651

38. Ghosh S, Das N, Mandal AK et al (2010) Mannosylated liposomal cytidine 5′ diphosphocholine prevent age related global moderate cerebral ischemia reperfusion induced mitochondrial cytochrome c release in aged rat brain. Neuroscience 171:1287–1299. https://doi.org/10.1016/j.neuroscience.2010.09.049

39. Fabel K, Dietrich J, Hau P et al (2001) Long-term stabilization in patients with malignant glioma after treatment with liposomal doxorubicin. Cancer 92:1936–1942

40. Koukourakis MI, Koukouraki S, Fezoulidis I et al (2000) High intratumoural accumulation of stealth liposomal doxorubicin (Caelyx) in glioblastomas and in metastatic brain tumours. Br J Cancer 83:1281–1286. https://doi.org/10.1054/bjoc.2000.1459

41. Zucchetti M, Boiardi A, Silvani A et al (1999) Distribution of daunorubicin and daunorubicinol in human glioma tumors after administration of liposomal daunorubicin. Cancer Chemother Pharmacol 44:173–176. https://doi.org/10.1007/s002800050964

42. Albrecht KW, de Witt Hamer PC, Leenstra S et al (2001) High concentration of daunorubicin and daunorubicinol in human malignant astrocytomas after systemic administration of liposomal daunorubicin. J Neuro-Oncol 53:267–271

43. Siegal T, Horowitz A, Gabizon A (1995) Doxorubicin encapsulated in sterically stabilized liposomes for the treatment of a brain tumor model: biodistribution and therapeutic efficacy. J Neurosurg 83:1029–1037. https://doi.org/10.3171/jns.1995.83.6.1029

44. Yang Y, Yan Z, Wei D et al (2013) Tumor-penetrating peptide functionalization enhances the anti-glioblastoma effect of doxorubicin liposomes. Nanotechnology 24:405101. https://doi.org/10.1088/0957-4484/24/40/405101

45. Soni V, Kohli DV, Jain SK (2005) Transferrin coupled liposomes as drug delivery carriers for brain targeting of 5-florouracil. J Drug Target 13:245–250. https://doi.org/10.1080/10611860500107401

46. Doi A, Kawabata S, Iida K et al (2008) Tumor-specific targeting of sodium borocaptate (BSH) to malignant glioma by transferrin-PEG liposomes: a modality for boron neutron capture therapy. J Neuro-Oncol 87:287–294. https://doi.org/10.1007/s11060-008-9522-8

47. Xiang Y, Liang L, Wang X et al (2011) Chloride channel-mediated brain glioma targeting of chlorotoxin-modified doxorubicine-loaded liposomes. J Control Release 152:402–410. https://doi.org/10.1016/j.jconrel.2011.03.014

48. Liu Y, Ran R, Chen J et al (2014) Paclitaxel loaded liposomes decorated with a multifunctional tandem peptide for glioma targeting. Biomaterials 35:4835–4847. https://doi.org/10.1016/j.biomaterials.2014.02.031

49. Shi K, Long Y, Xu C et al (2015) Liposomes combined an integrin $\alpha_v\beta_3$-specific vector with pH-responsible cell-penetrating property for highly effective antiglioma therapy through the blood–brain barrier. ACS Appl Mater Interfaces 7:21442–21454. https://doi.org/10.1021/acsami.5b06429

50. Liu Y, Mei L, Xu C et al (2016) Dual receptor recognizing cell penetrating peptide for selective targeting, efficient intratumoral diffusion and synthesized anti-glioma therapy. Theranostics 6:177–191. https://doi.org/10.7150/thno.13532

51. Ying X, Wen H, Lu W-L et al (2010) Dual-targeting daunorubicin liposomes improve the therapeutic efficacy of brain glioma in animals. J Control Release 141:183–192. https://doi.org/10.1016/j.jconrel.2009.09.020

52. Tian W, Ying X, Du J et al (2010) Enhanced efficacy of functionalized epirubicin liposomes in treating brain glioma-bearing rats.

Eur J Pharm Sci 41:232–243. https://doi.org/10.1016/j.ejps.2010.06.008

53. Du J, Lu W-L, Ying X et al (2009) Dual-targeting topotecan liposomes modified with tamoxifen and wheat germ agglutinin significantly improve drug transport across the blood–brain barrier and survival of brain tumor-bearing animals. Mol Pharm 6:905–917. https://doi.org/10.1021/mp800218q

54. Gao J-Q, Lv Q, Li L-M et al (2013) Glioma targeting and blood–brain barrier penetration by dual-targeting doxorubincin liposomes. Biomaterials 34:5628–5639. https://doi.org/10.1016/j.biomaterials.2013.03.097

55. Li X-T, Ju R-J, Li X-Y et al (2014) Multifunctional targeting daunorubicin plus quinacrine liposomes, modified by wheat germ agglutinin and tamoxifen, for treating brain glioma and glioma stem cells. Oncotarget 5:6497–6511. https://doi.org/10.18632/oncotarget.2267

56. Joshi S, Singh-Moon RP, Ellis JA et al (2015) Cerebral hypoperfusion-assisted intraarterial deposition of liposomes in normal and glioma-bearing rats. Neurosurgery 76:92–100. https://doi.org/10.1016/j.jdiacomp.2008.01.002.Postural

57. Lin Q, Mao KL, Tian FR et al (2016) Brain tumor-targeted delivery and therapy by focused ultrasound introduced doxorubicin-loaded cationic liposomes. Cancer Chemother Pharmacol 77:269–280. https://doi.org/10.1007/s00280-015-2926-1

58. Chen H, Qin Y, Zhang Q et al (2011) Lactoferrin modified doxorubicin-loaded procationic liposomes for the treatment of gliomas. Eur J Pharm Sci 44:164–173. https://doi.org/10.1016/j.ejps.2011.07.007

59. Qin Y, Chen H, Zhang Q et al (2011) Liposome formulated with TAT-modified cholesterol for improving brain delivery and therapeutic efficacy on brain glioma in animals. Int J Pharm 420:304–312. https://doi.org/10.1016/j.ijpharm.2011.09.008

60. Zhang Y, Calon F, Zhu C et al (2003) Intravenous nonviral gene therapy causes normalization of striatal tyrosine hydroxylase and reversal of motor impairment in experimental parkinsonism. Hum Gene Ther 14:1–12. https://doi.org/10.1089/10430340360464660

61. Xia C-F, Boado RJ, Zhang Y et al (2008) Intravenous glial-derived neurotrophic factor gene therapy of experimental Parkinson's disease with Trojan horse liposomes and a tyrosine hydroxylase promoter. J Gene Med 10:306–315. https://doi.org/10.1002/jgm.1152

62. Chen Z-L, Huang M, Wang X-R et al (2016) Transferrin-modified liposome promotes α-mangostin to penetrate the blood–brain barrier. Nanomedicine 12:421–430. https://doi.org/10.1016/j.nano.2015.10.021

63. Kuo Y-C, Tsao C-W (2017) Neuroprotection against apoptosis of SK-N-MC cells using RMP-7- and lactoferrin-grafted liposomes carrying quercetin. Int J Nanomedicine 12:2857–2869. https://doi.org/10.2147/IJN.S132472

64. Kuo Y-C, Wang C-T (2014) Protection of SK-N-MC cells against β-amyloid peptide-induced degeneration using neuron growth factor-loaded liposomes with surface lactoferrin. Biomaterials 35:5954–5964. https://doi.org/10.1016/j.biomaterials.2014.03.082

65. Sharma G, Modgil A, Layek B et al (2013) Cell penetrating peptide tethered bi-ligand liposomes for delivery to brain in vivo: biodistribution and transfection. J Control Release 167:1–10. https://doi.org/10.1016/j.jconrel.2013.01.016

66. Sharma G, Modgil A, Zhong T et al (2014) Influence of short-chain cell-penetrating peptides on transport of doxorubicin encapsulating receptor-targeted liposomes across brain endothelial barrier. Pharm Res 31:1194–1209. https://doi.org/10.1007/s11095-013-1242-x

67. Tremmel R, Uhl P, Helm F et al (2016) Delivery of copper-chelating trientine (TETA) to the central nervous system by surface modified liposomes. Int J Pharm 512:87–95. https://doi.org/10.1016/j.ijpharm.2016.08.040

68. Klatzo I (1987) Blood-brain barrier and ischaemic brain oedema. Z Kardiol 76(Suppl 4):67–69

69. Klatzo I (1985) Brain oedema following brain ischaemia and the influence of therapy. Br J Anaesth 57:18–22

70. Kuroiwa T, Ting P, Martinez H, Klatzo I (1985) The biphasic opening of the blood-brain barrier to proteins following temporary middle cerebral artery occlusion. Acta Neuropathol 68:122–129

71. Spatz M (2010) Past and recent BBB studies with particular emphasis on changes in ischemic brain edema. Acta Neurochir Suppl 106:21–27

72. Overgaard K (2014) The effects of citicoline on acute ischemic stroke: a review. J Stroke Cerebrovasc Dis 23:1764–1769. https://doi.

org/10.1016/j.jstrokecerebrovasdis.2014.
01.020

73. Lee JH, Engler JA, Collawn JF, Moore BA
(2001) Receptor mediated uptake of peptides
that bind the human transferrin receptor. Eur
J Biochem 268:2004–2012

74. Oh S, Kim BJ, Singh NP et al (2009) Synthesis and anti-cancer activity of covalent conjugates of artemisinin and a transferrin-receptor
targeting peptide. Cancer Lett 274:33–39.
https://doi.org/10.1016/j.canlet.2008.08.
031

75. Zidan AS, Aldawsari H (2015) Ultrasound
effects on brain-targeting mannosylated liposomes: in vitro and blood-brain barrier transport investigations. Drug Des Devel Ther
9:3885–3898. https://doi.org/10.2147/
DDDT.S87906

76. Liu Y, Lu W (2012) Recent advances in brain
tumor-targeted nano-drug delivery systems.
Expert Opin Drug Deliv 9:671–686.
https://doi.org/10.1517/17425247.2012.
682726

77. Miyata S, Kawabata S, Hiramatsu R et al
(2011) Computed tomography imaging of
transferrin targeting liposomes encapsulating
both boron and iodine contrast agents by
convection-enhanced delivery to F98 rat glioma for boron neutron capture therapy. Neurosurgery 68:1380–1387. https://doi.org/
10.1227/NEU.0b013e31820b52aa

78. Pardridge WM, Eisenberg J, Yang J (1987)
Human blood-brain barrier transferrin receptor. Metabolism 36:892–895

79. Jefferies WA, Brandon MR, Williams AF,
Hunt SV (1985) Analysis of lymphopoietic
stem cells with a monoclonal antibody to the
rat transferrin receptor. Immunology
54:333–341

80. Jefferies WA, Brandon MR, Hunt SV et al
(1984) Transferrin receptor on endothelium
of brain capillaries. Nature 312:162–163

81. Huwyler J, Wu D, Pardridge WM (1996)
Brain drug delivery of small molecules using
immunoliposomes. Proc Natl Acad Sci U S A
93:14164–14169

82. Huwyler J, Yang J, Pardridge WM (1997)
Receptor mediated delivery of daunomycin
using immunoliposomes: pharmacokinetics
and tissue distribution in the rat. J Pharmacol
Exp Ther 282:1541–1546

83. Schnyder A, Krähenbühl S, Drewe J, Huwyler
J (2005) Targeting of daunomycin using biotinylated immunoliposomes: pharmacokinetics, tissue distribution and *in vitro*
pharmacological effects. J Drug Target

13:325–335. https://doi.org/10.1080/
10611860500206674

84. Huwyler J, Cerletti A, Fricker G et al (2002)
By-passing of P-glycoprotein using Immunoliposomes. J Drug Target 10:73–79. https://
doi.org/10.1080/10611860290007559

85. Zhang Y, Jeong Lee H, Boado RJ, Pardridge
WM (2002) Receptor-mediated delivery of an
antisense gene to human brain cancer cells. J
Gene Med 4:183–194

86. Zhang Y, Zhu C, Pardridge WM (2002) Antisense gene therapy of brain cancer with an
artificial virus gene delivery system. Mol
Ther 6:67–72

87. Chaudhary B, Khaled YS, Ammori BJ, Elkord
E (2014) Neuropilin 1: function and therapeutic potential in cancer. Cancer Immunol
Immunother 63:81–99. https://doi.org/10.
1007/s00262-013-1500-0

88. Pardridge WM (1995) Transport of small
molecules through the blood-brain barrier:
biology and methodology. Adv Drug Deliv
Rev 15:5–36. https://doi.org/10.1016/
0169-409X(95)00003-P

89. Wagner E, Curiel D, Cotten M (1994) Delivery of drugs, proteins and genes into cells
using transferrin as a ligand for receptor-mediated endocytosis. Adv Drug Deliv Rev
14:113–135. https://doi.org/10.1016/
0169-409X(94)90008-6

90. Guo J, Schlich M, Cryan JF, O'Driscoll CM
(2017) Targeted drug delivery via folate
receptors for the treatment of brain cancer:
can the promise deliver? J Pharm Sci
106:3413–3420. https://doi.org/10.1016/
j.xphs.2017.08.009

91. Meeuwsen EJ, Melis RJF, Van Der Aa GCHM
et al (2012) Effectiveness of dementia follow-up care by memory clinics or general practitioners: randomised controlled trial. BMJ
344:e3086. https://doi.org/10.1136/BMJ.
E3086

92. Agrawal M, Ajazuddin TDK et al (2017)
Recent advancements in liposomes targeting
strategies to cross blood-brain barrier (BBB)
for the treatment of Alzheimer's disease. J
Control Release 260:61–77. https://doi.
org/10.1016/j.jconrel.2017.05.019

93. Zlokovic BV (2011) Neurovascular pathways
to neurodegeneration in Alzheimer's disease
and other disorders. Nat Rev Neurosci
12:723–738. https://doi.org/10.1038/
nrn3114

94. Wang Y, Xia Z, Xu J-R et al (2012)
α-Mangostin, a polyphenolic xanthone derivative from mangosteen, attenuates β-amyloid
oligomers-induced neurotoxicity by

inhibiting amyloid aggregation. Neuropharmacology 62:871–881. https://doi.org/10.1016/j.neuropharm.2011.09.016

95. Mash DC, Pablo J, Buck BE et al (1991) Distribution and number of transferrin receptors in Parkinson's disease and in MPTP-treated mice. Exp Neurol 114:73–81

96. Hwang O, Baker H, Gross S, Joh TH (1998) Localization of GTP cyclohydrolase in monoaminergic but not nitric oxide-producing cells. Synapse 28:140–153. https://doi.org/10.1002/(SICI)1098-2396(199802)28:2<140::AID-SYN4>3.0.CO;2-B

97. Mc Carthy DJ, Malhotra M, O'Mahony AM et al (2014) Nanoparticles and the blood-brain barrier: advancing from in-vitro models towards therapeutic significance. Pharm Res 32:1161. https://doi.org/10.1007/s11095-014-1545-6

98. Vieira DB, Gamarra LF (2016) Getting into the brain: liposome-based strategies for effective drug delivery across the blood–brain barrier. Int J Nanomedicine 11:5381–5414

99. Hervé F, Ghinea N, Scherrmann J-M (2008) CNS delivery via adsorptive transcytosis. AAPS J 10:455–472. https://doi.org/10.1208/s12248-008-9055-2

100. Farkhani SM, Valizadeh A, Karami H et al (2014) Cell penetrating peptides: efficient vectors for delivery of nanoparticles, nanocarriers, therapeutic and diagnostic molecules. Peptides 57:78–94. https://doi.org/10.1016/j.peptides.2014.04.015

101. Antimisiaris S, Mourtas S, Papadia K (2017) Targeted si-RNA with liposomes and exosomes (extracellular vesicles): how to unlock the potential. Int J Pharm 525:293–312. https://doi.org/10.1016/j.ijpharm.2017.01.056

102. Joshi S, Singh-Moon R, Wang M et al (2014) Cationic surface charge enhances early regional deposition of liposomes after intracarotid injection. J Neuro-Oncol 120:489–497. https://doi.org/10.1007/s11060-014-1584-1

103. Joshi S, Singh-Moon RP, Wang M et al (2014) Transient cerebral hypoperfusion assisted intraarterial cationic liposome delivery to brain tissue. J Neuro-Oncol 118:73–82. https://doi.org/10.1007/s11060-014-1421-6

104. Chen H, Tang L, Qin Y et al (2010) Lactoferrin-modified procationic liposomes as a novel drug carrier for brain delivery. Eur J Pharm Sci 40:94–102. https://doi.org/10.1016/j.ejps.2010.03.007

105. Knudsen KB, Northeved H, Pramod Kumar EK et al (2015) In vivo toxicity of cationic micelles and liposomes. Nanomedicine 11:467–477. https://doi.org/10.1016/j.nano.2014.08.004

106. Narayan R, Singh M, Ranjan OP et al (2016) Development of risperidone liposomes for brain targeting through intranasal route. Life Sci 163:38–45. https://doi.org/10.1016/j.lfs.2016.08.033

107. Wakabayashi T, Natsume A, Hashizume Y et al (2008) A phase I clinical trial of interferon-beta gene therapy for high-grade glioma: novel findings from gene expression profiling and autopsy. J Gene Med 10:610–618. https://doi.org/10.1002/jgm.1160

108. Moller P, Lykkesfeldt J (2014) Positive charge, negative effect: the impact of cationic nanoparticles in the brain. Nanomedicine 9:1441–1443

109. Koren E, Torchilin VP (2012) Cell-penetrating peptides: breaking through to the other side. Trends Mol Med 18:385–393. https://doi.org/10.1016/j.molmed.2012.04.012

110. Trabulo S, Cardoso AL, Mano M, de Lima MCP (2010) Cell-penetrating peptides-mechanisms of cellular uptake and generation of delivery systems. Pharmaceuticals 3:961–993. https://doi.org/10.3390/ph3040961

111. Stalmans S, Bracke N, Wynendaele E et al (2015) Cell-penetrating peptides selectively cross the blood-brain barrier in vivo. PLoS One 10:1–22. https://doi.org/10.1371/journal.pone.0139652

112. Qin Y, Chen H, Yuan W et al (2011) Liposome formulated with TAT-modified cholesterol for enhancing the brain delivery. Int J Pharm 419:85–95. https://doi.org/10.1016/j.ijpharm.2011.07.021

113. Qin Y, Zhang Q, Chen H et al (2012) Comparison of four different peptides to enhance accumulation of liposomes into the brain. J Drug Target 20:235–245. https://doi.org/10.3109/1061186X.2011.639022

114. Zong T, Mei L, Gao H et al (2014) Synergistic dual-ligand doxorubicin liposomes improve targeting and therapeutic efficacy of brain glioma in animals. Mol Pharm 11:2346–2357. https://doi.org/10.1021/mp500057n

Chapter 3

Nanofibers and Nanostructured Scaffolds for Nervous System Lesions

Jose L. Gerardo Nava, Jonas C. Rose, Haktan Altinova, Paul D. Dalton, Laura De Laporte, and Gary A. Brook

Abstract

Strategies aimed at repairing the injured nervous system have as their main goal the reconnection of axons with their appropriate targets through bridging devices. In order to achieve this, such devices must provide cues to support directed axonal growth and good integration with the host tissue. Differences in the anatomy of the central nervous system (CNS) and the peripheral nervous system (PNS) as well as their specific tissue response after injury, where the protective environment in the CNS contrasts with the more permissive one in the PNS, require strategies to be tailored for these specific locations. This chapter focuses on the development of nanostructured scaffolds (including hydrogels) in the formulation of strategies intended to promote axon regeneration and functional tissue repair following traumatic spinal cord injury (SCI) and peripheral nerve injury (PNI). The reader will be presented with a general introduction to the central nervous system and the peripheral nervous system, the pathophysiological consequences of such injuries, their incidence, and how advances in the state of the art of bioengineering nanostructured scaffolds are contributing to this important aspect of tissue engineering and regenerative medicine.

Key words Electrospinning, Nanofibers, Nervous system, Hydrogels

1 Introduction

In regenerative medicine, the interface between an implanted material and biological tissues always requires great consideration in order to support infiltrating cells and to prevent inflammatory reactions and fibrosis. This is particularly important in the case of delicate central nervous system (CNS) tissue such as spinal cord (SC) as well as in the peripheral nervous system (PNS) tissue. Regenerative materials should ideally contain biomimetic functionalities to support specific cell adhesion, migration, and differentiation. Incorporation of neurotrophic factors, possibly by recapitulating developmental guidance molecule signaling, can enhance the native regenerative capacity of the injured SC [1]. These molecules can be administered via a drug delivery system

Javier O. Morales and Pieter J. Gaillard (eds.), *Nanomedicines for Brain Drug Delivery*, Neuromethods, vol. 157,
https://doi.org/10.1007/978-1-0716-0838-8_3, © Springer Science+Business Media, LLC, part of Springer Nature 2021

(DDS) or by paracrine mechanisms of either material-delivered stem cells or through genetically modified cells. Furthermore, regenerating axons need to be guided unidirectionally across the lesion site to promote the most efficient tissue repair. Guidance can be achieved by chemotactic or haptotactic mechanisms [2], similar to the natural, endogenous SC extracellular matrix (ECM) of the PNS. The revival of electrospinning technology has allowed the generation of polymeric fibers in the nanoscale range that can be tailored to simulate aspects of natural ECM. By incorporating biofunctional molecules, their influence on cell behavior and axonal guidance can be improved. For this reason, the use of nanofiber-based structures for guidance has become popular, and this has boosted the development of nerve-guiding structures that provide controlled topographical and biochemical cues that improve directional cell and process growth (aligned nanofibers) as well as controlled conduit porosity (random nanofibers). Hydrogels are also being explored as promising intervention strategies in spinal cord injury (SCI) in an attempt to introduce a permissive environment for regeneration while preventing further damage to spared axons. Hydrogel research in this area of regenerative medicine focuses on ways to achieve orientational cell and axonal growth within the hydrogel by modifying fibrous organization during or after delivery.

This chapter focuses on the development of nanostructured scaffolds (including hydrogels) in the formulation of strategies intended to promote axon regeneration and functional tissue repair following traumatic spinal cord injury (SCI) and peripheral nerve injury (PNI). The reader will be presented with a general introduction to the central nervous system (CNS) and the peripheral nervous system (PNS), the pathophysiological consequences of such injuries, their incidence, and how advances in the state of the art of bioengineering nanostructured scaffolds are contributing to this important aspect of tissue engineering and regenerative medicine.

2 The Nervous System

The nervous system is responsible for monitoring multiple aspects of our environment (both internal and external) and allowing appropriate responses to be made (voluntary or involuntary) for any particular situation. The human brain contains, according to some estimates, tens of billions of neurons [3] and is, with the support of the spinal cord, responsible for receiving, processing, and storing information, whereas the peripheral nervous system is responsible for the transmission of motor, sensory, and autonomic signals between the CNS and the rest of the body (an exception being the enteric nervous system of gut which has the capacity to function in the absence of CNS input). The function and connectivity of the neuronal networks of the adult CNS and of their axonal projections through the nerves of the PNS are critically supported by populations of glial and non-glial (e.g., vascular and connective tissue) cells. Glia of the CNS are the oligodendrocytes, astrocytes, microglia, and ependymal cells (as well as their progenitors or stem cells), whereas those of the PNS are Schwann cells and satellite cells. The non-glial, components of nervous tissues include cells associated with the vasculature (including pericytes and perivascular fibroblasts) and those forming protective sheaths (e.g., fibroblasts, arachnoid, and pial cells of the meninges as well as endoneurial and epineurial fibroblast, and perineurial cells). All these cell types play important roles in supporting neuronal function in health but may also play significant roles following traumatic injury. It is beyond the scope of this review to provide a detailed description of the roles of all these cells, but where appropriate, their roles in the pathophysiology of traumatic injury to the spinal cord and peripheral nerves have been highlighted.

2.1   Spinal Cord Injury

Spinal cord injury affects approximately three million people worldwide, with 250,000–500,000 new cases occurring each year [4]. The yearly incidence of SCI (reflecting the number of new cases in a population during a particular time frame) for particular nations ranges from the highest value of 49.1 per million (New Zealand) to the lowest value 8 per million (Spain) with road traffic accidents being the main cause of injury. The peak age of incidence is less than 30 years of age, and there is a high male-to-female ratio.

Substantial progress has been made in understanding the cascade of cellular and molecular events initiated by traumatic SCI, to the extent that the general outlook of developing an effective treatment has changed from pessimism to one of great optimism. Although human SCI is a highly heterogeneous condition, its consequences are known to depend on a number of parameters such as the type and severity of injury. Four main classes of SCI have been identified: contusion-, compression-, and transection-type lesions, as well as solid core-type lesions [5, 6].

The immediate and devastating loss of motor, sensory, and autonomic functions after severe traumatic SCI is due to loss of neurons and glia at the lesion site, the shearing or severance of descending and ascending nerve fiber tracts, and the initiation of detrimental local events including the breakdown of the blood-spinal cord barrier, associated with bleeding, ischemia, and edema, as well as excitotoxicity, inflammation, scarring, and cystic cavitation. These secondary degenerative events take place over a timescales of hours, days, weeks, and even months after the initial injury [7]. The cell body response of axotomized, long-distance CNS-projecting neurons is much weaker than that observed in PNS-projecting neurons following injury [8]. This weak and transient neuronal cell body response is believed to contribute to the generally poor regenerative response of damaged CNS tissues. A description of the cell body response is given in the section describing the pathophysiology of peripheral nerve injury (see below). However, spared axons in the brain and spinal cord can sprout and form new connections that may contribute to a compensatory recovery of function. Lesioned lumbar projecting corticospinal axons were found to sprout and innervate cervical motor neurons as well as short- and long-distance projecting propriospinal neurons (PSN). With time, contacts with the short-distance projecting PSN were lost, while those with the long-distance projecting PSN were maintained. The long-distance projecting PSN extended synaptic contacts to deafferented lumbar motoneurons [9, 10]. Such spontaneous sprouting of damaged and spared axons has been suggested to effectively bypass the axon-growth inhibitory environment of the lesion site and was suggested to be responsible for the recovery of treadmill stepping in an animal model of spinal cord hemisection injury [11]. Interventions that support such compensatory sprouting and reorganization mechanisms could bring enormous benefit to patients suffering from SCI. The failure of severely damaged spinal tissues to demonstrate any spontaneous, functionally significant, long-distance axon regeneration is due to an overall imbalance between local axon-growth-inhibiting and axon-growth-promoting mechanisms at the lesion site, including the relatively poor expression of neurotrophic factors, the presence of potent molecular and physical barriers that form as part of the scarring process, and the lack of appropriately oriented guidance cues across the injury [7, 12].

Two environmental factors have been found to be of particular importance in reducing sprouting and preventing axon regeneration: CNS myelin-associated axon-growth inhibitors and the lesion-induced re-expression chondroitin sulfate proteoglycan associated with scar formation.

2.1.1 Myelin-Associated Axon-Growth Inhibitors

Traumatic injury results in substantial oligodendrocyte degeneration with the concomitant release, into the lesion site and surrounding tissues, of a number of myelin-associated molecules, several of which exert potent inhibitory effects on axon growth. NOGO-A, myelin-associated glycoprotein (MAG) and oligodendrocyte-myelin glycoprotein (OMgp), have all been reported to act via a complex of receptor molecules including the GPI-linked cell surface NOGO receptor (NgR1) and lead to growth cone collapse through the activation of the small GTPase, RhoA [13, 14]. The experimental administration of antibodies or peptides to block or inactivate NgR1 has resulted in significant axon regeneration and collateral sprouting that have been associated with functional recovery [15, 16]. The marked success with antibodies that neutralize or block the effects of NOGO-A resulted in the extension of the preclinical studies into phase I and II clinical trials including the ongoing European NISCI phase II trial involving centers in Switzerland, Germany, Italy, Spain, and the Czech Republic, the European Multicentre Study about Spinal Cord Injury [16] (Website: https://nisci-2020.eu/index.php?id=1449).

2.1.2 Spinal Cord Scarring and Axon-Growth-Repulsive Molecules

The general term "spinal cord scar" is often used interchangeably with that of the astroglial scarring; however, this oversimplifies the situation somewhat since it is clear that the scarring process involves both astroglial and fibroadhesive, connective tissue components. Astrocytes, oligodendrocyte progenitors, microglia/macrophages, Schwann cells, leptomeningeal fibroblasts, and blood vessel-derived pericytes/fibroblasts have all been reported to take part in the cellular reorganization at the lesion site, where they play significant roles in the formation of scar tissue after spinal cord injury [7, 17]. Reactive astrocytosis, in response to injury, the local migration of fibroblast-like cells and inflammatory cells, involves proliferation, hypertrophy, increased expression of cytoskeletal proteins such as vimentin, nestin, and glial fibrillary acidic protein (GFAP), as well as a marked upregulation of chondroitin sulfate proteoglycans (CSPGs), a family of highly sulfated extracellular matrix (ECM)-related molecules [12, 18, 19]. CSPGs display axon-growth-repulsive properties and play major roles in guiding axon growth during development (e.g., [20–23]). The generation of a layer of tightly packed astrocytes and their processes, as well as the deposition of a new glia limitans, highlight the formation of a physical and molecular barrier (scar) around the lesion site. This barrier formation is widely acknowledged to exert both beneficial and detrimental effects in that the scar protects surrounding spinal tissues from further inflammation-mediated damage [24, 25] but prevents axon regeneration through raised levels of axon-growth-repulsive CSPGs [12, 17]. Local injection or infusion of the enzyme chondroitinase ABC results in degradation of the glycosaminoglycan side chains of CSPGs and the promotion of sprouting and axon regeneration with some degree of functional recovery [26, 27].

Although fibroblast-like cells were initially thought to be derived from damaged leptomeninges and damaged blood vessels [28, 29], the recent use of transgenic animals has facilitated a more detailed understanding of the development of the fibroadhesive component scar after SCI. A subtype of vessel-associated pericytes has been reported to proliferate rapidly and contribute to the ECM-producing stromal or fibroblast-like population of cells within the connective tissue scar [30]. The involvement of fibroblast-type cells, particularly those associated with large diameter blood vessels, were also reported by Soderblom and colleagues to contribute to connective tissue scar formation [31]. However, it remains unclear if the lesion-induced changes in the perivascular fibroblasts, identified by Soderblom and colleagues, also included the type A pericytes that had previously been identified by Görlitz and colleagues. Transgenic mice have also been used to show SCI-induced ependymal cell proliferation and differentiation into oligodendrocyte progenitor cells and astrocytes that formed the innermost layer of reactive astrocytes that lined the lesion site [32]. However, this observation has been contradicted by more recent investigations that described ependymal cell responses to more physiologically relevant sized lesions of the spinal cord: either complete crush-type injuries or extensive penetrating injuries that were close to, but did not involve the central canal [33]. Only minimal- or no involvement of ependymal progenitors could be observed in the formation of reactive astroglia or of any other cell type after injury. This probably highlights the complexity of changes that are induced by spinal cord injury and the fact that different type/severities of injury can lead to scarring responses of differing cell compositions.

2.1.3 Lesion-Induced Inflammation After Spinal Cord Injury

As mentioned earlier, reactive astrocytes form a barrier that delineates the lesion site and effectively isolates the region of ongoing excitotoxic and free radical-induced degeneration (filled with activated, pro-inflammatory, and phagocytic neutrophils and macrophages) from the adjacent areas of spared CNS tissue. It has been suggested that reactive astrocytes regulate the number of migrating cells into the lesion site, whereas their associated re-expression of CSPGs influences macrophage spatial localization and activation [34]. The immune response to injury starts as early as 3 h after the insult with an early wave of neutrophil invasion that may peak at approximately 1 day after injury, followed by a second wave of microglial and monocyte-derived macrophage infiltration and activation, peaking at 7 and 60 days after injury and remaining present for at least 180 days [7, 35]. The phenotype of the macrophage response has been classified into an early (M1), pro-inflammatory, tissue-destructive, and myelin-clearing phenotype and a later (M2) anti-inflammatory or tissue-repairing phenotype [36]. Strategies that shift the relative polarization of macrophages at the lesion site to a regeneration supporting M2 phenotype are likely to promote neuroprotection and functional tissue repair (e.g., [37, 38]).

2.2 Peripheral Nerve Injury

Severe PNI, like severe SCI, results in the immediate loss of function. However, in contrast to the permanent functional deficit associated with severe SCI, substantial tissue repair and functional recovery are possible following PNI [39]. Injuries to the PNS have been reported to occur in approximately 2.8–5% of all trauma patients [40, 41], with the radial nerve being the most commonly injured nerve in the upper limb and the peroneal nerve being the commonly injured nerve of the lower limb [40].

Simple or complete transection injuries to the PNS, in which a gap of 1–2 mm may occur, can be surgically repaired by reconnecting individual nerve fascicles with tensionless sutures [42]. However, lesion-induced gaps of the PNS that are deemed too large for end-to-end reconstruction requires the gap between the nerve stumps to be bridged. The most commonly adopted surgical approach for this is to repair the gap by the transplantation of autologous nerves (usually sensory nerves), harvested from elsewhere in the patient's body [43]. Although autologous nerve transplantation is regarded as the current "gold standard" strategy for repairing large gaps in the PNS, it has some disadvantages such as the need for a second surgical site, the comorbidity, and risk of infection associated with harvesting the donor nerve(s), as well as the limitation of the amount of donor nerve(s) that can be obtained. Furthermore, only 40–50% of patients receiving autografts have been regarded as a success by the return of useful function [42].

2.2.1 Neuronal Cell Body Response

As mentioned earlier, the cell body response after PNI is stronger and longer lasting in PNS-related neuronal cell bodies than in CNS neuronal cell bodies. Experimental studies have shown that the cell body response and neuronal survival after peripheral nerve injury depend on the distance between the injury and neuronal cell body: greater survival correlating with longer distances and reduced survival with shorter distances [44, 45]. Over decades, many of the cellular and molecular events that influence neuronal survival, regenerative axonal growth, and target reinnervation after PNI have been identified (e.g., [46, 47]). The traumatic shearing of the axon induces retrograde signaling to the neuronal cell body over a time scale of seconds to weeks after injury [48]. The neuronal cell bodies of spinal cord and brain stem nuclei alpha motor neurons, as well as of sensory neurons in dorsal root ganglia, respond to axotomy by undergoing expansion and chromatolysis, in which cytoplasmic Nissl substance (i.e., rough endoplasmic reticulum-associated ribosomes) dissipates and the nucleus migrates to a peripheral location [8]. Axotomy induces signaling from Schwann cells, through the release of ciliary neurotrophic factor (CNTF), leukemia inhibitory factor (LIF), or neuronal interleukin-6 (IL-6) to activate signal transduction systems involving phosphorylation, dimerization, and translocation of signal transducer and activator of transcription (STAT-3) to the nucleus [49]. The phosphorylation of c-Jun assists in the shift of gene expression which switches neuronal cell function from one of the neurotransmission to one of growth with increased expression of regeneration-associated genes such as GAP-43, alpha-1 tubulin, and actin [47]. The increased expression of these (and many more) regeneration-associated proteins assists in growth cone motility and the extension of regenerating axonal processes [8, 47].

Degeneration at the site of injury and in the distal nerve stump is an active process that is required for subsequent axon regeneration to take place [50]. Schwann cells and macrophages both contribute to the clearance of dead cells and myelin debris during the process of Wallerian degeneration [51]. Macrophages are recruited into the distal nerve stump in response to chemokines such as monocyte chemoattractant protein-1 (MCP-1), macrophage inflammatory protein-1α (MIP-1α), interleukin-1β (IL-1β), and toll-like receptor 4 (TLR-4) [52, 53]. This recruitment process takes place as the perineurium breaks down, thus rendering the blood-nerve barrier more permeable to cell infiltration [54]. The process of inflammation during Wallerian degeneration in the lesioned PNS is more complete and substantially faster than in the lesioned CNS [55], with the pro-inflammatory, M1 subtype dominating during the early phase, being essential for myelin clearing and the stimulation of Schwann cell proliferation, and a later (7–14 days after injury) dominance of the anti-inflammatory, M2 subtype [56]. The loss of axonal contact induces Schwann cells to proliferate, reduce myelin-associated gene expression, and increase neurotrophins and cell adhesion molecule expression that is important for supporting axon regeneration [57]. The proliferating Schwann cells extend their processes and become highly aligned within the basal lamina of the endoneurial tubes, forming the so-called bands of Büngner. The alignment of Schwann cells within the bands of Büngner provides the neurotrophic and tropic cues that are pivotal for supporting and directing regenerating axons to the target end organs within the periphery [58–60]. Schwann cells are also important for the remyelination of regenerated large diameter axons for the restoration of rapid nerve conduction velocities; however, regenerated nerves typically have thinner myelin sheets, shorter internodal lengths, and relatively decreased functional capacity [61].

It is widely acknowledged that implantable biomaterials and scaffolds that support directional axon regeneration and tissue repair are likely to impart more favorable functional outcomes than those that have no orientational cues. The following sections provide an overview of key technological areas that are being pursued to develop and control their topographical diversity and functionalization in regenerative applications for CNS and PNS traumatology.

3   Electrospinning

Electrospinning is a fiber-forming technology that has been extensively researched since 1996 [62]. It produces nano- and micro-scale fibers using the electrostatic drawing of polymer-based fluids and has been widely adopted in research partly due to the low cost and simplicity of setting up the equipment [63]. The submicron diameter filaments generated by electrospinning have long been compared to collagen fibrils [64, 65], and there has been a sustained interest in developing electrospun materials for tissue engineering and regenerative medicine [66]. The majority of electrospinning research uses solutions, where both synthetic and/or biologically derived polymers are dissolved to achieve the macromolecular entanglements necessary to generate a fiber. In this configuration, solvent evaporation is required to produce the fiber. There are numerous excellent review articles on solution electrospinning [67–69]. Electrospinning can also be performed from the polymer melt—in this instance the fiber forms by the cooling and solidification of the electrified jet [70, 71].

In the simplest of configurations, electrospinning results in a nonwoven fibrous mesh (Fig. 1a, b) [72]. There are discussions about the cell invasiveness of electrospun meshes [79], although they are well-known as being excellent substrates upon which cells can grow [80–82]. For applications in neural tissue repair, where an oriented structure is of interest, there are three main approaches to fabricate directional substrates: (1) rotating collectors, (2) dual collectors, and (3) direct writing.

Rotating collectors: In this common form of electrospinning oriented substrates [83–86], the collector is designed so that it rotates at a high speed (Fig. 1c) [87]. This results in a "mostly" oriented electrospun material (Fig. 1d) [87]; however, there are numerous fibers that are deposited in non-desired directions. This approach has been augmented through the use of auxiliary electrodes so that the general orientation is improved [88, 89]. Similarly, sharp-tipped, fast-rotating, collectors (disks) are more effective in orienting the electrospun fibers [90]; however, the quantity of oriented material is reduced. The use of rotating collectors has also been used to manufacture tubular biomaterials, from both the solution [91] and the melt [92, 93].

Dual collectors: When two collectors are placed below the spinneret, separated by air or an insulating material, electrospun fibers individually bridge the gap between collector [75]. This approach, while mainly performed with polymer solutions, is also applicable to melt electrospinning [94]. As shown in Fig. 1g [76], these individual fibers are suspended between the collectors and can be later transferred to another substrate [95–97] for 2.5D in vitro culture. Such suspended fibers are much improved in orientation compared to rotating collectors (Fig. 1h) [76]; however, they are much more difficult to handle, and a secondary substrate is essential for later handling in vitro. Importantly, this principle of collecting suspended fibers has been adopted for 3D in vitro culture (Fig. 1i) [76].

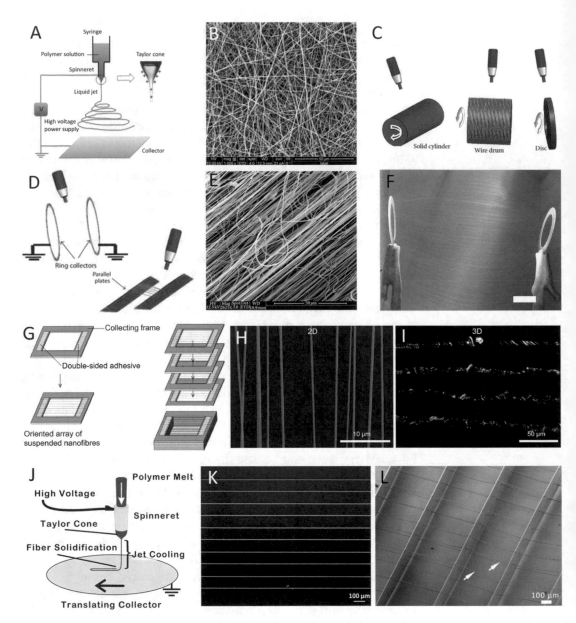

Fig. 1 Schematics of electrospinning configurations and collected material. (**a**) The simplest configuration is shown, where (**b**) randomly deposited, nonwoven fibers are collected. (**c**) A rotating collector is the basis for several fiber orientation strategies, shown in (**d**). In another approach for orienting fibers, (**e**) dual collectors result in (**f**) fibers suspended midair. Using the principle shown in (**e**), multiple frames (**g**) have been used to collect (**h**) suspended, oriented fibers that can then be suspended in a cell containing matrix (**i**). Applying (**j**) direct-writing principles to electrospinning, a series of oriented fibers (**k**) can also be accurately deposited and even (**l**) suspended (arrowed) between two support structures. (Figure (**a**) reproduced from [72], (**b**) from [73], (**c–e**) from [74], (**f**) from [75], (**g–i**) from [76], (**j**) from [77], (**k**) from [78], and (**l**) is unpublished data (courtesy of Mr. Andrei Hrynevich), all with permission)

Direct writing onto collectors: Less widely adopted than the two aforementioned orientation strategies, but significantly more controllable, is the direct writing of the electrospun fiber onto the collector (Fig. 1j). With polymer solutions, this is often termed "near-field electrospinning," since small collector distances (the distance between the spinneret and collector) are required to prevent significant electrical instabilities [98, 99] . For melts, larger collector distances [94] can be used in the direct-writing approach, when low conductivity polymers are electrospun. This approach has so far only resulted in a highly oriented single-fiber array (Fig. 1k) [78], and the development of such oriented structures into a truly 3D scaffold is still required; however, there are direct-write techniques being developed that allow melt electrospinning writing in air, across structures (Fig. 1l).

These three principles of electrospun fiber orientation cover the majority of the different fabrication approaches used within neural tissue engineering, for both the PNS and CNS. These configurations are applicable to nerve guides, in vitro studies of oriented fibers, and guidance substrates that are used within different neural tissue engineering paradigms.

4 Development of Bioengineered Nanofiber-Based Repair Strategies for Use in Spinal Cord Injury

The interest in generating an oriented matrix/scaffold for spinal cord repair has been demonstrated in numerous papers [100–103] and has been implemented in clinical trials (NCT02138110) (https://clinicaltrials.gov/ct2/show/study/NCT02138110). For this purpose, matrices and scaffolds have been aligned using numerous physical phenomena to induce orientation, including magnetic fields, templating, and uniaxial freezing. Oriented scaffolds have two general forms: (1) those formed by the creation of channels within a matrix or (2) bundled filamentous structures that form an oriented implant. There has also been substantial evidence of regeneration with such matrices/scaffolds in small animal models [104, 105] as well as the clinic, suggesting that there is potential in using materials as a regenerative/inflammatory substrate within the spinal cord. While many researchers consider a scaffold from a "regenerative substrate perspective," others describe how the scaffold alters the local inflammatory reaction and subsequent healing.

In the context of electrospinning, oriented nanofibers produce a substrate in vitro that could provide some insight as to the regenerative guidance within the spinal cord. Alternatively, nanofiber meshes have been used and placed over the dura, so that the blood-brain barrier has a membrane template that it can use to reseal itself. Such approaches can be combined with drug delivery strategies.

4.1 Substrates for In Vitro Analysis

In its simplest form, oriented electrospun fibers can be adhered to a substrate and cells seeded onto the surface. In a seminal paper by Schnell et al. [95], PCL fibers and PCL/collagen fibers were oriented by deposition on dual collectors and transferred to a non-cell adhesive star-polyethylene glycol (star-PEG) substrate. Cell from the PNS (i.e., Schwann cells, dorsal root ganglia (DRG), and olfactory ensheathing glia (OEG)) were oriented by the nanofibers and extended processes over greater distances with PCL/collagen nanofibers.

In a different study, oriented electrospun substrates were treated with poly-lysine, and oligodendrocyte culture resulted in myelination of the oriented fibers, mimicking the process by which these cells envelope and myelinate axons in the living body [106].

4.2 Implantable Scaffolds

Likely due to the easier implantation and clear requirement for entubulation for the peripheral nerve, there is far less in vivo implantation of electrospun nanofibers into the spinal cord. While nerve guides for the spinal cord have been performed in both full transection [107–110] and hemisection [111–113] models, there is significant effort (particularly postsurgical husbandry) required for this tissue compared to the peripheral nerve injury models.

In vitro studies performed on oriented electrospun using cells derived from the CNS demonstrate clear directed outgrowth [95, 97]. However, replicating such experiments within an in vivo spinal cord injury model is considerably more challenging. So far four in vivo studies have been performed in spinal cord transection models with variable outcomes. While not identical in injury location, type, materials, fiber diameter, or manufacturing configuration, in general the implanted materials were well integrated and did not exhibit a noticeably negative inflammatory response [114–117].

From a manufacturing perspective, two research groups used a rotating collector to manufacture a sheet of oriented nanofibers, which was then later rolled into a tube [116, 117]. Another study used dual collectors to manufacture oriented fibers that were then embedded within a collagen matrix [114, 115].

Hurtado et al. demonstrated a significant relationship between oriented and non-oriented electrospun poly-L-lactic acid (PLLA) fibers in a full transection model (3 mm gap at spinal cord level T8 in adult Sprague Dawley rats) [116]. Importantly, this study was extensive, with numerous time points (1, 2, and 4 weeks), with controls of PLLA sheet, including random and oriented electrospun fibers. In this scenario, the greatest extent of axon regeneration was supported by the implanted oriented electrospun fibers.

In a much smaller study, Chew and colleagues investigated oriented and random electrospun collagen fibers in a unilateral hemisection injury model at spinal level C3 [117] in the same species/strain/age as the Hurtado study. The diameter of the type I collagen fibers, however, was much smaller (210 ± 90 nm) than used in the PLLA study (between 1.2 and 1.6 µm). For this limited study, there was no statistical difference in regeneration between the oriented and random nanofiber groups; however, it is important to note the smaller number of animals used in this study by Chew and colleagues [117]. Even though a difference between oriented and non-oriented fiber groups was not observed, the authors concluded that "these findings clearly demonstrate the potential of electrospun collagen scaffolds for SCI repair" [117].

A second in vivo study was performed by Chew and colleagues that involved expanding the number of groups and increasing the number of experimental animals [115]. A hemi-section injury model was again performed, however, at spinal level C5. In addition, the rotating mandrel approach was substituted by aligning fibers using a dual collector configuration, while a different polymer (poly[ε-caprolactone-co-ethyl ethylene phosphate]) (PCLEEP) was embedded within a type I collagen hydrogel. Interestingly, for all groups investigated, the nanofiber-embedded collagen hydrogel supported greater neurite outgrowth than the collagen hydrogels alone. When the neurotrophin (NT-3) was included in such fibrous hydrogels, it did not have an impact on the length of neurites that regenerated into the scaffolds.

This approach was further modified in a third study by Chew and colleagues, to incorporate and deliver small noncoding RNAs. Here, PCLEEP was again used. It was aligned using a dual collector method, and oligonucleotides were included within the electrospun polymer solution. These oriented electrospun fibers were embedded within type I collagen, again including NT-3. The non-viral drug/gene delivery system "effectively directed neurite extensions and supported remyelination within the lesion sites" [114]. In summary, the development of oriented fiber/hydrogel composites for spinal cord implantation is achieving repeated positive outcomes while allowing the incorporation of additional therapeutic strategies.

4.3 Artificial Dura Mater

While not the focus of the in vivo study, a randomly electrospun fiber membrane was used to cover the exposed spinal cord after implanting a hydrogel scaffold in a small injury lesion model in the rat [118]. Here, the purpose was to reduce cell infiltration (fibroblasts and skeletal muscle cells) into the implantation site via the tissue space generated by the laminectomy. The efficacy of achieving such reduced cell infiltration was not described; however, nanofibrous materials could be combined in the future with in situ gelling hydrogels (such as HAMC) used to deliver therapeutics to the spinal cord.

5   Development of ECM-Related Molecules and Hydrogels with Controlled Nanostructures and Properties for Use in Spinal Cord Injury

To address the high sensitivity of the spinal cord tissue to mechanical trauma, any surgical intervention strategy should be minimally invasive to circumvent impairment of spared tissue with residual functionality [119]. This can be achieved via injection of the bioengineered material and subsequent solidification through biocompatible cross-linking mechanisms or shear thinning [120]. In addition, injectable materials provide a variety of advantages, as they can associate with irregular surfaces of lesion sites, be delivered with therapeutic factors, or combined with supporting cells. Notwithstanding the great progress over the last decades regarding the development of injectable materials intended to promote repair after SCI, a major challenge that remains is the implementation of in situ controlled mechanisms to guide the regenerating axons across the injury site. An overview of such hydrogel-based strategies for SCI is shown in Fig. 2.

Hydrogels have aroused great attention in the field of tissue engineering, as they mimic the ECM of soft tissues through their high water content and tunable physico-mechanical properties [121, 122]. Hydrogels can be subdivided into natural and synthetic hydrogels; however, most hydrogels are biohybrid materials. Generally, natural hydrogels, such as fibrin, Matrigel®, collagen, or gelatin, possess cell binding/signaling motifs, whereas synthetic hydrogels require the addition of biofunctionalities. One advantage of synthetic hydrogels is that their biofunctional domains can be tailored in a cell specific and highly reproducible manner, which cannot be achieved with natural materials. Moreover, synthetic materials provide control over mechanical properties, which have to emulate the microenvironment of the spinal cord (e.g., E-modulus between 100 and 1000 Pa) [120]. Apart from the material, the mechanism of hydrogel cross-linking at the injury site requires high biocompatibility. Physically bound hydrogels, which solidify upon an environmental change, such as pH, temperature, or the presence of ions, mostly contain mild gelation reactions [123]. However, the durability of these materials is often low, providing limited support throughout the complete regeneration process [124]. In contrast, chemically cross-linked hydrogels can be more stable with tunable degradation kinetics but these often rely on cytotoxic gelation reactions, which can further harm the injured tissue. Therefore, more biocompatible enzymatic or chemical cross-linking reactions have been developed. In addition, bio-orthogonal cross-linking reactions have emerged, which do not interfere with biological systems.

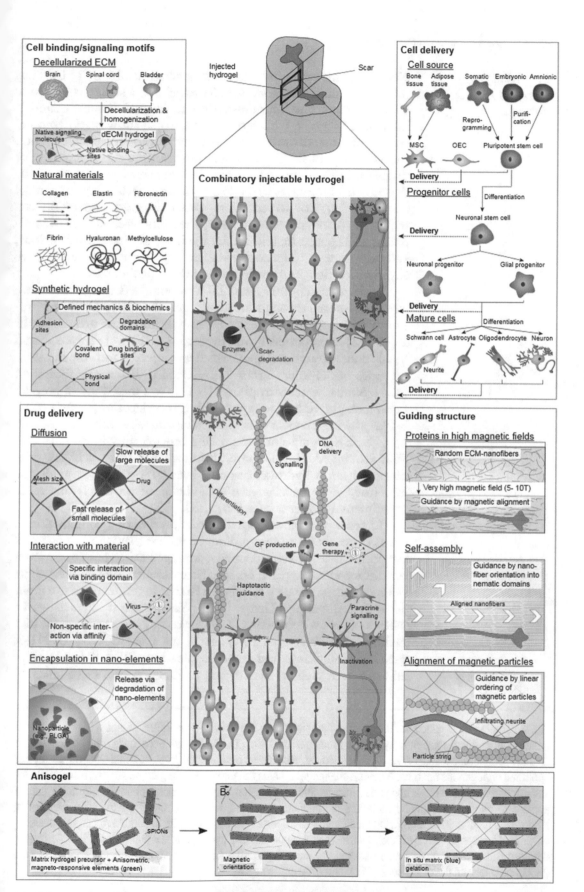

Fig. 2 Regenerative SCI repair concepts. Cell binding/signaling motifs can be delivered by decellularizing and homogenizing tissue, purifying natural materials, or mimicking biological cues in synthetic hydrogels. To

5.1 Injectable Hydrogels for Drug Delivery

Neurotrophins can reduce axonal degeneration after SCI [125] but mostly lack sufficient bioavailability to achieve a strong therapeutic effect. For sustained administration to injured tissues, material-integrated DDSs have been developed, which allow spatiotemporal control of drug release [126]. Apart from prolonged delivery times, the advantage of DDSs in comparison with the direct injection of drugs lies in their ability to deliver bioactive molecules with agent-dependent differences in release kinetics, achieving a parallel and/or serially connected effect. Other drug administration systems that are independent of the applied biomaterial, such as minipumps [127], may be rejected by the host, can have malfunctions, and are prone to infections, whereas regenerative materials have already been designed to circumvent these problems.

Initial material-integrated DDSs comprised drugs that were freely distributed throughout a hydrogel, allowing release via diffusion. Within one of the first approaches, fibroblast growth factor 2 (FGF-2) and epithelial growth factor (EGF) were mixed within highly concentrated collagen, which was injected into the intrathecal space of SC injured rats and resulted in less cavity formation [124]. Similarly, neurotrophin-3 (NT-3) was mixed into a polylactide-block-poly(ethylene glycol)-block-polylactide (PLA-b-PEG-b-PLA) hydrogel and released in therapeutically relevant doses for up to 6 days, significantly improving functionality in rodents after SCI. Prolonged release of chondroitinase ABC from a highly concentrated fibrin hydrogel has also been reported to degrade the inhibitory glycosaminoglycan side chains of proteoglycans in an experimental model of SCI [26, 128].

In order to achieve tailored release profiles of a combination of drugs or to obtain a more prolonged release that is independent of the drugs' hydrodynamic diameter, advanced DDSs have been developed. One strategy relies on specific molecular interactions, where bio-recognition sites can be attached to the hydrogel matrix to interact with neurotrophic factors [122]. Since heparin interacts with a wide range of growth factors [129], its conjugation to a regenerative hydrogel has been used to obtain a sustained delivery of NT-3, β-NGF, and brain-derived neurotrophic factor (BDNF) [130]. Heparin has been chemically activated to function as a cross-linker to

Fig. 2 (continued) enhance the therapeutic capacity, drugs are delivered via diffusion, specific or nonspecific interaction, or encapsulation in hydrolyzing nanoparticles (e.g., PLGA). Furthermore, cells may be delivered to support regeneration and nerve reconnection, as well as to differentiate into neurons, which can replace apoptotic endogenous neurons. A central challenge remains the formation of a guiding structure in situ. So far, self-assembly of peptide amphiphiles into nanofibers can form monodomains, or magnetic fields can be applied to align paramagnetic materials or magnetic particles, to form strings, which guide neurites by haptotactic mechanisms. Controlled unidirectional structures are obtained with injectable Anisogel, consisting of magneto-responsive rod-shaped elements and a surrounding hydrogel to fix their orientation. Combining different SCI repair concepts may possess the highest potential for regeneration. Mesenchymal stem cells (MSC), olfactory ensheathing cells (OEC), and growth factor (GF)

connect amine-terminated star-PEGs and bind to the cell-adhesive peptide, cyclic RGD, forming a FGF-2 delivering hydrogel for CNS injuries [131].

Another mechanism for drug delivery relies on combining sequentially degrading materials, such as fast degrading nanoparticles that are distributed within a slower degrading hydrogel. Therefore, a hybrid material composed of hyaluronan and methylcellulose (HAMC) was combined with poly(lactide-co-glycolide) (PLGA) nanoparticles, which can be loaded with drugs and degraded via hydrolysis, depending on the lactic to glycolic acid ratio [132]. The safety of injecting HAMC/PLGA into the intrathecal space of the uninjured and injured rat spinal cords was demonstrated by the absence of any significant increase in local inflammation and lack of any detrimental effects on astroglial scarring, cavity formation, or locomotor behavior [133]. Encapsulating NT-3 inside the PLGA nanoparticles prolonged release in vivo for up to 28 days, resulting in significant axonal growth and improved functional recovery [134]. Additional delivery of the antibody to NogoA also enhanced performance in several functional tasks after SCI [135]. By inducing covalent cross-links in the HAMC hydrogels, the rate of drug release could be reduced further [136, 137]. Alternatively, PLGA microspheres, loaded with glial-derived neurotrophic factor (GDNF), have been combined with a hydrogel composed of fibrinogen and alginate [138]. In another approach, microtubules were loaded with BDNF within an agarose hydrogel, which gelled in vivo in spinal cord injured rats, reducing local inflammatory reactions and enhancing the penetration of regenerating axons into the lesion site [139].

In comparison with direct drug delivery, cells delivered into or around the lesion can be genetically modified to function as small bioreactors. They can locally synthesize regenerative signals via paracrine mechanisms and distribute them throughout the tissues. Different types of nucleic acids (e.g., DNA, RNA, siRNA) can also be delivered, which encode for either growth-promoting and/or pro-angiogenic proteins [140] or block the expression of factors that would inhibit tissue regeneration (e.g., RNAi) (recently reviewed [141]). By delivering switchable gene cassettes to cells, such manipulated donor cells are capable of expressing multiple therapeutic factors at different time points with controlled amounts being made available over a prolonged period of time [142]. Importantly, manipulation of the genome in vivo can lead to permanent changes of the host cellular gene expression profiles and may introduce insertion mutations, inducing abnormal cell development.

For the delivery of genetic material, viral or nonviral carriers can be utilized. Highly efficient viral vectors have been developed over millions of years of evolution [143]. Nonviral gene transfer strategies, on the other hand, require protection of the nucleic acids to prevent endogenous enzymatic degradation. Therefore, nonviral nucleic acids are complexed by lipids, biomolecules, nanoparticles, or polymers that facilitate transport of the genetic load across the cell membrane for cytoplasmic or nuclear release [144, 145]. Criteria for deciding the vector delivery strategy include the triggered immune response, targeted cell population (type, mitotic, etc.), duration of expression (cytoplasmic vs. genome integrated), and stability of the vector. Recent advances in genetic engineering, such as zinc finger nucleases (ZFNs), transcription activator-like effector nucleases (TALENs), or clustered regularly interspaced palindromic repeats (CRISPR), enable genomic editing to introduce genetic modifications into almost any cell type and organism with high precision [146].

Regenerative materials can stabilize gene carriers by physical entrapment or (non)specific interactions, which determine the release kinetics, ranging from days to months [147]. As viral vectors have an innate ability to bind ECM proteins, these vectors have been successfully entrapped in collagen or fibrin hydrogels [148, 149]. On a more synthetic basis, viral vectors can be simply mixed, as in thermoresponsive poloxamer, pluronic F127, or incorporated via (non)specific interactions to knockout Lingo-1 expression, which has been proposed to contribute to inhibitory mechanisms to nerve regeneration [150]. For example, electrostatic interactions between viral vectors and IKVAV-Fmoc peptides have been utilized for sustained gene delivery [151]. On the other hand, lentivirus, incorporated in heparin-chitosan nanoparticles, inside a PEG hydrogel, has demonstrated better incorporation when compared to PEG modified with heparin or chitosan [152]. Nonviral vectors, such as plasmid or lipoplexes, have been delivered via fibrin and HA/PEG hydrogels, resulting in a prolonged delivery and high transfection efficiency [153, 154]. Alternatively, cationic lysine-based peptides have been coupled to a PEG hydrogel to interact with anionic lipoplexes [155]. To tune the carrier-matrix interaction more precisely, specific interactions, such as DNA-targeted antibodies or avidin/biotin, can be exploited [156, 157], or the vectors can be covalently conjugated to the material [158].

5.2 Injectable Hydrogels for Cell Delivery

Cells can be delivered to replace (injury-induced) lost cells or to support the regenerative process by paracrine signaling. The most commonly delivered cell types are stem or progenitor cells, which have the ability to proliferate and differentiate into nondividing cells such as neurons. A recent clinical study has shown that a single injection of human CNS stem cells into six patients, directly rostral and caudal to the injury epicenter, led to a functional recovery in four patients with no reported safety concerns related to the cells [159]. Other donor cell types in clinical trials are mesenchymal stromal cells, adipose-/bone-derived mesenchymal stem cells (MSCs), bone marrow nucleated cell, and bone marrow stem cells [160–163]. Cell types of increasing interest that are being investigated in animal studies include nasal olfactory ensheathing cells

(OECs), Schwann cells, embryonic neurons, embryonic stem cells, and oligodendrocyte progenitor cells (OPCs) [162–165]. As an alternative to embryonic or adult stem cells, host-derived induced pluripotent stem cells (iPSCs) can be applied to avoid immunological rejection. In this case, differentiation strategies have to be carefully chosen to prevent tumor formation [166].

A major challenge in such cell delivery strategies remains the low survival rate of donor cells (ranging from 1% to 20%) and their limited engraftment into the host tissue [167]. This is in part induced by the inflammatory microenvironment of the injury site and cannot be circumvented by simply supplying more cells, as these can become necrotic [168, 169]. These hurdles can be addressed by injecting cells in combination with a supporting hydrogel or biomaterial [170–172]. Different covalent chemistries, such as Michael-type addition [173] and azide-alkyne cycloaddition [174], can be applied to tailor the mechanical properties of the hydrogel via their cross-linking density, which also influences cell survival and their differentiation and thus the quantity of the desired cell type [175]. For example, ultrasoft hydrogels (<1 kPa) have demonstrated to favor neuronal differentiation and remyelination [173, 176]. Other hydrogels that are suitable for cell delivery due to their soft gelation have been prepared via physical bonds, based on protein-protein interaction [171]. Hydrogels, formed via the specific reversible interaction between a proline-rich (P1) and a WW peptide domain, supported the growth of NSCs in their progenitor status, while differentiation into neuronal and glial cells could be induced, depending on the applied molecular signals. To render the hydrogel less prone to biodegradation and increase the stability and efficiency of cell delivery in vivo, a second physical cross-linking step has been introduced after transplantation using network reinforcing thermoresponsive poly (N-isopropylacrylamide) (PNIPAM) [177].

To further enhance proliferation of delivered stem cells and induce and control their differentiation, additional tropic signals can be co-delivered within the material [178]. When NSC embryoid bodies were delivered in a fibrin hydrogel, supplemented with NT-3 and platelet-derived growth factor (PDGF), neuronal differentiation inside the hydrogel was enhanced by 8 weeks, resulting in significantly greater functional recovery in an experimental model of SCI [179]. In an alternative approach, brain-derived NSCs were combined with HAMC hydrogels, which were modified with PDGF to promote oligodendrocyte differentiation. This cell-hydrogel-growth factor combination led to a significant reduction in cavity formation and also improved functional recovery in an experimental model of SCI [180]. To enhance even further the paracrine signaling capacity of delivered cells and supply differentiation factors, the donor cells can also be genetically modified [141]. As an example, viral transduction of primary Schwann cells with a Cre-Lox system has been used to excise a constitutively expressing GDNF gene upon activation of Cre expression through tetracyclin family antibiotics, allowing for temporally controlled expression of GDNF [181].

5.3 Injectable Hydrogels for the Presentation of Cell Binding/ Signaling Motifs

For the design of an injectable hydrogel, neuron and glial-relevant cell binding and signaling motifs need to be identified and incorporated. As mentioned above, SCI induces changes in the local ECM environment from one of hyaluronic acid (HA) and other components, such as laminin, nidogen, tenascin, and low levels of growth factors and CSPGs to one that is subject to an acute inflammatory reaction and becomes generally hostile to axon regeneration and functional tissue repair [182, 183].

One strategy to promote tissue repair relies on thermally induced gelation of decellularized extracellular matrices (dECM), forming a nano-fibrillary hydrogel network [184]. Such dECM has been derived from the porcine brain, spinal cord, and urinary bladder, each providing a different set of molecular constituents and, thus, bioactive profiles [185]. Mouse neuroblastoma cells have been mixed with these dECM prior to gelation. After cultivation, the cells developed three-dimensional neurite networks, showing highest neurite length in brain-derived dECM. In a subsequent study, dECM was injected into a rodent spinal cord hemisection cavity, revealing enhanced neovascularization and axonal ingrowth [186]. However, cyst formation could not be prevented. This was attributed to rapid tissue degradation, which may have been boosted by massive macrophage infiltration. Apart from the rapid degradation, most methods of fabricating dECMs require harsh treatment steps, which can alter the bioactive molecules, and homogenize a material that is naturally heterogeneous [187]. In addition, the clinical application of such an approach may be hindered by the fact that dECMs show limited reproducibility and can cause immunological reactions, especially in the case of animal-derived materials [188].

Even though most naturally derived material-based strategies apply components of the native ECM, one of the first applied injectable materials for SCI was type I collagen, which has a low occurrence in the spinal cord [183]. Collagen forms a temperature-dependent nano-fibrillary network under physiological conditions depending on pH, and ionic strength. By injecting a collagen solution into a spinal cord mid-thoracic transection lesion, considerably more regenerative neurite ingrowth was achieved, as compared to the implantation of a preformed collagen hydrogel [189]. Other natural materials tested with in vivo nervous tissue injury models include agarose, methyl cellulose (MC), hyaluronic acid (HA), chitosan, fibronectin, and fibrin, all of which support an increase in regenerative neurite growth and reduced inflammation, however, with very limited or no associated functional improvements [120]. In an attempt to recapitulate the more complex neural ECM, natural materials have been further optimized. For example, fibrin hydrogels have been modified with additional peptides and proteins, utilizing their enzymatic cross-linking mechanism. The fibrin precursor fibrinogen is first catalyzed by thrombin, which removes the fibrinopeptides A and B, revealing the binding sites. These are then enzymatically cross-linked with activated factor XIII, which connects the ε-amino group of lysine (K) to the γ-carboxamide of glutamine (Q), forming an isopeptide bond [190]. By designing proteins or peptides with the same amino acid sequences that trigger fibrinogen conjugation, they can be covalently bound to fibrin [191]. The addition of four laminin-derived peptide sequences has revealed a synergistic effect between them, enhancing neurite outgrowth of DRGs by 77%, compared to fibrin-only hydrogel, and inducing an 85% increase of regenerated neurons in peripheral nerve tubes. One of the most prominent natural materials was developed in the Shoichet group and consists of hyaluronan and methylcellulose (HAMC) [192] . HAMC combines the thermal gelling properties of MC and shear-thinning properties of HA, enabling fast gelation in situ while being biocompatible and degradable via erosion. Injection of HAMC into the intrathecal space, which is located directly next to the spinal cord, led to a slightly improved functional recovery that was attributed to a decreased inflammatory response [193]. Other hydrogels capable of modulating reactive astrocytosis after SCI and reducing secondary neuronal damage include alginate/chitosan/genipin hydrogels that have also been developed, in which cross-linking in vivo was achieved by reacting with excess Ca^{2+} that was present at the injury site [194].

Hybrid materials have been developed to combine the high level of biocompatibility of natural materials with the flexible and robust molecular structure of synthetic materials. The interactions between both of these components, therefore, need to be chemically tuned. Recently, a chitosan-based material was rendered self-healing by dynamic covalent chemistry (Schiff-base linkage) between telechelic difunctional poly(ethylene glycol) and glycol chitosan [195]. The hydrogel created a mechanical environment with an E-modulus of 1.5 kPa that was capable of inducing neuronal differentiation of NSCs. Injecting this hydrogel with encapsulated NSC spheroids resulted in a superior degree of neuronal recovery of ethanol-exposed zebrafish embryos. Another approach is based on tethering enzymatically cross-linkable peptides to HA-vinyl sulfone (HA-VS) via Michael-type addition using the thiol group of a cysteine [196]. Two different peptides (K and Q peptide) were employed, which together were cross-linked by factor XIIIa, mimicking the fibrin gelation mechanism. An additionally imparted matrix metalloproteinase (MMP)-sensitive peptide sequence allowed cell-induced degradation [197]. Incorporated cortical neurons in vitro formed three-dimensional neuronal networks, which exhibited quick and long-lasting electrical activity.

**5.4 Injectable
Hydrogels that Provide
Guidance**

Neurotrophic stimulation of hydrogel-infiltrating cells can increase neuronal outgrowth, decrease inflammation, and even promote a limited degree of functional recovery. However, control of the direction of regenerating axons should promote a more rapid and possibly more coherent axonal regrowth that is less based on the plasticity of remaining functional spinal tissues. According to Geller and Fawcett [2], there is little point in stimulating axon regeneration if the axons wander randomly throughout the lesion site. Recent in vivo and ex vivo studies of developing neurons have clearly revealed the strong effect of mechanical gradients on neuronal guidance and sprouting [198]. Therefore, implementation of chemotactic or/and haptotactic mechanisms can guide infiltrating cells and regenerating axons across the injury site. Despite this widely accepted fact, a very limited number of injectable/orienting regenerative hydrogels have been generated because the orientation of such materials needs to be controlled in situ after injection. One example of a natural polymer that can form local heterogeneous structures is self-assembling collagen type-1 nanofibers [199]. By decreasing the gelation temperature, the global order increases, leading to longer oriented fibers in local domains. Such micromechanics have been demonstrated to affect cell morphology, proliferation, migration, and differentiation [200].

As synthetic examples, hydrogels, consisting of peptide amphiphiles (PAs), have been fabricated by self-assembly of supramolecular nanofibers [201]. PA hydrogels, tailored with the laminin-derived peptide sequence IKVAV, have been injected into compression injuries of the adult mouse spinal cord [202]. Although the PAs self-assembled into nonaligned nanofibers, they promoted descending motor and ascending sensory fiber growth through the lesion site. Astrogliosis and local cell death were reduced, the number of oligodendroglia at the injury site increased, and behavioral studies revealed significant functional improvement. Further manipulation of the PA hydrogel elasticity by altering the PA interactions led toward softer materials, revealing an accelerated neuronal polarity due to more dynamic cell-material interactions [203]. To render PA hydrogels directional or anisotropic, monodomain regions of aligned, supramolecular nanofibers were formed by heating, cooling, and collecting the PA solution in salt-containing medium [201]. The supramolecular aggregates (7–8 nm) undergo an entropy-driven dehydration during heating, enabling a closer, irreversible interaction and fusion between fibrous aggregates during cooling. The resulting nanofiber bundles have a diameter of approximately 40 nm and are ordered in microscopic monodomain regions, which can be aligned by flow [204]. Cardiomyocytes, encapsulated into these aligned fibrous scaffolds, revealed electrochemical communication that followed the axis of the fibers. RGD-, IKVAV-, or VFDNFVLK-(from TenascinC) modified PAs that underwent the thermal annealing process and salt-induced assembly demonstrated substantial alignment and enhanced neurite elongation in vitro [205, 206]. Co-delivered NPCs differentiated into neurons without the addition of differentiation promoting growth factors. Furthermore, the NPC-loaded liquid PA solution was applied to an injured rat spinal cord by injecting and retracting the needle following the longitudinal axis of the tissue. The liquid solidified upon contact with endogenous Ca^{2+}, resulting in locally aligned cell growth. Anisotropic PA hydrogels are fully synthetic and can be modified with specific adhesion sequences or function as growth factor delivery systems [207]. Yet, control of the nanofiber direction is limited to the direction of the injected hydrogel flow [205].

An alternative route to control the direction of the fibrils of a hydrogel is to magnetically align the matrix itself or structural elements inside the matrix. To produce oriented collagen and fibrin scaffolds, Tranquillo and colleagues utilized the paramagnetic properties of these materials to align nanofibrils with high-strength (5–10 T) magnetic fields [208, 209]. Interestingly, only nanofibers of larger diameters (i.e., 460 and 510 nm) were capable of aligning axonal growth, whereas aligned nanofibrils of around 150 nm showed no influence in comparison with isotropic hydrogels. Magnetically aligned collagen scaffolds have been transplanted into rats, leading to improved recovery after SCI by haptotactically guiding infiltrating axons [210]. However, such scaffolds have, so far, not been applied as an injectable material yet.

Finally another approach to create injectable and aligned collagen hydrogels relies on incorporating magnetic nanoscale particles (MNPs) into the liquid collagen solution [211]. By applying a static magnet (54–433 mT according to distance), the spherical MNPs assemble into chain-like structures inside the hydrogel, whereas the MNP movement locally aligns the collagen nanofibers, which are fixed by a temperature shift to solidify the hydrogel. Primary neurons cultured within such scaffolds displayed significant orientation that matched that of the MNPs and collagen nanofibers. For improved biocompatibility, MNPs have been coated with ECM proteins before assembly into chain-like structures inside Matrigel [212]. In such designs, only the MNPs strings themselves function as the guidance structures for cell orientation. Due to MNP topography, individual fibroblasts and PC12 neuron-like cells extended processes that grew both parallel to and perpendicular to the direction of the particle strings. Despite these encouraging results, these structures are defined by the amount, size, and assembly of the MNPs, which are known to be cytotoxic and neurite growth inhibiting at high concentrations [213]. In order to address these limitations, a minimally invasive hybrid hydrogel system was developed to form an oriented structure in situ after injection. The hydrogel is called an Anisogel and consists of modifiable rod-shaped microgels or short fibers that are receptive to external magnetic fields, and a surrounding nerve-supportive hydrogel, which cross-links in situ and fixes the aligned elements [214–217]. The unidirectional magnetic assembly of microgels or fibers was achieved by supplementing them with small amounts of superparamagnetic iron oxide nanoparticles below the cytotoxic range. Oriented microgels or fibers within a fibrin or PEG-base hydrogel functioned as guiding substrates and induced unidirectional growth of fibroblasts, neurites of DRGs, and primary nerve cells, with only a minimal amount of structural guiding elements required (<2 vol.% microgel/hydrogel) [218].

6   Development of Bioengineered Nanofiber-Based Structures for Use in Peripheral Nerve Injury

The rather simple anatomy of the peripheral nerve along with its high regenerative capacity simplifies the development of repair strategies. However, an "off-the-shelf" nerve guide, capable of bridging peripheral nerve defects larger than 30 mm, has yet to be achieved. To date, the autologous nerve graft is still the gold standard for the repair of gaps larger than 30 mm. The relatively recent revival of electrospinning technology, however, has allowed the generation of nanoscaled fibers and has boosted the development of nerve guides by providing controlled topographical cues that strongly influence directed cell growth (i.e., aligned nanofibers) as well as controlled conduit porosity (i.e., random nanofibers). As mentioned earlier, the combination of topographical cues of synthetic and naturally occurring polymers as well as biofunctionalization, and the introduction of axon-growth-supporting cells, has allowed the development of scaffolds and intervention strategies with great clinical potential. In this section we present a short description of the state of the art of peripheral nerve guide (PNG) development.

6.1   Current FDA/CE-Approved Conduits and Scaffolds that Are Commercially Available for Peripheral Nerve Injury: Relative Advantages and Disadvantages of these Materials

The use of conduits for the repair of PNI dates back to the second half of the nineteenth century, where different tubular tissues, such as decalcified bone and vessels from human and animal origin, were used to bridge large nerve defects [219]. Currently, surgeons have a number of bioengineered devices at their disposal, all of which have received approval by the US Food and Drug Administration (FDA) and Conformité Européenne (CE) for clinical use to bridge PNS lesions (Table 1). The majority of these products are simple hollow conduits, prepared from natural and/or synthetic polymers such as collagen (NeuraGen®, NeuroMatrix™), polyglycolic acid (PGA, Neurotube®), polyvinyl alcohol (PVA, SaluTunnel™), poly DL-lactide-ε-caprolactone (PLLC, Neurolac®), chitosan (Reaxon®), or porcine   small   intestinal   submucosa   (AxoGuard™   Nerve

Table 1
Available FDA- and or CE-approved bioengineered nerve guides

Device name	Company	Material	Structure	Diameter, mm	Length, mm
Avance®	Axogen Inc.	Cleansed and decellularized ECM	Native PNS	1–5	15–70
Neurotube®	Neuroregen, L.L.C.	PGA	Hollow tube	2.3–8	20–40
NeuraGen®	Integra Lifesciences Corp.	Type I collagen	Hollow tube	1.5–7	20–30
NeuroMatrix™	Collagen Matrix, Inc.	Type I collagen	Hollow tube	2–6	25
AxoGuard™ Nerve Connector	Cook Biotech, Inc.	Porcine small intestine submucosa	Hollow tube	1.5–7	10
Neurolac®	Polyganics Bv	Poly(DL-lactide-ε-caprolactone)	Hollow tube	1.5–10	30
SaluTunnel™	Salumedica L.C.C.	Polyvinyl alcohol	Hollow tube	2–10	63.5
NeuraGen 3D®	Integra Lifesciences Corp.	Collagen type I, glycosaminoglycan (chondroitin-6-sulfate)	Collagen tube filled with collagen and GAG	1.5–7	63
Reaxon®	Medovent GmbH	Chitosan	Hollow tube	2.1–6.0	30
Nerbridge™	Toyobo Co. Ltd.	PGA, medical-grade collagen	PGA tube filled with collagen	0.5–4.0	55

For a comprehensive description of some of these devices, the reader is referred to the following reviews [220, 221].

Connector). However, over recent years, conduits containing luminal components have been developed and approved. NeuraGen 3D®, an improved version of NeuraGen®, uses type I collagen conduit filled with a collagen hydrogel mixed with the glycosaminoglycan (GAG) chondroitin-6-sulfate. Nerbridge™ is a combination of a PGA conduit filled with medical-grade collagen hydrogel. Although hollow conduits have shown to be successful in bridging small gaps in animal models of PNI as well as in human PNI, none are recommended for their use in defects larger than 30 mm. The new devices or nerve guides containing fillings are expected to improve regenerative performance, and it has been suggested that Nerbridge, for example, be used in lesion-induced gaps of up to 50 mm. Recently, Avance® has been the first allograft material to receive FDA approval. The Avance® nerve graft (AxoGen Inc.) has been harvested from cadaveric sources and undergoes a series of processes to remove cellular components and axon-growth-inhibitory molecules, such as CSPGs, that are present in normal nerve and upregulated after injury [222]. It is available in different dimensions and has been designed to bridge gaps up to 70 mm. The limited clinical studies using the Avance® allogenic graft show superior performance compared to simple hollow conduits when used in nerve defects of sensory, motor, and mixed nerves, especially for the repair of the inferior alveolar nerve, bridging gaps up to 70 mm [223–225]. However, while these results are promising, the performance of the most recently approved devices, in particular of the nerve guides containing a luminal filling, still need to be documented, and comprehensive comparative studies with the gold standard autograft are required.

6.2 Nanofiber-Based Conduits for Peripheral Nerve Repair

For the development of bioengineered nerve guides, nature has been used as the main source of inspiration with PNR and the autografted nerve acting as the main blueprint. Tubular structures containing oriented guidance cues, as similes of the endoneurial basal lamina, have been engineered to induce a positive response of surrounding regenerating tissues. The main goal of these approaches has been to promote Schwann cell migration into the nerve guides as a means of supporting directed axonal regeneration. To achieve this, a series of physical and chemical cues, alone or in combination, have been proposed, such as patterning and stretching as well as by introducing molecular signals, axon-growth-promoting ECM molecules and cells. In peripheral nerve regeneration as in SCI, electrospinning has been the primary method of generating nanostructured substrates. As mentioned earlier, a great number of natural or synthetic polymers have been successfully electrospun into nanofibers and their biocompatibility studied, including their ability to influence neural cell and axonal behavior [226]. Over the past 5 years, there has been a significant increase in the number of nanofiber-based nerve guides being evaluated in preclinical in vivo studies, shedding light onto their potential use in the clinic.

Hollow conduits are the simplest form of structures used to bridge nerve injuries, and similar constructs, based on the use of nanofibers, have already been developed [227–229]. By creating mats of nanofibers and rolling them around a rod, several devices have been generated to investigate the benefits of incorporating nanofibers in the scaffold design. Randomly oriented fibers have generally been used in conduit design to tailor porosity of the wall and permit the exchange of nutrients/metabolites while limiting the infiltration or migration of highly proliferative, scar-inducing cells such as fibroblasts [221]. However, special attention has been placed on the incorporation of aligned nanofibers as a way to influence directed axonal regeneration and improved Schwann cell migration. The positive effects of substrate topography have been demonstrated by studies using nanofibers from materials lacking cell-specific binding sites [95, 97, 230]. By electrospinning poly (acrylonitrile-co-methylacrylate) (PAN-MA) onto a rotating drum, Kim and colleagues generated sheets of densely packed longitudinally aligned nanofibers. Ten to 12 thin sheets of aligned PAN-MA nanofibers were stacked and placed in a polysulfone tube to bridge a 17 mm gap of the adult rat sciatic nerve. Their results showed good tissue repair as indicated by detection of large numbers of S100+ Schwann cells throughout the construct that had migrated from the proximal and distal nerve stumps. This is in contrast to the poor Schwann cell migration observed in the conduits composed of randomly oriented nanofibers. The aligned nanofibers also facilitated axonal regeneration through the conduit, resulting in de novo reinnervation of target muscles and good functional recovery in the grid walk test as well the recovery of the compound action potentials of motor and sensory nerve [231]. In a later study, the same group showed that by reducing the number of sheets placed into the conduit (using either one or three sheets), greater numbers of regenerating axons and higher nerve conduction velocities (NCVs) were associated with only a single intraluminal sheet. This supported the notion that only minimal amount of properly positioned topographical cues is needed to support successful nerve regeneration [232, 233]. These encouraging results also suggest that the possibility of developing an "off-the-shelf" nerve guide from purely synthetic polymers is an achievable goal. Nerve guides intended for the bridging of large gaps, however, should incorporate biochemical cues such as cell adhesion motifs, growth factors, and/or axon-growth-promoting cells [221, 234]. Naturally occurring polymers such as collagen, gelatin, fibronectin, chitosan, and fibroin derived from silk have been successfully electrospun and are regarded as the ideal polymers because they incorporate both physical and biochemical cues [235–237]. The nanofibrous structures obtained from these materials mimic the more natural ECM milieu in which cells grow. However, the substrates produced with these polymers show, in most cases, only poor mechanical and structural stability. In order to improve the physical properties of such constructs, they are blended with synthetic polymers such as PCL, poly (L-lactide-co-caprolactone) (PLLACL), polylactic acid (PLA), and polydioxanone (PDO) [237].

Most of the experimental strategies proposed to date are introduced as nanostructured nerve guides generated from functionalized synthetic polymers as mentioned before by addition of biochemical elements using different methods, such as blending, adsorption, and coaxial electrospinning. Polymer blends are an easy way to produce electrospun nanofibers presenting biological cues on their surface with minimal manipulation. Presence of the molecule externally is important, but presence of the active molecule within the polymer fiber, especially in the case of degradable polymers, could be beneficial, as motifs are guaranteed to be available during degradation [238]. However, there is poor control over the location of the introduced cues (on the surface or within the fibers) and the electrospinning process itself or the solvents used to generate the blends can destroy the motifs [237]. A hollow porcine collagen type I nerve guide with a core of sheets of loosely packed oriented PCL- or C/PCL nanofibers suspended in a gelatin hydrogel has been evaluated by Kriebel and colleagues in vivo using the sciatic nerve resection injury model with a 15 mm nerve defect. The results at the 3-month post-lesion/implantation time revealed similar axon-growth-supporting performances of both PCL and C/PCL nanofiber types in histological and behavioral analysis; however, electrophysiological recordings showed a positive trend toward the C/PCL blend [239]. The continuous presence of growth factors known to be important for peripheral nervous system repair and axonal regeneration, such as NGF, BDNF, CNTF, and GDNF, is highly desired during the regenerative process, and strategies intended to deliver these factors within the nerve guide have also been explored [39, 221, 240, 241]. Coaxial electrospinning has allowed the use of composite nanofibers that serve not only as structural building blocks but at the same time renders them as drug delivery systems by producing fibers with cores rich in desired molecules such as growth factors, DNA, and siRNA sequences. Sheets of aligned PLGA nanofibers with PEG-NGF cores have been wound onto a stainless steel bar in order to produce hollow nerve guides, the tissue-repairing properties of which were evaluated in a 13 mm sciatic nerve defect in adult rats. In vitro release studies had shown a burst of NGF liberation within the first 4 days followed by a more steady release for up to 30 days. The early burst of NGF release has been attributed to the presence of NGF on the surface of the fibers (that occurs during the electrospinning process), and the slower, longer-term release of NGF was believed to be due to growth factor diffusion through pores on the surface of the PLGA shell. By 12 weeks after implantation into the sciatic nerve lesion model, the PLGA/PEG+NGF nerve guide showed significantly better regenerative performance when compared to PLGA-only fibers. The support of axon regeneration, by the PLGA/PEG-NGF nanofibers, was sufficiently strong that axon counts, myelin sheath thickness, nerve conduction velocities (NCV), and stimulus-evoked compound muscle action potentials (CMAP) were all similar to those observed in the control, autograft implantation group [241].

The high success rate of the autograft in repairing large gaps in experimental models of PNI is not only attributed to its oriented topography but also to the presence of supporting cells that protect, guide, and provide the ideal environment for axonal regeneration [39]. In nerve guides pre-seeded with cells, the interactions between nanofibers, especially those with defined orientation, and the incorporated cells are intended to mimic the typical alignment of cell in the bands of Büngner within the autograft. Schwann cells, the native glial cells of the peripheral nerve, are commonly used as the cell population of choice in seeded nanofiber-based scaffolds, due to their ability to support axonal regeneration after PNI [228, 242]. The use of OEGs is also a popular choice for PNI repair strategies, as these cells support axon growth from newly formed receptor neurons in the olfactory epithelium to the olfactory bulb throughout adulthood [243]. However, newly discovered populations of stem cells such as those from human exfoliated deciduous teeth (SHED) as well as iPSCs generated artificially from adult cells have added to the component that has been introduced into a number of different nanofibers constructs to improve tissue repair in vivo [227, 244]. While the introduction of a cellular components into nanostructure-based therapeutical strategies has the potential to improve regeneration, complicating issues similar to those observed following tissue grafts need to be overcome, including immunocompatibility and rejection in the case of allografted and xenografted cells, as well as the need of large numbers of cells that would be required for the seeding of scaffolds when using cells harvested from the same patient. Stem cell technologies such as iPSCs or the harvesting of stem cells from postnatal sources such as the collection of human umbilical vein endothelial cells (HUVECs) after birth has allowed the creation of imunnocompatible cell sources with the potential for personalized medicine; however, the controlled differentiation into the appropriate cell types required for a specific task is still in development, as is the fine tuning of the nanostructured devices that will act as their vehicles for implantation in tissue engineering and regenerative medicine (as described in the next section).

6.3 Advances in Nano-Scaffold Designs Intended to Control/Improve Cell-Substrate Interactions for Peripheral Nerve Repair

For the improvement of bioengineered nerve guides, researchers are focusing on two major fronts: biomaterial polymers and structural design, both following a biomimetic principle. As mentioned above, materials presenting macromolecules or peptide sequences on the surface of nanofibers are being extensively explored. One such peptide sequences is the RGD (arginine-glycine-aspartate) sequence, a signaling motif responsible for cell adhesion to ECM components following recognition by cell surface integrin receptors. RGD sequences have been successfully covalently bound to chitosan nanofibers through a PEG linker and show improved fibroblast adhesion properties [245]. In another study a blend of poly(serinol hexamethylene urea)-RGD PSHU-RGD and PCL was used to generate bioactive surfaces which performed as well as laminin-coated surfaces in in vitro assays in terms of supporting neurite outgrowth from PC12 cells [246]. By incorporating the RGD sequence into the PSHU element, the presence of the RGD sequence is not only on the surface of the fibers but also within them, guaranteeing motifs to be present as fibers degrade. The combined PSHU-RGD/PCL mixture was later successfully electrospun using a modified parallel collectors system. Rods of sucrose (200–500 µm) were placed perpendicular between two parallel wire collectors, and two syringes, placed on opposite sides of the arrangement, were used as polymer sources. Nanofibers were electrospun parallel to the sucrose rods until a defined sheet was obtained. The sheet was then removed from the parallel wires and rolled to obtain a 1.2 mm diameter structure. The sucrose rods were then dissolved in water leaving behind microchannels. The construct of aligned nanofiber-walled micro-channels was evaluated using human neural stem cells. Cells in the PSHU-RGD/PCL construct showed alignment along the channel/nanofiber direction in comparison with pure PCL constructs [247]. Following the concept of the effectiveness of minimal guidance cues, Hodde and colleagues proposed the embedding of stacked layers of low density, oriented PCL nanofibers into fibrin hydrogels as a possible 3D substrate for guided tissue repair. Schwann cells that were embedded in the nanofiber-containing hydrogels redistributed themselves to be almost entirely associated with the oriented nanofibers rather than remaining in the 3D fibrin hydrogel. Upon contact with the non-functionalized PCL nanofibers, the Schwann cells adopted an elongated morphology that followed the long axis of the fibers, establishing rapid and maximal process outgrowth after just 1 day in tissue culture [76]. The efficacy of such constructs to support directed axon regeneration and tissue repair in an animal model of PNI has yet to be demonstrated. Similar to the situation of SCI (mentioned above), self-assembling peptide nanofibrous hydrogels have also been used in experimental peripheral nerve repair [204, 248, 249]. In a study published by Li and colleagues, peptide amphiphiles with and without RGD functional motifs were passed through a 40 µm mesh screen, introducing sheer stress in the self-assembling peptide solution, in order to generate a highly aligned nanofibrous hydrogel within a hollow PLGA nerve guide. The fibers that contained the RGD sequence were shown to induce alignment of Schwann cells in vitro as well as in vivo and supported tissue repair that was comparable to implanted autologous nerve as shown through similar sciatic functional index (SFI) scores and hind paw withdrawal latency [204]. New trends in therapies for nerve regeneration have shown that electrical stimulation is a good complementary technique to enhance the rate of axonal regeneration following PNI, and electrical stimulation in nerve conduits has already been described [250, 251]. In order to increase the electrical conductivity of nerve guides, carbon nanotubes have been mixed with polymers such as PLLA and PLLA/PLGA in order to generate electrospun nanofiber-based substrates. The carbon nanotubes improved electrical conductivity of the material, and single-walled carbon nanotubes (SWCNTs) have also showed to be biocompatible [243, 252]. More recently Kabiri and colleagues used blends of SWCNTs and PLLA to generate nerve guide scaffolds that were based on oriented nanofiber technology with electrically conductive properties [243]. The constructs were produced with unseeded and seeded sheets of nanofibers with OEG that after reaching confluency were rolled to form a hollow cylindrical construct with four to five layers of cell nanofibers on its outer wall. Axonal regeneration was demonstrated through the rather short (i.e., 8 mm) rat sciatic nerve defect using the pre-seeded constructs, and improved SFI scores in comparison with hollow silicone conduits and non-cell-seeded scaffolds were shown. However no electrical stimulation was performed throughout the study, and it can only be assumed that such experiments will be repeated in the presence of electrical stimulation in the near future. In another study, a hollow nerve guide made of randomly oriented poly-L-lactic acid-co-ε-caprolactone PLCL nanofibers received a polypyrrole (PPy) coating, also intended to increase scaffold conductivity [253]. Such conductive nerve guides were implanted into a 15 mm sciatic nerve gap and electrically stimulated by applying a voltage of 100 mV for 1 h at 1, 3, and 7 days postimplantation, while the animal was under anesthesia. The regenerative performance of the stimulated conduit was evaluated at 4 and 8 weeks postimplantation, and parameters of NCV, CMAPs, as well as the SFI all showed a statistically significant improvement when compared with the non-stimulated conduit. However, tissue repair was still not as good as that observed following the implantation of the autograft.

7   Concluding Comments

Nanostructured scaffolds and injectable hydrogels for the repair of SCI and PNI have undergone a tremendous development throughout the last two decades. Current state-of-the-art biomaterials have the ability to mimic aspects of the endogenous nervous tissue microenvironments by tailoring of the molecular, mechanical, and physical properties (with nanofiber/fibril diameters being in the submicron range). The addition of trophic factors and/or pharmacological agents may directly enhance the regenerative capacity of the implanted scaffold for injured spinal cord tissues or may support repair indirectly by blocking or interfering with host axon-growth inhibitory/scarring mechanisms. The supplementary delivery of cells (i.e., stem cells, progenitors, or growth-promoting glia) can vitalize the implanted or injected materials and further support regenerating axons as well as improve their maturation and functional properties. The most promising achievements with regard to the reduction of scarring and enhanced functional recovery have been obtained by combining cell delivery, drug or gene delivery, and material design [120]. Incorporation of nano-features (in particular highly oriented nanostructures), which can, for example, enhance the material conductivity, can further improve neuroregeneration by electric excitation [254]. Therefore, carbon nanotubes [255, 256] or graphene nanoparticles [257] which show no adverse side effects have been proposed to have great promise in future tissue engineering and regenerative medicine approaches. Notwithstanding that much progress has been made over recent decades, a number of substantial challenges still lie ahead. These include reduction of fibroadhesive, as well as glial scarring, the creation of chemotactic or mechanotactic gradients in situ, and improved implant-host integration. Clearly, there is still some way to go before large cavitating lesions or fibroadhesive scars within the spinal cord or large gaps within peripheral nerves can be successfully bridged in a highly reproducible and controllable manner. To tackle this issue, light-induced patterning [122, 212, 258] and additive manufacturing techniques [259] might represent some of the most advanced systems. Progress in nano- and microfluidics may provide the basis for establishing biochemical gradients while injecting biomaterials [260]. Implementation and engineering of novel systems with existing regenerative materials will create a new toolbox for directed cell stimulation and regenerative tissue growth. The support of successful axon regeneration across large spinal cord and peripheral nerve injuries clearly represents a major step forward in regenerative medicine; however, an equally important determinant in the promotion of functional recovery involves the re-establishment of appropriate synaptic connectivity between regenerating axons and their target neurons or effector structures. The mechanisms that control such events in the lesioned adult nervous system remain largely unknown and uncontrolled. Nonetheless, it seems reasonable to take the optimistic view that continued multidisciplinary progress will be translated into clinically effective strategies for both CNS and PNS injured patients.

References

1. Harel NY, Strittmatter SM (2006) Can regenerating axons recapitulate developmental guidance during recovery from spinal cord injury? Nat Rev Neurosci 7:603–616

2. Geller HM, Fawcett JW (2002) Building a bridge: engineering spinal cord repair. Exp Neurol 174:125–136

3. Azevedo FAC et al (2009) Equal numbers of neuronal and nonneuronal cells make the human brain an isometrically scaled-up primate brain. J Comp Neurol 513:532–541

4. Singh A et al (2014) Global prevalence and incidence of traumatic spinal cord injury. Clin Epidemiol 6:309–331

5. Bunge RP et al (1993) Observations on the pathology of human spinal cord injury. A review and classification of 22 new cases with details from a case of chronic cord compression with extensive focal demyelination. Adv Neurol 59:75–89

6. Bunge RP, Puckett WR, Hiester ED (1997) Observations on the pathology of several types of human spinal cord injury, with emphasis on the astrocyte response to penetrating injuries. Adv Neurol 72:305–315

7. Schwab ME, Bartholdi D (1996) Degeneration and regeneration of axons in the lesioned spinal cord. Physiol Rev 76:319–370

8. Richardson PM et al (2009) Responses of the nerve cell body to axotomy. Neurosurgery 65: A74–A79

9. Bareyre FM et al (2004) The injured spinal cord spontaneously forms a new intraspinal circuit in adult rats. Nat Neurosci 7:269–277

10. Ghosh A et al (2010) Rewiring of hindlimb corticospinal neurons after spinal cord injury. Nat Neurosci 13:97–104

11. Courtine G et al (2008) Recovery of supraspinal control of stepping via indirect propriospinal relay connections after spinal cord injury. Nat Med 14:69–74

12. Silver J, Miller JH (2004) Regeneration beyond the glial scar. Nat Rev Neurosci 5:146–156

13. Fournier AE, GrandPre T, Strittmatter SM (2001) Identification of a receptor mediating Nogo-66 inhibition of axonal regeneration. Nature 409:341–346

14. McKerracher L, Rosen KM (2015) MAG, myelin and overcoming growth inhibition in the CNS. Front Mol Neurosci 8:51

15. Li S, Strittmatter SM (2003) Delayed systemic Nogo-66 receptor antagonist promotes recovery from spinal cord injury. J Neurosci 23:4219–4227

16. Starkey ML, Schwab ME (2012) Anti-Nogo-A and training: can one plus one equal three? Exp Neurol 235:53–61

17. Hackett AR, Lee JK (2016) Understanding the NG2 glial scar after spinal cord injury. Front Neurol 7:199

18. Busch SA, Silver J (2007) The role of extracellular matrix in CNS regeneration. Curr Opin Neurobiol 17:120–127

19. Brook GA et al (1999) Astrocytes re-express nestin in deafferented target territories of the adult rat hippocampus. Neuroreport 10:1007–1011

20. Snow DM, Steindler DA, Silver J (1990) Molecular and cellular characterization of the glial roof plate of the spinal cord and optic tectum: a possible role for a proteoglycan in the development of an axon barrier. Dev Biol 138:359–376

21. Dou CL, Levine JM (1994) Inhibition of neurite growth by the NG2 chondroitin sulfate proteoglycan. J Neurosci 14:7616–7628

22. Morgenstern DA, Asher RA, Fawcett JW (2002) Chondroitin sulphate proteoglycans in the CNS injury response. Prog Brain Res 137:313–332

23. Galtrey CM, Fawcett JW (2007) The role of chondroitin sulfate proteoglycans in regeneration and plasticity in the central nervous system. Brain Res Rev 54:1–18

24. Faulkner JR et al (2004) Reactive astrocytes protect tissue and preserve function after spinal cord injury. J Neurosci 24:2143–2155

25. Okada S et al (2006) Conditional ablation of Stat3 or Socs3 discloses a dual role for reactive astrocytes after spinal cord injury. Nat Med 12:829–834

26. Bradbury EJ et al (2002) Chondroitinase ABC promotes functional recovery after spinal cord injury. Nature 416:636–640

27. Massey JM et al (2006) Chondroitinase ABC digestion of the perineuronal net promotes functional collateral sprouting in the cuneate nucleus after cervical spinal cord injury. J Neurosci 26:4406–4414

28. Pasterkamp RJ et al (1999) Expression of the gene encoding the chemorepellent semaphorin III is induced in the fibroblast component of neural scar tissue formed following injuries of adult but not neonatal CNS. Mol Cell Neurosci 13:143–166

29. Niclou SP et al (2003) Meningeal cell-derived semaphorin 3A inhibits neurite outgrowth. Mol Cell Neurosci 24:902–912

30. Göritz C et al (2011) A pericyte origin of spinal cord scar tissue. Science 333:238–242

31. Soderblom C et al (2013) Perivascular fibroblasts form the fibrotic scar after contusive spinal cord injury. J Neurosci 33:13882–13887

32. Barnabé-Heider F et al (2010) Origin of new glial cells in intact and injured adult spinal cord. Cell Stem Cell 7:470–482

33. Ren Y et al (2017) Ependymal cell contribution to scar formation after spinal cord injury is minimal, local and dependent on direct ependymal injury. Sci Rep 7:41122

34. Rolls A et al (2008) Two faces of chondroitin sulfate proteoglycan in spinal cord repair: a role in microglia/macrophage activation. PLoS Med 5:e171

35. Beck KD et al (2010) Quantitative analysis of cellular inflammation after traumatic spinal cord injury: evidence for a multiphasic inflammatory response in the acute to chronic environment. Brain 133:433–447

36. Kigerl KA, McGaughy VM, Popovich PG (2006) Comparative analysis of lesion development and intraspinal inflammation in four strains of mice following spinal contusion injury. J Comp Neurol 494:578–594

37. Rapalino O et al (1998) Implantation of stimulated homologous macrophages results in partial recovery of paraplegic rats. Nat Med 4:814–821

38. Kwon MJ, Yoon HJ, Kim BG (2016) Regeneration-associated macrophages: a novel approach to boost intrinsic regenerative

capacity for axon regeneration. Neural Regen Res 11:1368–1371

39. Deumens R et al (2010) Repairing injured peripheral nerves: bridging the gap. Prog Neurobiol 92:245–276

40. Noble J et al (1998) Analysis of upper and lower extremity peripheral nerve injuries in a population of patients with multiple injuries. J Trauma 45:116–122

41. Robinson LR (2000) Traumatic injury to peripheral nerves. Muscle Nerve 23:863–873

42. Lee SK, Wolfe SW (2000) Peripheral nerve injury and repair. J Am Acad Orthop Surg 8:243–252

43. Pabari A et al (2010) Modern surgical management of peripheral nerve gap. J Plast Reconstr Aesthet Surg 63:1941–1948

44. Goldberg JL, Barres BA (2000) The relationship between neuronal survival and regeneration. Annu Rev Neurosci 23:579–612

45. Novikov L, Novikova L, Kellerth JO (1997) Brain-derived neurotrophic factor promotes axonal regeneration and long-term survival of adult rat spinal motoneurons in vivo. Neuroscience 79:765–774

46. Makwana M, Raivich G (2005) Molecular mechanisms in successful peripheral regeneration. FEBS J 272:2628–2638

47. Shin JE, Cho Y (2017) Epigenetic regulation of axon regeneration after neural injury. Mol Cells 40:10–16

48. Yudin D et al (2008) Localized regulation of axonal RanGTPase controls retrograde injury signaling in peripheral nerve. Neuron 59:241–252

49. Schmitt AB et al (2003) Identification of regeneration-associated genes after central and peripheral nerve injury in the adult rat. BMC Neurosci 4:8

50. Saxena S, Caroni P (2007) Mechanisms of axon degeneration: from development to disease. Prog Neurobiol 83:174–191

51. Perry VH, Brown MC, Gordon S (1987) The macrophage response to central and peripheral nerve injury. A possible role for macrophages in regeneration. J Exp Med 165:1218–1223

52. Karanth S et al (2006) Nature of signals that initiate the immune response during Wallerian degeneration of peripheral nerves. Exp Neurol 202:161–166

53. Boivin A et al (2007) Toll-like receptor signaling is critical for Wallerian degeneration and functional recovery after peripheral nerve injury. J Neurosci 27:12565–12576

54. Brosius LA, Barres BA (2014) Contrasting the glial response to axon injury in the central and peripheral nervous systems. Dev Cell 28:7–17

55. Vargas ME, Barres BA (2007) Why is Wallerian degeneration in the CNS so slow? Annu Rev Neurosci 30:153–179

56. Stratton JA, Shah PT (2016). Macrophage polarization in nerve injury: do Schwann cells play a role? Neural Regen Res 11(1): 53–57

57. Sulaiman W, Gordon T (2013) Neurobiology of peripheral nerve injury, regeneration, and functional recovery: from bench top research to bedside application. Ochsner J 13:100–108

58. Ide C (1983) Nerve regeneration and Schwann cell basal lamina: observations of the long-term regeneration. Arch Histol Jpn 46:243–257

59. Son YJ, Thompson WJ (1995) Schwann cell processes guide regeneration of peripheral axons. Neuron 14:125–132

60. Thompson DM, Buettner HM (2006) Neurite outgrowth is directed by Schwann cell alignment in the absence of other guidance cues. Ann Biomed Eng 34:161–168

61. Gaudet AD, Popovich PG, Ramer MS (2011) Wallerian degeneration: gaining perspective on inflammatory events after peripheral nerve injury. J Neuroinflammation 8:110

62. Reneker DH, Chun I (1996) Nanometre diameter fibres of polymer, produced by electrospinning. Nanotechnology 7:216–223

63. Venugopal J et al (2008) Nanotechnology for nanomedicine and delivery of drugs. Curr Pharm Des 14:2184–2200

64. Li W-J et al (2002) Electrospun nanofibrous structure: a novel scaffold for tissue engineering. J Biomed Mater Res 60:613–621

65. Chew SY et al (2006) The role of electrospinning in the emerging field of nanomedicine. Curr Pharm Des 12:4751–4770

66. Hutmacher D et al (2008) Scaffold design and fabrication. In: van Blitterswijk C et al (eds) Tissue engineering. Elsevier, Amsterdam, pp 403–454

67. Pham QP, Sharma U, Mikos AG (2006) Electrospinning of polymeric nanofibers for tissue engineering applications: a review. Tissue Eng 12:1197–1211

68. Reneker DH, Yarin AL (2008) Electrospinning jets and polymer nanofibers. Polymer (Guildf) 49:2387–2425

69. Sun B et al (2014) Advances in three-dimensional nanofibrous macrostructures via electrospinning. Prog Polym Sci 39:862–890

70. Hutmacher DW, Dalton PD (2011) Melt electrospinning. Chem Asian J 6:44–56

71. Brown TD, Dalton PD, Hutmacher DW (2016) Melt electrospinning today: an opportune time for an emerging polymer process. Prog Polym Sci 56:116–166

72. Li F, Zhao Y, Song Y (2010) Core-shell nanofibers: nano channel and capsule by coaxial electrospinning. In: Kumar A (ed) Nanofibers. InTech, London

73. K.S. Athira, Pallab S. & Chatterjee K. (2014) Fabrication of poly(caprolactone) nanofibers by electrospinning. J Polym Biopolym Phys Chem 2, 62–66

74. Sahay R, Thavasi V, Ramakrishna S (2011) Design modifications in electrospinning setup for advanced applications. J Nanomater 2011: 317673

75. Dalton P.D., Klee D. & Möller M. (2005) Electrospinning with dual collection rings. Polymer (Guildf) 46, 611–614

76. Hodde D. et al. (2016) Characterisation of cell-substrate interactions between Schwann cells and three-dimensional fibrin hydrogels containing orientated nanofibre topographical cues. Eur J Neurosci 43, 376–87

77. Brown T.D. et al. (2014) Melt electrospinning of poly(ε-caprolactone) scaffolds: phenomenological observations associated with collection and direct writing. Mater Sci Eng C Mater Biol Appl 45, 698–708

78. Hochleitner G. et al. (2016) Fibre pulsing during melt electrospinning writing. BioNanoMat 173–4159–171

79. Ekaputra A.K. et al. (2008) Combining electrospun scaffolds with electrosprayed hydrogels leads to three-dimensional cellularization of hybrid constructs. Biomacromolecules 9, 2097–103

80. Neal RA et al (2009) Laminin nanofiber meshes that mimic morphological properties and bioactivity of basement membranes. Tissue Eng Part C Methods 15:11–21

81. Xu C et al (2004) In vitro study of human vascular endothelial cell function on materials with various surface roughness. J Biomed Mater Res A 71:154–161

82. Min B et al (2004) Formation of nanostructured poly(lactic-co-glycolic acid)/chitin matrix and its cellular response to normal human keratinocytes and fibroblasts. Carbohydr Polym 57:285–292

83. Lee S-H, Yoon J-W, Suh MH (2002) Continuous nanofibers manufactured by electrospinning technique. Macromol Res 10:282–285

84. Lee CH et al (2005) Nanofiber alignment and direction of mechanical strain affect the ECM production of human ACL fibroblast. Biomaterials 26:1261–1270

85. Mathew G et al (2006) Preparation and anisotropic mechanical behavior of highly-oriented electrospun poly(butylene terephthalate) fibers. J Appl Polym Sci 101:2017–2021

86. Sun Z et al (2012) The effect of solvent dielectric properties on the collection of oriented electrospun fibers. J Appl Polym Sci 125:2585–2594

87. Góra A et al (2011) Melt-electrospun fibers for advances in biomedical engineering, clean energy, filtration, and separation. Polym Rev 51:265–287

88. Arras MML et al (2012) Electrospinning of aligned fibers with adjustable orientation using auxiliary electrodes. Sci Technol Adv Mater 13:35008

89. Carnell LS et al (2008) Aligned mats from electrospun single fibers. Macromolecules 41:5345–5349

90. Wang HB et al (2009) Creation of highly aligned electrospun poly-L-lactic acid fibers for nerve regeneration applications. J Neural Eng 6:16001

91. Kang YK et al (2008) Development of thermoplastic polyurethane vascular prostheses. J Appl Polym Sci 110:3267–3274

92. Brown TD et al (2012) Design and fabrication of tubular scaffolds via direct writing in a melt electrospinning mode. Biointerphases 7:13

93. Jungst T et al (2015) Melt electrospinning onto cylinders: effects of rotational velocity and collector diameter on morphology of tubular structures. Polym Int 64:1086–1095

94. Dalton PD et al (2007) Electrospinning of polymer melts: phenomenological observations. Polymer (Guildf) 48:6823–6833

95. Schnell E. et al. (2007) Guidance of glial cell migration and axonal growth on electrospun nanofibers of poly-epsilon-caprolactone and a collagen/poly-epsilon-caprolactone blend. Biomaterials 28, 3012–25

96. Klinkhammer K et al (2009) Deposition of electrospun fibers on reactive substrates for in vitro investigations. Tissue Eng Part C Methods 15:77–85

97. Gerardo-Nava J et al (2009) Human neural cell interactions with orientated electrospun nanofibers in vitro. Nanomedicine (Lond) 4:11–30

98. Sun D et al (2006) Near-field electrospinning. Nano Lett 6:839–842

99. Chang C, Limkrailassiri K, Lin L (2008) Continuous near-field electrospinning for large area deposition of orderly nanofiber patterns. Appl Phys Lett 93:123111

100. Flynn L., Dalton P.D. & Shoichet M.S. (2003) Fiber templating of poly(2- hydroxyethyl methacrylate) for neural tissue engineering. Biomaterials 24, 4265–72

101. Stokols S, Tuszynski MH (2004) The fabrication and characterization of linearly oriented nerve guidance scaffolds for spinal cord injury. Biomaterials 25:5839–5846

102. King VR et al (2003) Mats made from fibronectin support oriented growth of axons in the damaged spinal cord of the adult rat. Exp Neurol 182:383–398

103. Prang P et al (2006) The promotion of oriented axonal regrowth in the injured spinal cord by alginate-based anisotropic capillary hydrogels. Biomaterials 27:3560–3569

104. Yoshii S et al (2004) Restoration of function after spinal cord transection using a collagen bridge. J Biomed Mater Res 70A:569–575

105. Stokols S, Tuszynski MH (2006) Freeze-dried agarose scaffolds with uniaxial channels stimulate and guide linear axonal growth following spinal cord injury. Biomaterials 27:443–451

106. Lee S et al (2012) A culture system to study oligodendrocyte myelination processes using engineered nanofibers. Nat Methods 9:917–922

107. Dalton PD, Shoichet MS (2001) Creating porous tubes by centrifugal forces for soft tissue application. Biomaterials 22:2661–2669

108. Tsai EC et al (2004) Synthetic hydrogel guidance channels facilitate regeneration of adult rat brainstem motor axons after complete spinal cord transection. J Neurotrauma 21:789–804

109. Tsai EC et al (2006) Matrix inclusion within synthetic hydrogel guidance channels improves specific supraspinal and local axonal regeneration after complete spinal cord transection. Biomaterials 27:519–533

110. Nomura H et al (2008) Delayed implantation of intramedullary chitosan channels containing nerve grafts promotes extensive axonal regeneration after spinal cord injury. Neurosurgery 63:127–143

111. Dalton P et al (2008) Tissue engineering of the nervous system. In: van Blitterswijk C et al (eds) Tissue engineering. Elsevier, Amsterdam, pp 611–647

112. Bamber NI et al (2001) Neurotrophins BDNF and NT-3 promote axonal re-entry into the distal host spinal cord through Schwann cell-seeded mini-channels. Eur J Neurosci 13:257–268

113. Cai P et al (2009) Survival of transplanted neurotrophin-3 expressing human neural stem cells and motor function in a rat model of spinal cord injury. Neural Regen Res 4:485–491

114. Nguyen LH et al (2017) Three-dimensional aligned nanofibers-hydrogel scaffold for controlled non-viral drug/gene delivery to direct axon regeneration in spinal cord injury treatment. Sci Rep 7:42212

115. Milbreta U et al (2016) Three-dimensional nanofiber hybrid scaffold directs and enhances axonal regeneration after spinal cord injury. ACS Biomater Sci Eng 2:1319–1329

116. Hurtado A et al (2011) Robust CNS regeneration after complete spinal cord transection using aligned poly-L-lactic acid microfibers. Biomaterials 32:6068–6079

117. Liu T et al (2012) Nanofibrous collagen nerve conduits for spinal cord repair. Tissue Eng Part A 18:1057–1066

118. Li HY et al (2013) Host reaction to poly (2-hydroxyethyl methacrylate) scaffolds in a small spinal cord injury model. J Mater Sci Mater Med 24:2001–2011

119. Pakulska MM, Ballios BG, Shoichet MS (2012) Injectable hydrogels for central nervous system therapy. Biomed Mater 7:24101

120. Macaya D, Spector M (2012) Injectable hydrogel materials for spinal cord regeneration: a review. Biomed Mater 7:12001

121. Lee KY, Mooney DJ (2001) Hydrogels for tissue engineering. Chem Rev 101:1869–1879

122. Peppas NA et al (2006) Hydrogels in biology and medicine: from molecular principles to bionanotechnology. Adv Mater 18:1345–1360

123. Taylor DL, In Het Panhuis M (2016) Self-healing hydrogels. Adv Mater 28:9060–9093

124. Jimenez Hamann MC, Tator CH, Shoichet MS (2005) Injectable intrathecal delivery system for localized administration of EGF and FGF-2 to the injured rat spinal cord. Exp Neurol 194:106–119

125. Sayer FT, Oudega M, Hagg T (2002) Neurotrophins reduce degeneration of injured ascending sensory and corticospinal motor axons in adult rat spinal cord. Exp Neurol 175:282–296

126. Katz JS, Burdick JA (2009) Hydrogel mediated delivery of trophic factors for neural repair. Wiley Interdiscip Rev Nanomed Nanobiotechnol 1:128–139

127. Ramer MS, Priestley J, V & McMahon S.B. (2000) Functional regeneration of sensory axons into the adult spinal cord. Nature 403:312–316

128. Hyatt AJT et al (2010) Controlled release of chondroitinase ABC from fibrin gel reduces the level of inhibitory glycosaminoglycan chains in lesioned spinal cord. J Control Release 147:24–29

129. Ruoslahti E, Yamaguchi Y (1991) Proteoglycans as modulators of growth factor activities. Cell 64:867–869

130. Sakiyama-Elbert SE, Hubbell JA (2000) Controlled release of nerve growth factor from a heparin-containing fibrin-based cell ingrowth matrix. J Control Release 69:149–158

131. Freudenberg U et al (2009) A star-PEG-heparin hydrogel platform to aid cell replacement therapies for neurodegenerative diseases. Biomaterials 30:5049–5060

132. Panyam J, Labhasetwar V (2003) Biodegradable nanoparticles for drug and gene delivery to cells and tissue. Adv Drug Deliv Rev 55:329–347

133. Baumann MD et al (2010) Intrathecal delivery of a polymeric nanocomposite hydrogel after spinal cord injury. Biomaterials 31:7631–7639

134. Elliott DI, Tator CH, Shoichet MS (2015) Sustained delivery of bioactive neurotrophin-3 to the injured spinal cord. Biomater Sci 3:65–72

135. Elliott DI, Tator CH, Shoichet MS (2016) Local delivery of neurotrophin-3 and anti-NogoA promotes repair after spinal cord injury. Tissue Eng Part A 22:733–741

136. Führmann T et al (2015) Click-crosslinked injectable hyaluronic acid hydrogel is safe and biocompatible in the intrathecal space for ultimate use in regenerative strategies of the injured spinal cord. Methods 84:60–69

137. Pakulska MM et al (2015) Hybrid crosslinked methylcellulose hydrogel: a predictable and tunable platform for local drug delivery. Adv Mater 27:5002–5008

138. Ansorena E et al (2013) Injectable alginate hydrogel loaded with GDNF promotes functional recovery in a hemisection model of spinal cord injury. Int J Pharm 455:148–158

139. Jain A et al (2006) In situ gelling hydrogels for conformal repair of spinal cord defects, and local delivery of BDNF after spinal cord injury. Biomaterials 27:497–504

140. De Laporte L et al (2010) Patterned transgene expression in multiple-channel bridges after spinal cord injury. Acta Biomater 6:2889–2897

141. Walthers CM, Seidlits SK (2015) Gene delivery strategies to promote spinal cord repair. Biomark Insights 10:11–29

142. Naidoo J, Young D (2012) Gene regulation systems for gene therapy applications in the central nervous system. Neurol Res Int 2012:595410

143. Lentz TB, Gray SJ, Samulski RJ (2012) Viral vectors for gene delivery to the central nervous system. Neurobiol Dis 48:179–188

144. Yin H et al (2014) Non-viral vectors for gene-based therapy. Nat Rev Genet 15:541–555

145. Yao L et al (2012) Non-viral gene therapy for spinal cord regeneration. Drug Discov Today 17:998–1005

146. Gersbach CA, Perez-Pinera P (2014) Activating human genes with zinc finger proteins, transcription activator-like effectors and CRISPR/Cas9 for gene therapy and regenerative medicine. Expert Opin Ther Targets 18:835–839

147. De Laporte L, Shea LD (2007) Matrices and scaffolds for DNA delivery in tissue engineering. Adv Drug Deliv Rev 59:292–307

148. Doukas J et al (2001) Matrix immobilization enhances the tissue repair activity of growth factor gene therapy vectors. Hum Gene Ther 12:783–798

149. Kidd ME, Shin S, Shea LD (2012) Fibrin hydrogels for lentiviral gene delivery in vitro and in vivo. J Control Release 157:80–85

150. Wu H-F et al (2013) The promotion of functional recovery and nerve regeneration after spinal cord injury by lentiviral vectors encoding Lingo-1 shRNA delivered by Pluronic F-127. Biomaterials 34:1686–1700

151. Rodriguez AL et al (2016) Tailoring minimalist self-assembling peptides for localized viral vector gene delivery. Nano Res 9:674–684

152. Thomas AM et al (2014) Heparin-chitosan nanoparticle functionalization of porous poly (ethylene glycol) hydrogels for localized lentivirus delivery of angiogenic factors. Biomaterials 35:8687–8693

153. Lei P, Padmashali RM, Andreadis ST (2009) Cell-controlled and spatially arrayed gene delivery from fibrin hydrogels. Biomaterials 30:3790–3799

154. Wieland JA, Houchin-Ray TL, Shea LD (2007) Non-viral vector delivery from PEG-hyaluronic acid hydrogels. J Control Release 120:233–241

155. Shepard JA et al (2011) Gene therapy vectors with enhanced transfection based on hydrogels modified with affinity peptides. Biomaterials 32:5092–5099

156. Zhang L-H et al (2011) Anti-DNA antibody modified coronary stent for plasmid gene delivery: results obtained from a porcine coronary stent model. J Gene Med 13:37–45

157. Segura T et al (2005) Crosslinked hyaluronic acid hydrogels: a strategy to functionalize and pattern. Biomaterials 26:359–371

158. Padmashali RM, Andreadis ST (2011) Engineering fibrinogen-binding VSV-G envelope for spatially- and cell-controlled lentivirus delivery through fibrin hydrogels. Biomaterials 32:3330–3339

159. Cummings BJ et al (2005) Human neural stem cells differentiate and promote locomotor recovery in spinal cord-injured mice. Proc Natl Acad Sci U S A 102:14069–14074

160. Satti HS et al (2016) Autologous mesenchymal stromal cell transplantation for spinal cord injury: a phase I pilot study. Cytotherapy 18:518–522

161. Jarocha D et al (2015) Continuous improvement after multiple mesenchymal stem cell transplantations in a patient with complete spinal cord injury. Cell Transplant 24:661–672

162. Dasari VR, Veeravalli KK, Dinh DH (2014) Mesenchymal stem cells in the treatment of spinal cord injuries: a review. World J Stem Cells 6:120–133

163. Kakabadze Z et al (2016) Phase 1 trial of autologous bone marrow stem cell transplantation in patients with spinal cord injury. Stem Cells Int 2016:6768274

164. Falkner S et al (2016) Transplanted embryonic neurons integrate into adult neocortical circuits. Nature 539:248–253

165. Arboleda D et al (2011) Transplantation of predifferentiated adipose-derived stromal cells for the treatment of spinal cord injury. Cell Mol Neurobiol 31:1113–1122

166. Herberts CA, Kwa MSG, Hermsen HPH (2011) Risk factors in the development of stem cell therapy. J Transl Med 9:29

167. Marquardt LM, Heilshorn SC (2016) Design of injectable materials to improve stem cell transplantation. Curr Stem Cell Rep 2:207–220

168. Roberts T, De Boni U, Sefton MV (1996) Dopamine secretion by PC12 cells microencapsulated in a hydroxyethyl methacrylate—methyl methacrylate copolymer. Biomaterials 17:267–275

169. Sontag CJ et al (2014) Injury to the spinal cord niche alters the engraftment dynamics of human neural stem cells. Stem cell Rep 2:620–632

170. Elliott DI et al (2014) Cell and biomolecule delivery for tissue repair and regeneration in the central nervous system. J Control Release 190:219–227

171. Wong Po Foo CTS et al (2009) Two-component protein-engineered physical hydrogels for cell encapsulation. Proc Natl Acad Sci U S A 106:22067–22072

172. Führmann T et al (2016) Injectable hydrogel promotes early survival of induced pluripotent stem cell-derived oligodendrocytes and attenuates longterm teratoma formation in a spinal cord injury model. Biomaterials 83:23–36

173. Li X et al (2013) Engineering an in situ crosslinkable hydrogel for enhanced remyelination. FASEB J 27:1127–1136

174. Madl CM, Katz LM, Heilshorn SC (2016) Bio-orthogonally crosslinked, engineered protein hydrogels with tunable mechanics and biochemistry for cell encapsulation. Adv Funct Mater 26:3612–3620

175. Aguado BA et al (2012) Improving viability of stem cells during syringe needle flow through the design of hydrogel cell carriers. Tissue Eng Part A 18:806–815

176. Leipzig ND, Shoichet MS (2009) The effect of substrate stiffness on adult neural stem cell behavior. Biomaterials 30:6867–6878

177. Cai L, Dewi RE, Heilshorn SC (2015) Injectable hydrogels with in situ double network formation enhance retention of transplanted stem cells. Adv Funct Mater 25:1344–1351

178. Willerth SM, Sakiyama-Elbert SE (2008) Cell therapy for spinal cord regeneration. Adv Drug Deliv Rev 60:263–276

179. Johnson PJ et al (2010) Tissue-engineered fibrin scaffolds containing neural progenitors enhance functional recovery in a subacute model of SCI. Soft Matter 6:5127–5137

180. Mothe AJ et al (2013) Repair of the injured spinal cord by transplantation of neural stem cells in a hyaluronan-based hydrogel. Biomaterials 34:3775–3783

181. Wu-Fienberg Y et al (2014) Viral transduction of primary Schwann cells using a Cre-lox system to regulate GDNF expression. Biotechnol Bioeng 111:1886–1894

182. Condic ML, Lemons ML (2002) Extracellular matrix in spinal cord regeneration: getting beyond attraction and inhibition. Neuroreport 13:A37–A48

183. Volpato FZ et al (2013) Using extracellular matrix for regenerative medicine in the spinal cord. Biomaterials 34:4945–4955

184. Badylak SF, Taylor D, Uygun K (2011) Whole-organ tissue engineering: decellularization and recellularization of three-dimensional matrix scaffolds. Annu Rev Biomed Eng 13:27–53

185. Medberry CJ et al (2013) Hydrogels derived from central nervous system extracellular matrix. Biomaterials 34:1033–1040

186. Tukmachev D et al (2016) Injectable extracellular matrix hydrogels as scaffolds for spinal cord injury repair. Tissue Eng Part A 22:306–317

187. Frantz C, Stewart KM, Weaver VM (2010) The extracellular matrix at a glance. J Cell Sci 123:4195–4200

188. Keane TJ et al (2012) Consequences of ineffective decellularization of biologic scaffolds on the host response. Biomaterials 33:1771–1781

189. Joosten EA, Bär PR, Gispen WH (1995) Collagen implants and cortico-spinal axonal growth after mid-thoracic spinal cord lesion in the adult rat. J Neurosci Res 41:481–490

190. Ryan EA et al (1999) Structural origins of fibrin clot rheology. Biophys J 77:2813–2826

191. Schense JC et al (2000) Enzymatic incorporation of bioactive peptides into fibrin matrices enhances neurite extension. Nat Biotechnol 18:415–419

192. Gupta D, Tator CH, Shoichet MS (2006) Fast-gelling injectable blend of hyaluronan and methylcellulose for intrathecal, localized delivery to the injured spinal cord. Biomaterials 27:2370–2379

193. Austin JW et al (2012) The effects of intrathecal injection of a hyaluronan-based hydrogel on inflammation, scarring and neurobehavioural outcomes in a rat model of severe spinal cord injury associated with arachnoiditis. Biomaterials 33:4555–4564

194. McKay CA et al (2014) An injectable, calcium responsive composite hydrogel for the treatment of acute spinal cord injury. ACS Appl Mater Interfaces 6:1424–1438

195. Tseng T-C et al (2015) An injectable, self-healing hydrogel to repair the central nervous system. Adv Mater 27:3518–3524

196. Broguiere N, Isenmann L, Zenobi-Wong M (2016) Novel enzymatically cross-linked hyaluronan hydrogels support the formation of 3D neuronal networks. Biomaterials 99:47–55

197. Moisse K, Strong MJ (2006) Innate immunity in amyotrophic lateral sclerosis. Biochim Biophys Acta 1762:1083–1093

198. Koser DE et al (2016) Mechanosensing is critical for axon growth in the developing brain. Nat Neurosci 19:1592–1598

199. Jones CAR et al (2014) The spatial-temporal characteristics of type I collagen-based extracellular matrix. Soft Matter 10:8855–8863

200. Jones CAR et al (2015) Micromechanics of cellularized biopolymer networks. Proc Natl Acad Sci U S A 112:E5117–E5122

201. Zhang S et al (2010) A self-assembly pathway to aligned monodomain gels. Nat Mater 9:594–601

202. Tysseling-Mattiace VM et al (2008) Self-assembling nanofibers inhibit glial scar formation and promote axon elongation after spinal cord injury. J Neurosci 28:3814–3823

203. Sur S et al (2013) Tuning supramolecular mechanics to guide neuron development. Biomaterials 34:4749–4757

204. Li A et al (2014) A bioengineered peripheral nerve construct using aligned peptide amphiphile nanofibers. Biomaterials 35:8780–8790

205. Berns EJ et al (2014) Aligned neurite outgrowth and directed cell migration in self-assembled monodomain gels. Biomaterials 35:185–195

206. Berns EJ et al (2016) A tenascin-C mimetic peptide amphiphile nanofiber gel promotes neurite outgrowth and cell migration of neurosphere-derived cells. Acta Biomater 37:50–58

207. Jiao Y et al (2014) BDNF increases survival and neuronal differentiation of human neural precursor cells cotransplanted with a nanofiber gel to the auditory nerve in a rat model of neuronal damage. Biomed Res Int 2014:356415

208. Dubey N, Letourneau PC, Tranquillo RT (2001) Neuronal contact guidance in magnetically aligned fibrin gels: effect of variation in gel mechano-structural properties. Biomaterials 22:1065–1075

209. Kriebel A et al (2014) Three-dimensional configuration of orientated fibers as guidance structures for cell migration and axonal growth. J Biomed Mater Res B Appl Biomater 102:356–365

210. Han Q et al (2009) Linear ordered collagen scaffolds loaded with collagen-binding brain-derived neurotrophic factor improve the recovery of spinal cord injury in rats. Tissue Eng Part A 15:2927–2935

211. Antman-Passig M, Shefi O (2016) Remote magnetic orientation of 3D collagen hydrogels for directed neuronal regeneration. Nano Lett 16:2567–2573

212. Kim J, Staunton JR, Tanner K (2016) Independent control of topography for 3D patterning of the ECM microenvironment. Adv Mater 28:132–137

213. Pisanic TR et al (2007) Nanotoxicity of iron oxide nanoparticle internalization in growing neurons. Biomaterials 28:2572–2581

214. Abdolrahman Omidinia-Anarkoli, Sarah Boesveld, Urandelger Tuvshindorj, Jonas C. Rose, Tamás Haraszti, Laura De Laporte, (2017) An injectable hybrid hydrogel with oriented short fibers induces unidirectional growth of functional nerve cells. Small 13 (36):1702207

215. Jonas C. Rose, David B. Gehlen, Tamás Haraszti, Jens Köhler, Christopher J. Licht, Laura De Laporte, (2018) Biofunctionalized aligned microgels provide 3D cell guidance to mimic complex tissue matrices. Biomaterials 163:128–141

216. Christopher Licht, Jonas C. Rose, Abdolrahman Omidinia Anarkoli, Delphine Blondel, Marta Roccio, Tamás Haraszti, David B. Gehlen, Jeffrey A. Hubbell, Matthias P. Lutolf, Laura De Laporte, (2019) Synthetic 3D PEG-anisogel tailored with fibronectin fragments induce aligned nerve extension. Biomacromolecules 20 (11):4075–4087

217. Jonas C. Rose, Maaike Fölster, Lukas Kivilip, Jose L. Gerardo-Nava, Esther E. Jaekel, David B. Gehlen, Wilko Rohlfs, Laura De Laporte, (2020) Predicting the orientation of magnetic microgel rods for soft anisotropic biomimetic hydrogels. Polym Chem 11 (2):496–507

218. Rose JC et al (2017) Nerve cells decide to orient inside an injectable hydrogel with minimal structural guidance. Nano Lett 17:3782. https://doi.org/10.1021/acs.nanolett.7b01123

219. Konofaos P, Ver Halen JP (2013) Nerve repair by means of tubulization: past, present, future. J Reconstr Microsurg 29:149–164

220. Kehoe S., Zhang X.F. & Boyd D. (2012) FDA approved guidance conduits and wraps for peripheral nerve injury: a review of materials and efficacy. Injury 43, 553–72

221. Lackington W.A., Ryan A.J. O'Brien F.J. (2017) Advances in nerve guidance conduit-based therapeutics for peripheral nerve repair. ACS Biomater Sci Eng doi:10.1021/acsbiomaterials.6b00500, 3, 1221

222. Zuo J, Hernandez YJ, Muir D (1998) Chondroitin sulfate proteoglycan with neurite-inhibiting activity is up-regulated following peripheral nerve injury. J Neurobiol 34:41–54

223. Brooks DN et al (2012) Processed nerve allografts for peripheral nerve reconstruction: a multicenter study of utilization and outcomes in sensory, mixed, and motor nerve reconstructions. Microsurgery 32:1–14

224. Zuniga JR (2015) Sensory outcomes after reconstruction of lingual and inferior alveolar nerve discontinuities using processed nerve allograft--a case series. J Oral Maxillofac Surg 73:734–744

225. Salomon D, Miloro M, Kolokythas A (2016) Outcomes of immediate allograft reconstruction of long-span defects of the inferior alveolar nerve. J Oral Maxillofac Surg 74:2507–2514

226. Tian L, Prabhakaran MP, Ramakrishna S (2015) Strategies for regeneration of components of nervous system: scaffolds, cells and biomolecules. Regen Biomater 2:31–45

227. Lv Y et al (2015) In vivo repair of rat transected sciatic nerve by low-intensity pulsed ultrasound and induced pluripotent stem cells-derived neural crest stem cells. Biotechnol Lett 37:2497–2506

228. Xie J et al (2014) Nerve guidance conduits based on double-layered scaffolds of electrospun nanofibers for repairing the peripheral nervous system. ACS Appl Mater Interfaces 6:9472–9480

229. Jiang X et al (2014) Nanofibrous nerve conduit-enhanced peripheral nerve regeneration. J Tissue Eng Regen Med 8:377–385

230. Corey JM et al (2007) Aligned electrospun nanofibers specify the direction of dorsal root ganglia neurite growth. J Biomed Mater Res A 83:636–645

231. Kim Y-T et al (2008) The role of aligned polymer fiber-based constructs in the bridging of long peripheral nerve gaps. Biomaterials 29:3117–3127

232. Clements IP et al (2009) Thin-film enhanced nerve guidance channels for peripheral nerve repair. Biomaterials 30:3834–3846

233. Ngo T-TB et al (2003) Poly(L-lactide) microfilaments enhance peripheral nerve regeneration across extended nerve lesions. J Neurosci Res 72:227–238

234. Ichihara S, Inada Y, Nakamura T (2008) Artificial nerve tubes and their application for repair of peripheral nerve injury: an update of current concepts. Injury 39(Suppl 4):29–39

235. Wang W et al (2009) Effects of Schwann cell alignment along the oriented electrospun

chitosan nanofibers on nerve regeneration. J Biomed Mater Res A 91:994–1005

236. Mottaghitalab F et al (2013) A biosynthetic nerve guide conduit based on silk/SWNT/ fibronectin nanocomposite for peripheral nerve regeneration. PLoS One 8:e74417

237. Sell SA et al (2010) The use of natural polymers in tissue engineering: a focus on electrospun extracellular matrix analogues. Polymers (Basel) 2:522–553

238. Koh HS et al (2008) Enhancement of neurite outgrowth using nano-structured scaffolds coupled with laminin. Biomaterials 29:3574–3582

239. Kriebel A et al (2017) Cell-free artificial implants of electrospun fibres in a three-dimensional gelatin matrix support sciatic nerve regeneration in vivo. J Tissue Eng Regen Med 11:3289. https://doi.org/10. 1002/term.2237

240. Koh HS et al (2010) In vivo study of novel nanofibrous intra-luminal guidance channels to promote nerve regeneration. J Neural Eng 7:46003

241. Wang C-Y et al (2012) The effect of aligned core-shell nanofibres delivering NGF on the promotion of sciatic nerve regeneration. J Biomater Sci Polym Ed 23:167–184

242. Dimos JT et al (2008) Induced pluripotent stem cells generated from patients with ALS can be differentiated into motor neurons. Science 321:1218–1221

243. Kabiri M et al (2015) Neuroregenerative effects of olfactory ensheathing cells transplanted in a multi-layered conductive nanofibrous conduit in peripheral nerve repair in rats. J Biomed Sci 22:35

244. Beigi M-H et al (2014) In vivo integration of poly(ε-caprolactone)/gelatin nanofibrous nerve guide seeded with teeth derived stem cells for peripheral nerve regeneration. J Biomed Mater Res A 102:4554–4567

245. Wang Y-Y et al (2010) Cellular compatibility of RGD-modified chitosan nanofibers with aligned or random orientation. Biomed Mater 5:54112

246. Yun D et al (2014) Biomimetic poly(serinol hexamethylene urea) for promotion of neurite outgrowth and guidance. J Biomater Sci Polym Ed 25:354–369

247. Jenkins PM et al (2015) A nerve guidance conduit with topographical and biochemical cues: potential application using human neural stem cells. Nanoscale Res Lett 10:972

248. Wang X et al (2014) A novel artificial nerve graft for repairing long-distance sciatic nerve defects: a self-assembling peptide nanofiber scaffold-containing poly(lactic-co-glycolic acid) conduit. Neural Regen Res 9:2132–2141

249. Wu X et al (2017) Functional self-assembling peptide nanofiber hydrogel for peripheral nerve regeneration. Regen. Biomater 4:21–30

250. Haastert-Talini K, Grothe C (2013) Electrical stimulation for promoting peripheral nerve regeneration. Int Rev Neurobiol 109:111–124

251. Zhang Z et al (2007) Electrically conductive biodegradable polymer composite for nerve regeneration: electricity-stimulated neurite outgrowth and axon regeneration. Artif Organs 31:13–22

252. Gupta A et al (2015) Biocompatibility of single-walled carbon nanotube composites for bone regeneration. Bone Joint Res 4:70–77

253. Song J et al (2016) Polymerizing pyrrole coated poly (l-lactic acid-co-ε-caprolactone) (PLCL) conductive Nanofibrous conduit combined with electric stimulation for long-range peripheral nerve regeneration. Front Mol Neurosci 9:117

254. Schmidt CE et al (1997) Stimulation of neurite outgrowth using an electrically conducting polymer. Proc Natl Acad Sci U S A 94:8948–8953

255. Lovat V et al (2005) Carbon nanotube substrates boost neuronal electrical signaling. Nano Lett 5:1107–1110

256. Sang L et al (2016) Thermally sensitive conductive hydrogel using amphiphilic crosslinker self-assembled carbon nanotube to enhance neurite outgrowth and promote spinal cord regeneration. RSC Adv 6:26341–26351

257. Annabi N et al (2016) Highly elastic and conductive human-based protein hybrid hydrogels. Adv Mater 28:40–49

258. Luo Y, Shoichet MS (2004) A photolabile hydrogel for guided three-dimensional cell growth and migration. Nat Mater 3:249–253

259. Andersen MØ et al (2013) Spatially controlled delivery of siRNAs to stem cells in implants generated by multi-component additive manufacturing. Adv Funct Mater 23:5599–5607

260. Toh AGG et al (2014) Engineering microfluidic concentration gradient generators for biological applications. Microfluid Nanofluidics 16:1–18

Chapter 4

Self-Assembling Peptide Nanofibrous Scaffolds in Central Nervous System Lesions

Na Zhang, Liumin He, and Wutian Wu

Abstract

Central nervous system (CNS) lesion leads to the loss of tissues and functional defect. After injury, the local microenvironment typically contains potential inhibitory molecules and glial scars, which inhibit axonal regeneration. Self-assembling peptide nanofibrous scaffolds (SAPNSs) which have been shown to mimic the structure of natural extracellular matrix (ECM) demonstrate the effectiveness in minimizing secondary injury as well as promoting nerve regeneration. In this chapter, the application of SAPNSs and the functionalized self-assembling peptide (SAP) in different brain injury models, and, especially, the methods of SAP delivery, will be elaborately discussed. The use of peptide in CNS injuries is a promising approach for the reconstruction of the brain tissues and the rebuilding of the neuronal circuits.

Key words Self-assembling peptide nanofibrous scaffolds, Traumatic brain injury, Intracerebral hemorrhage, Extracellular Matrix, Tissue engineering

1 Introduction

Tissue-engineered scaffolds-based brain tissue reconstruction is promising in bridging structural gaps, re-knitting the injured brain, hence promoting neurite extension and providing a platform for cell migration and nerve regeneration [1]. In terms of tissue engineering applications, self-assembling proteins and peptides serving as potential scaffolds for tissue repair have gained interests in the past years. As a commonly used method in nanofiber scaffold fabrication, self-assembly demonstrated the spontaneous combination of molecules and components without human interface, by which different patterns or structures can be formed during this process [2]. Molecules are linked with weak non-covalent bonds (hydrogen bonds in particular), electrostatic interactions, hydrophobic interactions, and van der Waals interactions. Although these bonds are isolated relatively, when combined together, they dominate all structures of biological shapes such as fibrils, fibril network, gels, and membranes. These biological shapes could influence other

Javier O. Morales and Pieter J. Gaillard (eds.), *Nanomedicines for Brain Drug Delivery*, Neuromethods, vol. 157,
https://doi.org/10.1007/978-1-0716-0838-8_4, © Springer Science+Business Media, LLC, part of Springer Nature 2021

molecules, which are rather important in the living system [3]. Moreover, the breakdown products of self-assembly peptide (SAP) are nontoxic natural L-amino acids which could be absorbed by adjacent cells potentially. Because of its comparable dimension with natural collagen fibers, the nanofibrous scaffolds can imitate the structure of nature extracellular matrix (ECM) which is functional in supporting cells structures and providing cell-instructive cues due to the internal molecules such as proteins, glycosaminoglycan, polysaccharides, etc. The bioactive signals sequences inside ECM could be recognized by cells through transmembrane receptor, which is named as integrin and can interact with bioactive epitopes of ECM whereby activating signal transduction mechanisms. Subsequently, specific cellular functions including adhesion, migration, differentiation, and proliferation will be induced [4]. Due to these special properties, nanofibrous scaffolds have been extensively investigated as a promising platform for neural tissue engineering. It has been reported that self-assembling peptide nanofibrous scaffolds (SAPNSs) are able to self-assemble into nanofibrous morphology with bilayer beta-sheet structure and become hydrogel whose mechanical stiffness is analogous to brain tissue due to the special physiochemical properties [5]. Based on all these properties, self-assembly peptide nanofibrous scaffolds can serve as a powerful tool for CNS lesions.

1.1 Self-Assembling Peptide Nanofibrous Scaffolds

As an ionic self-complementary peptide, RADA16-I (Ac-(RADA) 4-CONH2) is able to form stable β-sheet structure and eventually becomes a hydrogel scaffold during self-assembly process whose water content is more than 99.5%. Although it is relatively soft, the movement of molecules and other nanoscale and microscale substances will be confined once the scaffold is formed. In vitro studies indicated that RADA16-I could promote the neurite outgrowth of PC12 cells after seeding on the three-dimensional (3D) hydrogel matrix for 10–14 days [6]. Furthermore, RADA16-I was reported to be able to enhance the differentiation of neurons derived from hippocampal slices [7]. In addition to serving as a 3D cell culture system, RADA16-I has also been widely used in animal models. In vivo studies indicated that RADA16-I could decrease inflammatory response and reduce apoptosis in traumatic brain injury model [1]. Moreover, the regeneration of peripheral axons across and beyond the 10 mm gap in peripheral nerve injury model can be activated by RADA16-I [8]. In another study, Schwann cells (SCs) were retrieved from human fetal sciatic nerves and were cultured within SAP nanofibrous scaffolds for 2 days and then implanted into the injured rat spinal cord. The experiment yielded reduced aster gliosis, and the infiltration of endogenous S-100 cells into the injury site was significantly increased 8 weeks after grafting, suggesting that the SAPNS RADA16-I is helpful to the recovery of spinal cord injury [9]. RADA16-I with its closely resembling structure of ECM showed great potential in the applications of regenerative medicine.

However, the applications of RADA16-I have been limited because of its main drawback caused by low pH value (pH = 3.5), which may have negative effects on the surrounding cells and host tissues after direct exposure. When it is exposed to an acidic environment with even lower pH value (e.g., pH = 1.0), RADA16-I was maintained the typical beta-sheet structure and self-assembled into long nanofiber, despite of the 10% decreased beta-sheet content. However, the typical beta-sheet structure of the peptide has been dramatically lost, and the peptide will self-assemble into various small-sized globular aggregates when it is exposed to either the neutral or alkali environment, suggesting that both the temperature and the pH value have effects on biophysical and morphological properties of SAPNS RADA16-I [10]. Owing to the abilities of adding functional motifs then further expanding the utility in tissue engineering, RADA16-I can be commercially synthesized with high purity and, more importantly, be custom-tailored for specific cell culture applications [11].

1.2 Functionalized Self-Assembling Peptide Nanofibrous Scaffolds

By incorporating functional motifs, the pH for maintaining the typical beta-sheet structure will be changed. These functionalized peptides self-assembled into nanofiber scaffolds in which cells can be fully embedded in 3D culture system. Gelain et al. reported that designer peptide nanofibrous scaffolds can produce 3D cultures for neural stem cells. In this study, RADA16-I was conjunct with bone marrow homing motifs BMHP1 and BMHP2. These two functionalized peptides not only enhanced survival of the neural stem cells significantly but also promoted differentiation toward cells expressing neuronal and glial markers [12]. A. Horii et al. also studied functionalized RADA16 by extending the original RADA16 at the C-terminus directly through solid phase synthesis with three short peptide motifs [osteogenic growth peptide OGP (ALKRQGRTLYGFGG), osteopontin cell adhesion motif DGR (DGRGDSVAYG), and a specifically designer 2 units of RGD sequence (PRGDSGYRGDS)] in order to closer mimic ECM, which enhanced differentiation, migration, and pre-osteoblast proliferation [13].

In our strategy, we prepared nanofiber hydrogels from two designer SAPs at the environment with neutral pH value. RADA16-I was appended with neurite outgrowth peptide IKVAV and cell adhesion peptide RGD, which were derived from laminin and fibronectin, respectively. The two functional SAPs, namely, RADA16-RGD (Ac-(RADA) 4-DGDRGDS) and RADA16-IKVAV (Ac-(RADA) 4-RIKVAV), were specially designed to have opposite net charges. The combination of both SAPs yielded a nanofiber hydrogel (-IKVAV/RGD). In the current study, RADA16-I was appended with IKVAV and RGD, which were

Table 1
Physicochemical properties for SAP used in this study

Code	Sequences	Net charge, pH 7.4	ζ, pH 7.4	MW, Da
RADA16-1	Ac-(RADA)$_4$-CONH$_2$	Neutral	0	1713
RADA16-RGD	Ac-(RADA)$_4$-DGDRGDS	–	−18.6	2416
RADA16-IKVAV	Ac-(RADA)$_4$-RIKVAV	+	19.7	2451

derived from laminin and fibronectin, respectively, as shown in Table 1. RADA16-IKVAV and RADA16-RGD exhibited zeta potentials of 19.7 and 0.18.6, at pH 7, respectively. The opposite charge polarities were in good agreement with our original design.

The combination of these two designer SAPs forms a 3D nanofiber hydrogel (Fig. 1). The two bioactive motifs were thereby incorporated in 3D hydrogel scaffold, yielding a synergistic promotion effect on nerve cell response. Constant neutral pH was maintained during the process of self-assembly so that bioactive molecules and cells could be fully embedded in 3D environment. In addition, the designed SAP hydrogel could be transferred directly to living tissues [14].

These designer peptide nanofiber scaffolds exhibit inherent advantages. Firstly, they can be designed easily using known, biologically active motifs. At the second place, they are commercially custom-synthesizable by using mature solid phase peptide synthesis technology. Finally, they are applicable in selection of an extensive repertoire of biological active motifs detected in some ECM components, etc.

2   Materials

For in vivo studies, we use liquid form of RADA16-I as 1% solution (w/v). RADA16-I solution can also be obtained in powder form [15]. The powder can be reconstituted readily as a solution by dissolving 10 mg in 1 ml of Milli-Q water in an Eppendorf tube, mixed, sonicated for 30 s, and then filtered. The solution can be well kept at room temperature for 1 month prior to use.

For the functionalized SAP RADA16-IKVAV/RGD, the two functionalized peptides RADA16-RGD and RADA16-IKVAV were custom-synthesized by American Peptide Company, Inc. with the purity above 95%. The peptide powders were dissolved in distilled water with a concentration of 1% (v/w), filter-sterilized with an Acrodisc Syringe Filter (0.2 μm HT Tuffryn membrane, Pall Corp., Ann Arbor, MI) for subsequent use. Both peptide solutions were carefully titrated to neutral liquor using 1 M Tris solution. The hydrogel formed by RADA16-RGD and RADA16-IKVAV was defined as -IKVAV/-RGD here (Fig. 2).

RADA16-IKVAV   RADA16-RGD   -IKVAV/-RGD

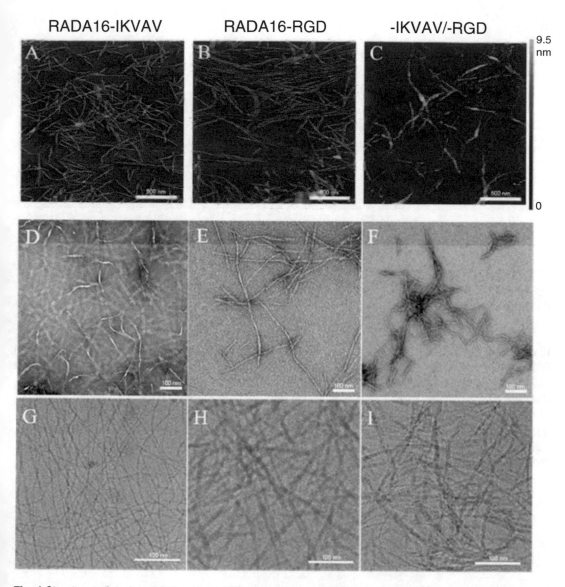

Fig. 1 Structures of designed SAPs. (**a–c**) AFM phase images of different designer SAPs. (**d–f**) TEM images of different designer SAPs by negative staining at room temperature. (**g–i**) Cryo-TEM images of different designer SAPs. The dimensions of AFM images were 2 × 2 μm². Scale bars of cryo-TEM images were 100 nm

3 Brain Injury Models and SAP Delivery Methods

3.1 Traumatic Brain Injury

Traumatic brain injury (TBI) leads to extensive loss of cerebral parenchyma as well as the formation of a cavity in the brain. TBI consists of a primary result, which is the mechanical displacement of tissue that occurs at the time of injury, and a secondary result, which occurs gradually and involves a number of cellular processes [16–18]. Because the primary outcome is not amenable by treatment, the secondary brain outcome has been the major focus of

a   b   c

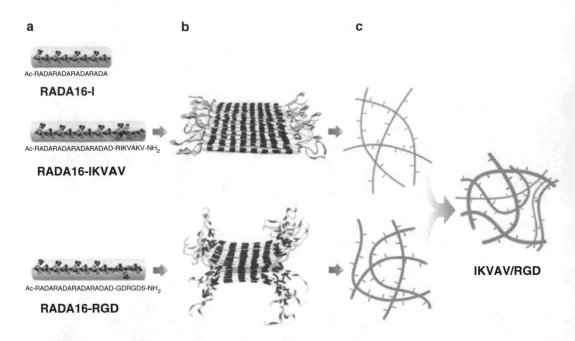

Ac-RADARADARADARADA

RADA16-I

Ac-RADARADARADARADAD-RIKVAKV-NH₂

RADA16-IKVAV

Ac-RADARADARADARADAD-GDRGDS-NH₂

RADA16-RGD

IKVAV/RGD

Fig. 2 Proposed molecular model for dynamic reassembly of self-assembling RADA16-IKVAV and RADA16-RGD as well as a combination of the two designer SAPs. (**a**) Amino acid sequence and molecular model of RADA16-I, RADA16-IKVAV, and RADA16-RGD. (**b**) Bilayer structures of RADA16-IKVAVA and RADA16-RGD. Color schemes for self-assembling "core" RADA16-I were blue (basic), white (hydrophobic), and red (acidic) whereas in the motifs green (polar), dark gray (hydrophobic), and light gray (hydrophilic residues). (**c**) Nanofibers were formed in aqueous RADA16-IKVAV and RADA16-RGD as well as in -IKVAV/-RGD hydrogel. This was composed by alternating basic residues (blue) with acid residues (red) and hydrophobic ones (white). Polar neutral residues (green) are present in added functional motifs

most studies to identify potential targets for treating TBI [16]. However, despite extensive research to develop new neuro-protective therapies, surgery and neurocritical care remain the only treatment options available for TBI patients. Therefore, searching for an alternative and clinically feasible treatment is rather urgent.

3.1.1 Surgery Procedure

The rats were anesthetized with ketamine (80 mg/kg) and xylazine (10 mg/kg) and then fixed in a stereotaxic apparatus. The rat head was horizontally secured by two lateral ear pins and an incisor bar. For craniotomy, a trephine (3 mm of outer diameter) was used to drill holes in the skull, and dura incision was performed to expose the forebrain. On the cortex 2 mm to the right of the bregma, a 1 mm × 2 mm rectangle of 2-mm-thick cortical tissue was cut and removed by a vacuum pump-aided aspirator. Then, the skull flap was restored and the scalp was sutured. For the bilateral injury model, the same-size injuries were performed in both the left and right hemispheres in the same setting. Animals were placed in a heated chamber and monitored after injury until anesthesia effects had worn off and the animal was returned to home cage. The symptoms such as arched back, erect hair, unconsciousness, and slow respiration were observed for all TBI rats, but they recovered to normal within the following 2 h.

3.1.2 Refill
of the Damaged Area

After traumatic brain injuries, there was bleeding in the lesioned civility. As RADA16-I has an immediate hemostatic property, 20 μL of 1% RADA16-I peptide solution was first administered to stop the bleeding. Thirty seconds after the surgical injury, the cavity was cleaned to remove coagulated blood. Subsequently, fresh RADA16-I was injected to fill in the damaged area. As for the control group, the same volume of saline was injected into the cavity after the bleeding was stopped by Gelfoam. After RADA16-I (or saline) administration, the dura was sutured with 11-0 surgical suture (Fig. 3) and covered by dura foam (dura graft implant; Johnson & Johnson, Raynham, MA). The skull window was closed with Parafilm, which was fixed in position by tissue glue.

The animals were sacrificed in 2 days, 2 weeks, or 6 weeks after injury for immunohistochemistry studies. Inflammatory-related antibodies and neuronal markers were tested, which includes rabbit anti-GFAP (1:1000; Chemicon, Temecula, California) for astrocytes; mouse anti-ED1 (1:1000; Serotec, Raleigh, North Carolina) for macrophages; and rabbit anti-IBA1 (1:1000; Osaka, Wako,

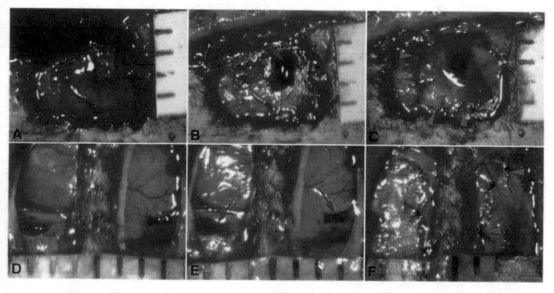

Fig. 3 (**a–c**) Experimental procedures performed unilaterally on the brain right hemisphere. (**a**) At distance of 2 mm to the right of the bregma, a rectangle of 1 mm × 2 mm was cut with a depth of 2 mm, and the cut tissue was removed with a vacuum pump-aided aspirator. (**b**) SAPNS was administered to fill the cavity after bleeding stopped (arrow). (**c**) The dura was sutured with 11-0 surgical suture (arrowheads). (**d–f**) Experimental procedures performed bilaterally. (**d**) Rectangles were cut as above on both sides of the brain. (**e**) SAPNS was administered to fill the right cavity (arrow), and saline was administered into the left cavity as control (arrow). (**f**) Suturing of the dura opening (arrowheads). Scale bar: 1 mm

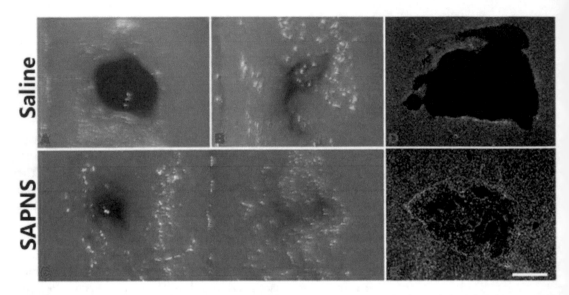

Fig. 4 Lesion sites with different treatments 6 weeks after surgery. (**a–c**) Surface of the injured brain. (**a**) A moderate-sized cavity in the saline-treated brain with unilateral injury. (**b**) The wound in the SAPNS-treated brain with unilateral injury has closed. (**c**) A sample of bilaterally injured brain with a small cavity in the saline-treated left hemisphere and SAPNS-filled lesion site in the right hemisphere. (**d**, **e**) Nissl and DAPI double staining showing lesion sites of saline- (**d**) and SAPNS- (**e**) treated brain. Panel (**e**) shows that the cavity formed by trauma was filled with SAPNS, which integrated very well with the host tissue with no obvious gaps. Scale bar: **a–c**, 1 mm; **d–e**, 500 μm

Japan) for microglia. Fluorescent TUNEL staining and fluorescent Nissl staining were carried out to detect the apoptosis and the morphology of the lesion regions, respectively. The data demonstrated that the SAPNS was a prospective biomaterial for the repair of the acutely injured brain. The present results showed that the TBI-induced cavity in the brain was closed by the treatment with SAPNS and the graft of SAPNS was well integrated with the host tissue with no obvious gaps (Fig. 4). The SAPNS-treated brain tissue had fewer astrocytes and macrophages around the lesion site as compare to the saline-treated brain tissue. SAPNSs are able to reconstruct the acute traumatically injured brain, facilitate support cells to migrate and survive within the lesion site, and reduce both the glia reaction and the inflammation of the injured brain tissue, suggesting that SAPNS is a good biomaterial candidate for brain tissue restoration in humans [1]. It has potential in the application of filling and reconstructing any kind of cavities in the brain, such as the removal of brain tumor. However, one possible weakness of this biomaterial is that we could not detect neurons or nanofibers in the grafted scaffold, which is similar as the other biomaterials in the previous studies [19, 20].

3.2 Intracerebral Hemorrhage

Intracerebral hemorrhage (ICH) is defined as non-vascular rupture hemorrhage caused by traumatic brain parenchyma, accounting for

10–15% of all strokes and affecting 10–30 cases per 100,000 adults [21–24]. Early mortality of ICH is high, and most survivors leave sequela such as movement disorders, cognitive deficits, and speech impediment [15]. Intracerebral hemorrhage occurs due to destructive changes in brain arteries like segmental lipohyalinosis, pseudoaneurysm, or rupture [25]. These changes cause pain to patients and impose great pressure on family and the society. Therefore, being fully aware of the severity of ICH and finding effective therapy become extremely urgent.

3.2.1 Surgery Procedure

The mouse ICH model was caused by intracranial injection of type IV collagenase (Sigma-Aldrich, St. Louis, USA). Corpus striatum was the target region. Briefly, a burr hole of 0.15 mm diameter was drilled along the right coronal suture at 2.0 mm lateral to the bregma. A 30 G needle was inserted into the right striatum with its tip at 0.26 mm anterior to the bregma, 2 mm lateral to the midline, and 3.75 mm underneath the dural surface. ICH was induced by a slow injection of 0.25 U collagenase IV in 1 μL saline into the right striatum over 10 min.

3.2.2 Intracranial Injection

The intracranial injection delivery method was applied in a type IV collagenase-induced intracerebral hemorrhage model. Three and a half hours after type IV collagenase injection, aspiration was carried out by gentle suction through a 1 ml syringe attached to a 23 G needle. The needle was placed at the same stereotactic coordinates as the collagenase injection. Four aspirations were performed over 15 min. Following aspiration, 10 μL RADA16-IKVAV/RGD solution, 10 μL RADA16-I solution, or 10 μL saline was injected into the lesion over 10 min, respectively. Finally, the burr hole was sealed with bone wax, and the incision was sutured.

Histomorphology analysis by Nissl staining suggested that collagenase injection induced intracerebral hemorrhage and considerable loss of neurons. However, the injected SAPNS filled in the voids and the boundary between SAPNS and the host fit tightly (Fig. 5).

Hematoma volume was measured 3 days after injury, and local delivery of RADA16-IKVAV/RGD did not reduce the amount of hematoma. Hematoma was significantly reduced by RADA16-I, which has been verified as a hemostatic hydrogel. In addition, different markers including IBA-1, CD11b, and NeuN have been used to detect inflammatory response and cell survival. Apoptosis was detected by TUNEL staining. Findings showed that local delivery of RADA16-IKVAV/RGD reduced acute brain injury by lowering the number of apoptotic cells, reducing inflammatory response as well as promoting cell survival. Three behavioral tests were done in weekly interval including catwalk test, rotarod test, and grip strength test. Results showed that 4 weeks later animals treated with RADA16-IKVAV/RGD showed better functional

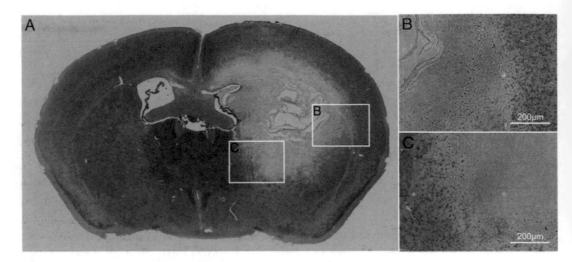

Fig. 5 The overall morphology of the injured brain was detected by Nissl staining. (**a**) There was almost no neuron survival inside the injured area. (**b, c**) The boundaries between materials and organizations were in perfect fit

recovery than that in RADA16-I group. Two months after surgery, nerve fibers inside RADA16-IKVAV/RGD were detected using its specific marker, antibody against neurofilament-200. Results showed that 2 months later RADA16-IKVAV/RGD could attach well with the host, and new nerve fibers have grown into the new SAP hydrogel. RADA16-I also could attach with the host very well; however, no fibers were detected inside the hydrogel, suggesting that RADA16-IKVAV/RGD could serve as a bridge for fibers to grow through [26].

3.3 Other Surgical Brain Injury and SAP Delivery Methods

The cortical vessel cut injury model was carried out to investigate the function of RADA16-I as a topical hemostatic and neuroprotective agent. The surgical procedure can be given as follows. The rats were anesthetized with ketamine (80 mg/kg) and xylazine (10 mg/kg) and then fixed in a stereotaxic apparatus. The head was horizontally secured by two lateral ear pins and an incisor bar. Following dura opening, a right cortical incision (2 mm in anterior–posterior length; 1 mm in depth; centered 2 mm lateral to the sagittal suture; and 1 mm anterior to the coronal suture) by a stereotactically guided micro-knife is made under an operating microscope. The incision is designed to transect a cortical vessel.

Another brain injury model we carried out is the knife wound injury made at the midbrain. Briefly, the rats were anesthetized and the head was fixed in a stereotaxic apparatus. The cortex was exposed and aspirated to reveal the rostral edge of the SC and the brachium of the SC on the left side. Animals received a complete transection of the brachium of the SC. The lesions are shallow in many CNS regeneration models. To confirm that there would be no regeneration around the cut, the cut depth of our model is 2 mm. The increased depth prevented axons from growing around the bottom of the cut, ensuring the growth of axons through the center of the lesion site rather than around it. In addition, special attention was paid to ensure the complete transection of the brachium of the SC from the lateral edge to the midline.

3.3.1 Patching Up the Wound

SAPNS RADA16-I has been proved to be functional in stopping bleeding. The decomposition products of RADA16-I was amino acids, which can serve as building blocks for tissues and be used to repair the site of injury. Since the pH value of 1% RADA16-I was about 3.5 and the ECM has a positive charge, fibers could absorb wound edges by electrostatic interactions to seal the wound. Further, RADA16-I was highly hydrated and exhibited a water content of about 99%. It is assembled prior to filling the gap irregularly and then assembled to form a nanofiber scaffold molecule [27]. Based on this property, RADA16-I has been applied to stop bleeding in the cortical vessel cut experiment. The animals were fitted in a head holder. The left lateral part of the cortex was exposed, and each animal received a transection of a blood vessel leading to the superior sagittal sinus (Fig. 6). With the aid of a sterile glass micropipette, 20 μL of 1% wt/vol RADA16-I solution was applied to the injury site by patching up the wound. For the control group, iced saline was used.

In the groups treated with RADA16-1, hemostasis was achieved in less than 10 s in both hamsters and rats. Control group hamsters ($n = 5$) and rats ($n = 5$) irrigated with saline bled for more than 3 min. This is the first time that nanotechnology has been used to stop bleeding in a surgical setting for animal models. A new class of hemostatic agent has been demonstrated, which does not rely on heat, pressure, platelet activation, adhesion, or desiccation to stop bleeding. As a synthetic and biodegradable peptide [28, 29], RADA16-I does not contain any blood products, collagens, or biological contaminants that may be present in human- or animal-derived hemostatic agents such as fibrin glue [30–33]. It can be applied directly onto/into a wound without consideration of material expanding, thus reducing the risk of secondary tissue damage as well as the problems caused by constricted blood flow.

With this discovery the ability to speedily achieve hemostasis will dramatically reduce the quantity of blood required during the surgery, as 50% of surgical time was spent on packing wounds to reduce or control bleeding. The RADA16-I solutions may represent a step change in technology and could revolutionize bleeding control during surgery and trauma [34].

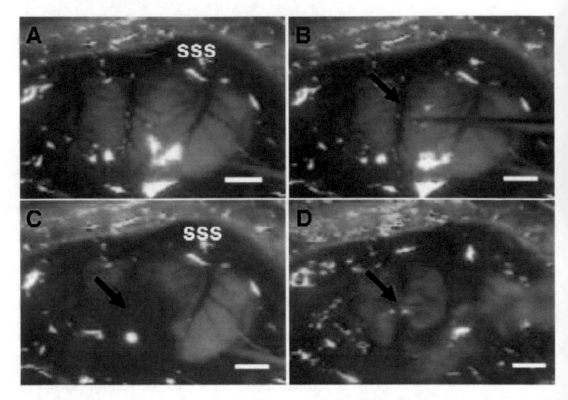

Fig. 6 Complete hemostasis in brain and femoral artery. The pictures are time-lapse images at each stage of the experiment for the brain of adult rat cortex hemostasis. Part of the overlying skull has been removed in an adult rat, and one of the veins of the superior sagittal sinus is transected and treated with 1% self-assembling RADA16-1. (**a**) The brain and veins of the superior sagittal sinus (SSS) are exposed; (**b**) cutting of the vein (arrow); (**c**) bleeding of the ruptured vein (arrow); (**d**) the same area 5 s after application of the self-assembling RADA16-1 to the location of the cut (arrow) as seen under the clear RADA16-1. Scale bars represent 1 mm

3.3.2 Other Combined Delivery Methods

Tissue engineering holds great potential in regenerative medicine, which usually combines cells, bioactive molecules, and scaffolds [13]. Cells encapsulated in 3D scaffold are often chosen to create a biomimic microenvironment for cell growth so that cells are active in regeneration process by secreting their own ECM and regulating cytokines and growth factors. Neural stem cells show the potential of unlimited self-renewal and the ability to differentiate into neurons, astrocytes, and oligodendrocytes, which are favorable in neural regeneration. Hence, besides the application of SAP hydrogel alone in treating brain injuries, new therapeutic strategies have been focused on using a functional self-assembling peptide 3D hydrogel with encapsulated neural stem cells to enhance the reconstruction of the injured brain.

Briefly, traumatic brain injury models were carried out in both left and right frontal cortex. RADA16-IKVAV peptide with or without encapsulated neural stem cells and RADA16 peptide without cells were injected into the damaged sites. After these cavities were filled in with the in situ peptide matrix, hemostasis was observed in seconds. The control groups were injected with neural stem cells only or normal saline by the same procedure. The wound defects were covered with the dura membrane (Duraform Sponge, Codman), and the skin was sutured with 4-0 surgical suture. Related studies successfully demonstrated that injectable RADA16-IKVAV SAP hydrogel possessed the ability to in situ support and bridge the damaged brain wounds. Furthermore, RADA16-IKVAV/NSCs could significantly improve the regeneration and repair of injured brain [35].

4 Conclusion

CNS lesions cause serious tissue loss and neurodegeneration; tissue engineering platforms show great potential in bridging the large injured gap and provide a biomimic microenvironment which functions as nature ECM. After injury, stopping bleeding was the first step, and SAP nanofibrous hydrogel RADA16-I shows its advantages in hemostasis within 10 s. In addition, RADA16-I could fill up the injured lesions by simple injection; hence the cavity in the brain was closed, and the SAP graft was well integrated to the host tissue without obvious gaps in the traumatic brain injury models. Furthermore, RADA16-I could reconstruct the acute traumatic brain injury, facilitate support cells to migrate and survive within the lesion site, and reduce the glia reaction. The injectable property of self-assembling peptide nanofibrous hydrogel also facilitates the application of RADA16-I in intracerebral hemorrhage models. After hematoma aspiration, SAP was injected to the lesion side. Besides reducing inflammation reaction, the SAP-treated group demonstrated a smaller cavity as compare to the aspiration-only group and saline group. Based on RADA16-I, functional motifs RGD and IKVAV were appended to the C-terminus. The combination of these two functionalized SAPs could form a 3D nanofiber hydrogel. The functionalized SAP demonstrated not only the ability of decreasing the inflammation reaction but also great potential in promoting nerve regeneration after hemorrhagic brain injury.

In recent years, the stem cell therapies combined with 3D matrix have become more and more popular in the field of neural tissue engineering. The property of biomimic ECM facilitates the proliferation and differentiation of neural stem cells into neurons in vivo. The implanted graft combined with the functioned neuronal cells makes up the loss of neurons after brain injury and further benefits the functional recovery and neural circuit reformation.

SAPNSs are good candidates for central nervous system lesions. SAPNSs provide a true biomimicking platform with their nanofibrous architecture, which can serve as powerful tools in CNS regeneration process [36–38]. Furthermore, SAPs are amendable by computational methods through which appropriate positions for incorporation of functional motifs can be predicted [39–41]. These methods may be helpful to fabricate scaffolds in the near future, as computation design of self-assembling protein nanomaterials can recently be achieved with atomic-level accuracy [42].

Despite its advantages, the limitations of SAPNSs such as forming uncontrollably macro-sized pores, unstable mechanization of its 3D structure [43], and the lack of its use in PNS nerve regeneration as PNS regeneration needs more strict scaffold structure [44] could not be ignored. Despite its shortcomings, SAPNSs still play a vital role in nerve regeneration and tissue repair. Much remains unexplored and great efforts need to make in the area of CNS regeneration.

References

1. Guo J, Leung KKG, Su H et al (2009) Self-assembling peptide nanofiber scaffold promotes the reconstruction of acutely injured brain. Nanomedicine 5:345–351

2. Whitesides GM, Grzybowski B (2002) Self-assembly at all scales. Science 295:5564

3. Wang T, Zhong X, Wang S et al (2012) Molecular mechanisms of RADA16-1 peptide on fast stop bleeding in rat models. Int J Mol Sci 13:15279–15290

4. Loo Y, Goktas M, Tekinay AB et al (2015) Self-assembled proteins and peptides as scaffolds for tissue regeneration. Adv Healthc Mater 4:2557–2586

5. Liu X, Pi B, Wang H et al (2015) Self-assembling peptide nanofiber hydrogels for central nervous system regeneration. Front Mater Sci 9:1–13

6. Holmes TC, de Lacalle S, Su X et al (2000) Extensive neurite outgrowth and active synapse formation on self-assembling peptide scaffolds. Proc Natl Acad Sci U S A 97:6728–6733

7. Semino CE, Kasahara J, Hayashi Y et al (2004) Entrapment of migrating hippocampal neural cells in three-dimensional peptide nanofiber scaffold. Tissue Eng 10:643–655

8. Wang X, Pan M, Wen J et al (2014) A novel artificial nerve graft for repairing long-distance sciatic nerve defects: a self-assembling peptide nanofiber scaffold-containing poly(lactic-co-glycolic acid) conduit. Neural Regen Res 9:2132–2141

9. Moradi F, Bahktiari M, Joghataei MT et al (2012) BD PuraMatrix peptide hydrogel as a culture system for human fetal Schwann cells in spinal cord regeneration. J Neurosci Res 90:2335–2348

10. Ye Z, Zhang H, Luo H et al (2008) Temperature and pH effects on biophysical and morphological properties of self-assembling peptide RADA16-I. J Pept Sci 14:152–162

11. Cunha C, Villa S, Silva O et al (2011) 3D culture of adult mouse neural stem cells within functionalized self-assembling peptide scaffolds. Int J Nanomedicine 6:943

12. Gelain F, Bottai D, Vescovi A et al (2006) Designer self-assembling peptide nanofiber scaffolds for adult mouse neural stem cell 3-dimensional cultures. PLoS One 1:e119

13. Horii A, Wang X, Gelain F et al (2007) Biological designer self-assembling peptide nanofiber scaffolds significantly enhance osteoblast proliferation, differentiation and 3-D migration. PLoS One 2:e190

14. Sun Y, Li W, Wu X et al (2016) Functional self-assembling peptide nanofiber hydrogels designed for nerve degeneration. ACS Appl Mater Interfaces 8:2348–2359

15. Hallevi H, Dar NS, Barreto AD et al (2009) The IVH Score: a novel tool for estimating intraventricular hemorrhage volume: clinical and research implications. Crit Care Med 37:969–9e1

16. Gajavelli S, Sinha VK, Mazzeo AT et al (2015) Evidence to support mitochondrial neuroprotection, in severe traumatic brain injury. J Bioenerg Biomembr 47:133–148

17. Katz DI, White DK, Alexander MP et al (2004) Recovery of ambulation after traumatic brain injury. Arch Phys Med Rehabil 85:865–869

18. Ucar T, Tanriover G, Gurer I et al (2006) Modified experimental mild traumatic brain injury model. J Trauma Inj Infect Crit Care 60:558–565

19. Zhang H, Hayashi T, Tsuru K et al (2007) Vascular endothelial growth factor promotes brain tissue regeneration with a novel biomaterial polydimethylsiloxane–tetraethoxysilane. Brain Res 1132:29–35

20. Deguchi K, Tsuru K, Hayashi T et al (2006) Implantation of a new porous gelatin–siloxane hybrid into a brain lesion as a potential scaffold for tissue regeneration. J Cereb Blood Flow Metab 26:1263–1273

21. Broderick JP, Brott T, Tomsick T et al (1993) Intracerebral hemorrhage more than twice as common as subarachnoid hemorrhage. J Neurosurg 78:188–191

22. Iniesta JA, Corral J, González-Conejero R et al (2003) Polymorphisms of platelet adhesive receptors: do they play a role in primary intracerebral hemorrhage. Cerebrovasc Dis 15:51–55

23. Labovitz DL, Halim A, Boden-Albala B et al (2005) The incidence of deep and lobar intracerebral hemorrhage in whites, blacks, and hispanics. Neurology 65:518–522

24. Qureshi AI, Tuhrim S, Broderick JP et al (2001) Spontaneous intracerebral hemorrhage. N Engl J Med 344:1450–1460

25. Hesami O, Kasmaei HD, Matini F et al (2015) Relationship between intracerebral hemorrhage and diabetes mellitus: a case-control study. J Clin Diagn Res 9:8–10

26. Zhang N, Luo Y, He L et al (2016) A self-assembly peptide nanofibrous scaffold reduces inflammatory response and promotes functional recovery in a mouse model of intracerebral hemorrhage. Nanomedicine 12:1205–1217

27. Nune M, Kumaraswamy P, Krishnan UM et al (2013) Self-assembling peptide nanofibrous scaffolds for tissue engineering: novel approaches and strategies for effective functional regeneration. Curr Protein Pept Sci 14:70–84

28. Schonauer C, Tessitore E, Barbagallo G et al (2004) The use of local agents: bone wax, gelatin, collagen, oxidized cellulose. Eur Spine J 13:S89–S96

29. Sabel M, Stummer W (2004) Haemostasis in spine surgery. the use of local agents: surgicel and surgifoam. Eur Spine J 13:S97–S101

30. Alam HB, Burris D, DaCorta JA et al (2005) Hemorrhage control in the battlefield: role of new hemostatic agents. Mil Med 170:63–69

31. Petersen B, Barkun A, Carpenter S et al (2004) Tissue adhesives and fibrin glues. Gastrointest Endosc 60:327–333

32. Hoffman M (2004) The cellular basis of traumatic bleeding. Mil Med 169:5–7

33. Bhanot S, Alex JC (2002) Current applications of platelet gels in facial plastic surgery. Facial Plast Surg 18:027–034

34. Ellis-Behnke RG, Liang Y-X, Tay DKC et al (2006) Nano hemostat solution: immediate hemostasis at the nanoscale. Nanomedicine 2:207–215

35. Cheng T-Y, Chen M-H, Chang W-H et al (2013) Neural stem cells encapsulated in a functionalized self-assembling peptide hydrogel for brain tissue engineering. Biomaterials 34:2005–2016

36. Yang Y-H, Khan Z, Ma C et al (2015) Optimization of adhesive conditions for neural differentiation of murine embryonic stem cells using hydrogels functionalized with continuous Ile--Lys-Val-Ala-Val concentration gradients. Acta Biomater 21:55–62

37. Silva GA, Czeisler C, Niece KL et al (2004) Selective differentiation of neural progenitor cells by high-epitope density nanofibers. Science 303:1352–1355

38. Burdick JA, Anseth KS (2002) Photoencapsulation of osteoblasts in injectable RGD-modified PEG hydrogels for bone tissue engineering. Biomaterials 23:4315–4323

39. Tamamis P, Kasotakis E, Mitraki A et al (2009) Amyloid-like self-assembly of peptide sequences from the adenovirus fiber shaft: insights from molecular dynamics simulations. J Phys Chem B 113:15639–15647

40. Hauser CAE, Deng R, Mishra A et al (2011) Natural tri- to hexapeptides self-assemble in water to amyloid beta-type fiber aggregates by unexpected alpha-helical intermediate structures. Proc Natl Acad Sci U S A 108:1361–1366

41. Tamamis P, Terzaki K, Kassinopoulos M et al (2014) Self-assembly of an aspartate-rich sequence from the adenovirus fiber shaft: insights from molecular dynamics simulations and experiments. J Phys Chem B 118:1765–1774

42. King NP, Sheffler W, Sawaya MR et al (2012) Computational design of self-assembling protein nanomaterials with atomic level accuracy. Science 336:1171

43. Ma PX (2008) Biomimetic materials for tissue engineering. Adv Drug Deliv Rev 60:184–198

44. Ma Z, Kotaki M, Inai R et al (2005) Potential of nanofiber matrix as tissue-engineering scaffolds. Tissue Eng 11:101–109

Chapter 5

The Use of Peptide and Protein Vectors to Cross the Blood-Brain Barrier for the Delivery of Therapeutic Concentration of Biologics

Mei Mei Tian and Reinhard Gabathuler

Abstract

The normal functioning of central nervous system is protected by the blood-brain barrier (BBB), which regulates the brain homeostasis and the transport of endogenous compounds. The BBB formed by the endothelial cells of the brain capillaries restricts access to brain cells of blood-borne compounds and allows only essential nutrients such as amino acids, glucose, and hormones to reach brain cells for their normal metabolism. The highly selective regulation of the brain homeostasis by the BBB presents a major obstacle in the incapacity of therapeutic compounds small and large to reach the brain. Diverse ranges of strategies are now being developed to enhance delivery of therapeutic compounds in the brain parenchyma. In this review, we will more specifically address new methods developed as physiological approaches to transport biologics across the BBB. Use of specific molecules such as protein and peptide vectors to facilitate the delivery of biologics to cross the BBB will be addressed. Additionally, their relevance to nanomedicines for brain drug delivery will also be summarized.

Key words Peptide-based delivery, Protein-based delivery, Biologics brain delivery, Antibodies, Blood-brain barrier, $\times B^3$, Melanotransferrin, p97

1 Introduction

Central nervous system (CNS) disorders affect as many as 1.5 billion people worldwide [1], accounting for an economic burden of more than $2 trillion in the EU and USA [2]. In 2014, the drug delivery technologies used in the CNS market was estimated to be about $10 billion, which is expected to grow to $22.5 billion in 2018 [3]. Better treatments are needed, but new CNS drugs have suffered from considerably lower success rates than those for non-CNS indications [4]. This situation is due mainly by the challenge related to drug brain delivery [5]. The largest obstacle to effective

---

Mei Mei Tian and Reinhard Gabathuler contributed equally to this work.

Javier O. Morales and Pieter J. Gaillard (eds.), *Nanomedicines for Brain Drug Delivery*, Neuromethods, vol. 157,
https://doi.org/10.1007/978-1-0716-0838-8_5, © Springer Science+Business Media, LLC, part of Springer Nature 2021

CNS delivery is the BBB. This problem has led the pharmaceutical companies to close research lab concerned with the discovery and development of new treatment for brain disorders and to more focus their research in cardiovascular and oncology indications.

2   Challenges for Brain Delivery Across the BBB

Specialized capillary endothelial cells form the BBB that protects the brain from harmful substances and provides the brain with the required nutrients for its proper function [6]. In contrast to the peripheral capillaries that allow relatively free exchange of substances between cells and tissues, the BBB strictly restricts the transport into the brain by both physical and metabolic barriers [7]. These barriers regulate brain homeostasis (electrolytes, glucose, nucleosides, and amino acids) through multiple efflux and uptake transporters, metabolic enzymes, low pinocytotic activity, and low paracellular permeability. The impermeability of the BBB results from tight junctions between capillary endothelial cells formed by cell adhesion molecules. Brain endothelial cells also possess few alternate transport pathways (e.g., fenestra, transendothelial channels, pinocytotic vesicles) and express high levels of active efflux transport proteins, including P-glycoprotein (P-gp, MDR1, or ABCB1). Even though the volume occupied by capillaries and endothelial cells is about 1% and 0.1% of the brain volume, respectively, the total surface area of brain microvasculature developed is approximately 20 m^2 and a total length of approximately 400 miles [7]. This high vascularization confers to the brain a very complex network of capillaries with an average distance of approximately 40 μm between the capillaries [8].

As BBB restricts diffusion of drugs with a molecular weight of greater than 500 Da [9] or a cross-sectional area of greater than 70 Å [10], the most important factors determining to what extend a small molecule will be delivered from blood into the CNS are lipid solubility, molecular mass, and charge. These structural characteristics allow for passive penetration of the molecules across the BBB. The "classical" neuropharmaceutical agents in the market or in clinical trials are typically less than 600 Da mw with a sufficient degree of solubility.

The overwhelming majority of small molecules, proteins, and peptides do not cross efficiently the BBB. It is key not only to deliver a therapeutic to the CNS but also to target it to adequate brain cells and to their correct intracellular compartment in order to reach therapeutic concentration for efficacy. By localizing drugs at their desired site of action, one can reduce toxicity and increase treatment efficacy. In response to the insufficiency in conventional delivery mechanisms, aggressive research efforts have recently focused on the development of new strategies to more effectively deliver drug molecules to the CNS.

3 Endogenous Mechanisms of BBB Penetration

Molecules successful in penetrating the brain cells use specific mechanism such as transporters and receptors to cross the BBB [6, 11–14]:

- Hydrophilic molecules such as amino acids, glucose, and other small molecules use transporters expressed at the luminal and basolateral side of the endothelial cells.

- Larger hydrophilic molecules such as hormones, insulin, transferrin for iron, and lipoproteins use specific receptors that are highly expressed on the luminal side of the endothelial cells for transcytosis across the BBB.

- Lipophilic molecules can diffuse passively across the BBB into the brain. But these molecules are exposed first to efflux pump (Pgp and others) highly expressed on the luminal side of the BBB and expulsed from the endothelial cells; if not they will be exposed to degrading enzymes localized in the cytoplasm of endothelial cells.

- Brain endothelial cells (BECs) are able to produce secretions for communication purposes, which they can in turn use to modify their own behavior. For example, the granulocyte-macrophage colony-stimulating factor (GM-CSF) and IL-6 released from BECs due to activation of innate immune system can enhance the transport of cell-free HIV-1 across the BBB by acting on the luminal surface of the BECs [14, 15].

Three different approaches are currently used to design drugs able to reach the CNS at therapeutic concentration: invasive, pharmacological, and physiological.

4 Current Therapeutic Approaches for BBB Penetration

4.1 Invasive Approaches

4.1.1 Convection-Enhanced Delivery

Convection-enhanced delivery (CED) uses bulk flow for a uniform distribution and diffusion of the drug throughout the target region [16]. Drugs, peptides, siRNA, and other molecules can potentially be delivered to brain tissue using CED [17, 18].

The mechanism of convection-enhanced delivery involves insertion of a small catheter into the target site within the brain. Through the catheter, the drug is actively pumped into the brain using continuous positive pressure generated by an infusion pump, which causes the drug to penetrate into the interstitial space.

CED causes minimal structural damage to the brain, with the exception of the catheter track. CED has been used in the clinical treatment of neurological diseases such as malignant brain tumors [19–22], neurodegenerative disorders [23–26], epilepsy [27], and stroke [9]. Clinically, CED has been used in the delivery of chemo-therapeutic agents to tumors in the treatment of gliomas, which are aggressive brain tumors with currently few effective treatments. Treatments had very limited survival rate and most were discontinued. Using CED the pressure due to continuous infusion will help a drug to migrate and diffuse in the CNS but also will push it to a path of less resistance, which is not necessarily where the drug needs to be going and be active. This explains the very low success rate of this approach.

4.1.2
Intracerebroventricular
(ICV) Infusion; Intracerebral
(IC) Injection; and
Intrathecal (IT) Injection

ICV, IC, and IT administration have been done with a lot of different molecules small and large with very low success rate specifically due to the undetermined quantification of drugs administered to specific sites in the brain and to the problem of migration of these different molecules in the brain parenchyma [28, 29]. These administrations appear to have a marked improvement in overcoming previously identified translational barriers, such as inhibition by preexisting neutralizing Abs, high peripheral organ biodistribution, and reduced efficiency of CNS transduction. It should be noted, however, that IT administration, especially for AAV9, did not completely restrict the vector within the CNS [30]. IT injections are routine non-surgical procedures that are often done in an outpatient setting with minimal risk. This approach has strong translational implications for lysosomal storage diseases, or any other approach in which a secreted gene product is utilized [31]. The efficiency of spinal cord transduction would also suggest feasible applications for motor neuron diseases such as spinal muscular atrophy, giant axonal neuropathy, and amyotrophic lateral sclerosis [30].

4.1.3   Polymer or
Microchip Systems That
Directly Release
Therapeutics After
Implantation

The Food and Drug Administration (FDA) have approved direct administration into the brain using carmustine-loaded wafers (Gliadel) for the treatment of malignant glioma, providing about 3 weeks gain in overall survival [32, 33]. A major concern with this intervention is that it can only be done at one occasion and drug release from the wafer is a finite process. Moreover, direct administration only results in a localized delivery, which may not be suitable in case of a large and infiltrative tumor or when multiple tumors are located in the brain. Microchip technology has existed since the 1990s and is mainly used for local chemotherapeutic delivery [34]. The potential advantage in overcoming some of the limitations of polymer technology is controlled release by time-dependent biodegradation [35] or electrochemical dissolution [34]. More importantly, it has been shown capable of being remotely controlled [36]. However, potential drawbacks of this technology are finite drug release and possible issues with electronics and magnetic fields [37].

4.1.4 Disruption of the BBB

Using Osmotic Disruption

Osmotic BBB disruption has been a mainstay for increasing permeability of the BBB, usually using high concentrations of mannitol introduced into the carotid artery. Major limitation that occurs with osmotic BBB disruption is from the lack of specificity and the potential toxicity problem due to brain entry of unwanted blood compounds [38].

Using Ultrasound

Focused ultrasound (FUS) is a transient drug delivery technique which employs transcranial delivery of low-frequency ultrasound waves that cause BBB disruptions in targeted areas of the brain [39]. The use of intravenously administered microbubbles in conjunction with this technique has further enhanced the safety by lowering the energy required for BBB disruption [40]. The microbubbles are small air-filled, lipid-, or protein-shelled bubbles which facilitate reversible disruption of the BBB [41]. Although FUS shows promise in preclinical phase, limitations to the technology do exist, such as signal attenuation and distortion from the skull [38]. By disrupting the BBB, potential leakage of bacteria, antibodies, and other toxic substances into the CNS remains high.

Using Bradykinin and Analogs

Bradykinin and its synthetic analogs (RMP-7, also known as cereport and lobradimil) are vasodilators shown to increase the BBB permeability by acting on the tight junctions between the brain capillary endothelial cells [42]. The use of chemical mediators to increase the permeability of BBB can be advantageous; however this technology is limited by transient affect and short therapeutic window with permeability peak at 15 min after treatment [43, 44]. This approach has been tested in clinical setting using carboplatin and a synthetic bradykinin analog (lobradimil) with minimal therapeutic benefit [45].

4.2 Pharmacological Approaches

Noninvasive approaches exploit natural properties of BBB such as molecular size (being less than 500 Da), charge (low hydrogen bonding capabilities), and lipophilicity (the more lipophilic, the better the transport) [46]. A number of pharmaceutical companies have experimented with chemical modification of drugs by addition of lipid groups to the polar ends in making the drug more lipophilic (lipidization) [47]. The main concerns with lipid modification are low selectivity and tissue retention [48]. Lipidization not only alters the permeability of drug to the targeted region but the entire body as well mainly due to binding of plasma protein which affects all pharmacokinetic parameters [49]. This creates an issue of getting sufficient drug to the target site without causing detrimental side effects. Furthermore, lipidization could potentially impact the rate of oxidative metabolism by enzymes such as cytochrome P450 [49]. Optimizing drugs is likely to end up with lipophilic compounds showing high nonspecific binding to brain tissue and, thus, low free concentration in the interstitial fluid to exhibit the pharmacological activity.

4.3 Physiological Approaches

Physiological approaches consist of modification of drugs to take advantage of native BBB nutrient transport systems or by conjugation to ligands of receptors expressed at the BBB which will piggy-back the drug across the BBB after receptor-mediated transcytosis.

The brain requires essential substances for its survival, i.e., glucose, insulin, growth hormone, LDL, etc. These substances are recognized by specific receptors resulting in active transport into the brain. Since the high perfusion of the brain is characterized by an average distance of 50 μm separating capillaries [50], the most effective way of delivering biopharmaceutical drugs is achieved by targeting these essential internalizing (uptake) transport receptors or transporters on these capillaries. Drugs can be modified to take advantage of native BBB nutrient transport systems or by conjugation to ligands to receptors expressed at the BBB which will piggy-back the drug across the BBB after receptor-mediated transcytosis.

4.3.1 Transporter-Mediated Transport

Carrier-Mediated Transporter

BBB express various endogenous saturable, bi-directional transporters for nutrients, vitamins, and minerals [51, 52]. These transporters include (a) the hexose transport systems for glucose (GLUT-1 and GLUT-3) and mannose, (b) amino acid transport by carriers such as cationic amino acid transporter (CAT-1) and large neutral amino acid transporter (LAT-1), (c) monocarboxylic acid transporter for lactate and short-chain fatty acids, (d) choline transporter for choline and thiamine, (e) amine transporter for mepyramine, (f) nucleoside transporter for purine, and (g) peptide transporter for small peptides such as thyrotropin-releasing hormone [12, 53, 54].

As the uptake rate across the BBB for carrier-mediated transport of endogenous ligand is about ten times greater than by transmembrane diffusion [54], these transporters are potential candidates for drug targeting. A number of different nutrient transport systems have been identified, with each capable of transporting a group of nutrients with similar structure. Consequently, only drugs that highly mimic the endogenous carrier substrates will be transported across the BBB into the brain. However, one can modify the structure of a drug such that it mimics a particular nutrient structurally and thus is able to use a carrier-mediated transporter expressed on the endothelial cells forming the BBB to increase transport [50]. Use of BBB transport proteins such as the choline transporter and the amino acid transporter has been done for few drugs. The most well-known example is the large neutral amino acid transporter (LAT-1)-mediated transport of L-Dopa, widely used in the treatment of Parkinson's disease [55].

Large Amino Acid Transporter 1 (LAT-1)

Levodopa (L-Dopa) is transported into the brain via LAT-1 transporters. L-Dopa is metabolically transformed to dopamine in the brain. Other transporters such as glucose transporter 1 (GluT-1), ascorbic acid transporters, and others have been successfully utilized as drug carriers to the brain [56–58]. Modification of small drugs by addition of amino acid or glucose to form a prodrug facilitates their transport by glucose or amino acid transporters and mimics nutrients necessary for the survival of brain cells.

GluT-1 and CD98hc Transporters

Using proteomic analysis of mouse brain capillary endothelial cells, multiple highly expressed proteins were identified [59]. Among these GluT-1 and CD98hc, antibodies against these targets were seen enriched in the brain after systemic administration. In particular, antibodies against CD98hc showed a significant accumulation in the brain and pharmacodynamic response using bispecific antibodies anti-CD98hc and BACE-1 with lower affinity to CD98hc [59]. CD98hc functions as an amino acid (AA) transporter (together with another subunit) and integrin signaling enhancer [60].

Glutathione Transporters

Glutathione is an endogenous tripeptide thiol with antioxidant-like properties and plays a central role in detoxification of intracellular metabolites [61]. Glutathione is highly expressed in the brain and cerebral vasculature and is actively transported across the BBB [62, 63]. Glutathione is considered safe and has been marketed as food ingredient, antioxidant, and part of supportive therapy in cancer and HIV treatments [64]. Although the molecular mechanism for transport remains unknown, it has been shown that glutathione can be used as a targeting ligand coupled to pegylated liposomes to enhance drug delivery to the brain (G-Technology) [65, 66]. The G-Technology is based on coating the surface of nanosized liposomes with glutathione, which has been shown to effectively deliver several drugs to the brain by various independent laboratories [64]. In particular, glutathione pegylated liposomal doxorubicin (2B3-101) has completed a Phase I/IIa clinical trial for various forms of brain cancer and is found to be safe, well tolerated, and active [67]. Glutathione pegylated liposomal methylprednisolone (2B3-201) targeting treatment of patients suffering from acute and chronic neuro-inflammatory disease has also completed a double-blind crossover Phase I study in healthy volunteers.

Using BBB transporter protein as drug delivery vector targeting the brain, several factors need to be considered including (1) kinetics available to transport physiological molecules, (2) structural binding configuration to the transporter, (3) drug manipulation without affecting the activity in vivo, and (4) target delivery within the brain.

Large molecules necessary for the normal function of the brain are delivered to the brain by specific receptors, which are highly expressed on the endothelial cells forming the BBB. Unlike transporter-mediated transport, receptor-mediated transport is characterized by highly specific active transport.

The receptor-mediated transcytosis occurs in three steps:

1. Receptor-mediated endocytosis of the compound at the luminal (blood) side

2. Movement through the endothelial cell cytoplasm

3. Exocytosis of the drug at the luminal (brain) side of the brain capillary endothelium

Receptors highly expressed at the BBB and used as targets for drug brain delivery:

- Transferrin receptor (TR)

- Insulin receptor (IR)

- Insulin-like growth factor receptor

- Leptin receptor

- LDL receptor

- LDL receptor-related protein (LRP)

- Others

Transferrin Receptors (TR)
Owing to the high expression of transferrin receptor on brain capillary endothelial surfaces and various malignant cells, transferrin family ligands are rationalized to be one of the well-known candidates for brain targeting [68, 69]. Drug targeting to the TR can be achieved by utilizing the endogenous ligand transferring (Tf) or by targeting the TR. However, Tf is not considered an optimal targeting ligand as the high concentration of endogenous Tf in the circulation creates highly competitive environment for the injected vector for TfR binding [68, 70]. Receptor-specific binding on the brain microvascular endothelial cells facilitating the transport of associated therapeutical agent to cross the BBB has garnered more popularity in the recent years [71].

HAIYPRH (T7), a transferrin receptor-specific peptide [72, 73], and OX26, a monoclonal antibody against the transferrin receptor [74, 75] have been explored for their ability to facilitate BBB transport. For instance, the brain uptake of brain-derived neurotropic factor (BDNF)-OX26 conjugate caused a 65–70% reduction in stroke volume in rat stroke model [71]. However, study on correlation between binding affinity of antibody against TR and TR with transport efficiency across the BBB has shed light on the need for optimization of targeting antibody [76]. High-affinity TR antibody does not necessarily translate into high transport efficiency, as high-affinity antibody remained mostly associated to the vasculature unable to detach from the TR. By contrast, antibody with 25% lower TR binding affinity resulted in ~50% higher antibody activity compare to high-affinity antibody counterpart [76]. Therefore, modest changes in affinity can significantly affect brain uptake across the BBB. More recently, Niewoehner et al. developed an anti-TR Fab to mediate BBB transcytosis of an attached immunoglobulin [77]. To test the therapeutic potential of this "brain shuttle," Niewoehner et al. [77] re-engineered a monoclonal antibody (mAb) against Aß, the toxic peptide that accumulates in Alzheimer's disease brain [78], by fusing the single-chain anti-TR Fab to either one or both C-terminus of the anti-Aß mAb (sFab and dFab, respectively). Notably, in a transgenic mouse model of Alzheimer's disease, the monovalent sFab fusion mediated effective uptake, transcytosis, and TfR recycling, and the presence of two Fab fragments on mAb31 (dFab) resulted in uptake followed by trafficking to lysosomes and an associated reduction in TfR levels. Treatment with mAb31-sFab for 3 months significantly reduced amyloid plaque burden even at a relatively low dose when compared to treatment with unmodified mAb31. Target engagement at the amyloid plaques was improved more than 50-fold for the sFab construct based on fluorescence intensity quantification using a labeled secondary antibody. Whereas the sFab construct showed extensive plaque decoration, the dFab was only detectable in the microvessels, indicating that the dFab construct targets and enters brain microvessels but fails to escape at the abluminal side [77]. Overall, the sFab anti-TfR brain shuttle module enhanced the delivery and potency of a plaque reducing Ab antibody and could potentially be expanded to the delivery of other therapeutic cargo.

A new format of bispecific antibody, termed dual-variable-domain immunoglobulin (DVD-Ig™), has been generated and described in 2007 [79]. With the DVD-Igs™ format, the target-binding variable domains of two mAbs can be combined via naturally occurring linkers or widely used glycine-serine linkers to create tetravalent, dual-targeting single agent [80]. With proper peptide linkages between the two variable domains in both HC and LC, the various motions within Fab region may provide dual-binding capabilities [81]. Design of DVD-Igs™ with dual-specific targeting for the TfR and anti-Ab has shown brain delivery and binding to specific targets in the brain. The levels and localization of DVD-Igs™, which were injected systemically, were assessed. Preliminary data show brain uptake and retention of DVD-Igs™ up to 96 h [82].

Insulin Receptor (IR)

Similar to TR, insulin receptor is also highly expressed on the endothelial cells making up the BBB [83, 84]. However, its endogenous ligand insulin has not been considered an optimal candidate for brain delivery due to short serum half-life of 10 minutes and possible implication of hypoglycemia through interference with natural balance of insulin [74]. Pardridge group have extensively documented the use of the insulin receptor for the targeted delivery of drugs to the brain using specific antibodies directed against the IR [85] such as using the 83-14 mouse mAb against the human insulin receptor (HIR) in rhesus monkey. Total uptake is 4% which corresponds to 0.04%/g brain tissue 3 h after IV injection [74]. Both chimeric antibody and a fully humanized form of the 83-14 antibody against HIR have been created [86]. Despite promising results, this field is still in its infancy, as recent work by Watts et al. found brain accumulation of IR-specific antibodies did not exceed the antibody control and performed well below that of TR-specific antibodies at doses up to 20 mg/kg [59, 87]. Expression level of IR on mouse brain endothelial cells was also found to be low, which may explain the low brain accumulation [59]. In the manner similar to TR, an antibody only recognizes one site on receptor; thus any alteration to an antibody in order to lower its affinity for targeted receptor may require high dosing leading to possible unwanted side effects outside the brain. Ideally, a vector with low affinity and high capacity binding to the receptor will likely make a good candidate for efficient transport across the BBB in order to avoid negative side effects associated with high dosing.

Low-Density Lipoprotein Receptor-Related Protein 1 (LRP1)

A type I transmembrane protein belonging to an ancient family of endocytic receptors, low-density lipoprotein (LDL) receptor family [88, 89], LRP1 is a multifunctional endocytosis receptor that mediates the internalization and degradation of ligands involved in diverse metabolic pathways [90, 91]. LRP1 is synthesized as 600 kDa precursor protein and processed into a large 515 kDa extracellular α-subunit and a smaller 85 kDa ß-subunit containing the transmembrane domain and cytoplasmic tail that remain non-covalently linked [92]. Like all the members of the LDL receptor family, the extracellular α-subunit consists of four ligand-binding domains (DI, DII, DIII, and DIV) and epidermal growth factor (EGF) repeats, which facilitate the interaction with more than 40 different ligands [93, 94]. The ß-subunit is made up of a transmembrane fragment and a shorter cytoplasmic tail containing YxxL and dileucine motifs that function as principal endocytosis signals and two NPxY motifs that serve as secondary endocytosis signals, as well as binding sites for signaling adapter proteins [90, 93, 95–97]. LRP is a multiligand lipoprotein receptor which interacts with a broad range of secreted proteins and resident cell surface molecules including apoE, α2M, tPA, PAI-1, APP, Factor VIII, and lactoferrin. LRP is expressed in many tissues including the CNS [98] and has been associated with a number of human diseases such as Alzheimer's disease [99, 100], multiple sclerosis [101, 102], cancer [103–106], and ischemic cardiomyopathy

[107]. The diverse function and association with diseases suggest that LRP1 is likely a multifunctional receptor playing important roles in various human diseases.

The LRP 1 has been exploited to target drugs to the brain in a similar fashion as TR and IR.

Polysorbate 80-Coated Nanoparticles

Nanoparticles coated with polysorbate 80 (Tween 80) have been used for brain delivery of active compounds including analgesic peptides (dalargin) and chemotherapeutics such as doxorubicin in animal models [108]. It is thought that polysorbate 80-coated polybutyl-cyanoacrylate nanoparticles adsorb apolipoproteins E and B from the bloodstream after IV injection [109] and therefore uses LRP for transcytosis across the BBB [110]. The precise mechanism of transcytosis however is still debatable.

Melanotransferrin

Melanotransferrin (MTf), also known as human melanoma-associated antigen p97, is a 97 kDa sialoglycoprotein belonging to the Tf family of proteins. MTf has been proposed to be one of the oldest members of the Tf family, dating back to 670 million years ago [111]. Similar to other members of Tf family, it possesses two metal-binding lobes; however, MTf uniquely exist in two different forms: a glycosyl-phosphatidyl-inositol (GPI)-anchored membrane-bound form and as a soluble secreted form in the serum [112, 113]. Recombinant soluble form of MTf (80 kDa) has been shown to actively transcytose across the BBB at a transport rate 10–15 times higher than that of either Tf or lactoferrin with the involvement of LRP [114]. MTf is being further developed by BiOasis Technologies Inc. under the name of $\times B^3$ for the use of transport of anti-cancer agent such as doxorubicin, paclitaxel, and biologics such as monoclonal antibodies or lysosomal enzymes for brain delivery [115, 116]. Recently, MTf has been shown to effectively deliver trastuzumab to the brain leading to a reduction of the number of preclinical human Her2+ breast cancer metastases in the brain by 68%, while the tumors which remained after treatment were 46% smaller compared to the control group. In comparison, trastuzumab alone had minimal to no effect on reducing the number or the size of metastases in the brain [117].

MTf has a clear potential as the endogenous protein is found at very low concentration in the blood of most normal individuals, and binding kinetics are consistent with MTf having a relatively medium to low affinity for its receptor [114], thereby minimizing competition with exogenous ligands of the receptor. As an autologous human protein, immune hypersensitivity and elimination by neutralizing antibodies after repeated treatments in clinical therapies are likely to be minimized. A second generation of this technology is now in development and consists of a family of peptides derived from the $\times B^3$ sequence. This family of peptides has shown efficient transport across the BBB and the ability to transport biologics to the brain. A 12-amino acid peptide has shown high transport level to the CNS and ability to transport various payloads such as enzymes, antibodies, and oligonucleotides to therapeutic levels in the brain.

Receptor-Associated Protein

Other proteins such as the receptor-associated protein (RAP), an antagonist and a ligand for both LRP and VLDLR, have been shown to be efficiently transported across the BBB into the brain parenchyma [118]. RAP is found in the endoplasmic reticulum (ER) where it plays the role of a chaperone for the LDL receptor family, which includes LRP1 and 2 facilitating its transport to the cell surface avoiding interaction with endogenous ligands [119], and exogenously introduced RAP can also act to prevent uptake of LRP ligands via binding to the receptor [120]. The application of RAP as a potential drug carrier to the brain is in development by Raptor Pharmaceutical [121].

Lentiviral Vector

The use of the lentivirus vector system to deliver the lysosomal enzyme, glucocerebrosidase, and a secreted form of GFP to the neurons and astrocytes in the CNS has been demonstrated by using the low-density lipoprotein receptor-binding domain of the apolipoprotein B with the targeted protein [122]. This transport was specific to the protein with the ApoB LDLR domain, as the control protein sGCm did not cross the BBB. Although the ApoB LDLR sequence is 38 aa, the length did not appear to greatly affect delivery or function of the recombinant protein. Unlike other retroviral genera, lentiviruses are capable of infecting both dividing and non-dividing cells, which makes it ideal for the mostly non-dividing cells in the postnatal brain [123, 124]. Lentiviruses posses numerous attractive qualities for CNS delivery, such as lack of viral gene expression, relatively large cloning capacity [125], lower probability of generating replication-competent retroviruses [126, 127], and ability to expend lentiviral tropism by pseudotyping [124]. The most common pseudotyping is mediated by the viral surface glycoprotein, VSV-G, which facilitates wide tropism through their interaction with LRP [128]. Through differential pseudotyping, targeted delivery within a brain region is possible. However, as most lentiviral vectors are of HIV origin, safety remains top concern. Even though no serious effects have risen, the inability to control where lentiviral vector integration happens remains an issue of concern [129].

Angiopep

The field of peptide shuttle was pioneered by Stephen Dowby and coworkers in the late 1990s [130, 131]; however, the brain transport capability by a peptide in a selective manner was not proven until 2007 with RVG29 [132] and soon after with Angiopep [131, 133]. Angiopeps were derived from alignment of amino acid sequence of aprotinin with LRP ligands bikunin, amyloid ß/

A4 protein precursor, and the Kunitz inhibitor-1 precursor [68]. Angiopep-2, a 19 aa peptide derived from the Kunitz domain, demonstrates the highest transcytosis rate [133] involving LRP1 [134]. The potential of this platform technology by Angiochem Inc. in brain-targeted therapy has been demonstrated with ANG1005, a paclitaxel-Angiopep-2 conjugate [135], where intra-peritoneal treatment increased average survival and prolonged life span of mice without any loss of cytotoxic effect from paclitaxel [135]. ANG1005 underwent Phase I/II clinical trials and was well tolerated and showed activity in heavily pretreated patients with advanced solid tumors, including those who had brain metastases and/or failed prior taxane therapy [136]. ANG1005 is now in two Phase II clinical trials for recurrent malignant gliomas and for advanced cancer and brain metastases.

It should be noted that the relevance of Angiopep-2 is less pronounced when applied with nanoparticle-mediated delivery to the brain, as Angiopep-2 did not significantly enhance the brain uptake of liposomes in vivo [137].

Vect-Horus Peptide

More recently, LDL receptor has been targeted by Vect-Horus for peptide-based brain delivery. A family of cyclic peptides were isolated by phage display biopanning and were shown to undergo receptor-mediated endocytosis involving LDL receptor without competition from LDL [138]. Chemical optimization of a cyclic 15-mer peptide showed improved biochemical parameters including LDL receptor binding affinity and in vitro blood stability. In vivo data also show preferential accumulation of peptide and/or its metabolites in LDL receptor-enriched tissues [139]. This offers a new approach to peptide-based CNS delivery of therapeutics.

4.3.3 The BBB Transmigrating Llama Single-Domain Antibodies

Discovered by Hamers-Casterman and coworkers [140], single-domain antibodies (sdAbs) are 12–15 kDa functional antibodies characterized by lack of light chains and CH1 domain [141, 142]. Produced as a humoral immune response, sdAb seems to be limited to only *Camelidae* species within mammals [143]. Using a Llama single-domain antibody (sdAb) phage display library [144, 145], a new antigen-ligand system was identified for transvascular brain delivery [146, 147]. The transport of two sdAbs, FC5 and FC44 across the human brain endothelial cells, was shown to be polarized, charge independent, and temperature dependent suggesting a receptor-mediated process [145]. FC5 was detected in clathrin-enriched fractions following internalization and was shown to target the early endosomes, bypass late endosomes/lysosomes, and remain intact after transcytosis. The clathrin-mediated internalization was associated with a cell surface $\alpha(2,3)$-sialoglycoprotein, which was later identified as transmembrane domain protein 30A (TMEM 30A) [145]. In vivo pharmacokinetics study of FC5 in rat demonstrated nearly identical plasma pharmacokinetics compare to control sdAb, with significantly higher concentration in CSF (10- to 25-fold) [148]. Compare to whole antibody, lower molecular weight, increased stability, and absence of complement system-triggered cytotoxicity associated with Fc make sdAbs a promising vector candidate for BBB drug delivery.

4.3.4   Receptor-Targeted Nanoparticles

Trojan Horses Liposomes

Trojan Horses Liposomes (THL) are considered by Pardridge et al. [149, 150] and ArmaGen Inc. for the delivery of non-viral plasmid DNA across the BBB for expression of interfering RNA (shRNA). The plasmid DNA is encapsulated in the interior of a 100 nm liposome. The surface of the liposome was conjugated with several thousand strands of 2000 Da PEG. The tips of 1–2% of the PEG strands were conjugated with a receptor (R)-specific mAb. TR mAbs-targeted THL with expression plasmid for tyrosine hydroxylase (TH) can treat Parkinson's disease (PD) induced in rats. This approach has been used to deliver shRNA against EGFR and resulted in knockdown of EGFR expression and increase survival of mice implanted intracerebrally with brain tumors [149].

Nanoparticles Coated with Transferrin or Transferrin Receptor Antibodies

Human serum albumin (HAS) nanoparticles covalently coupled to transferrin or transferrin receptor monoclonal antibodies have been used to transport loperamide across the BBB. Loperamide is a model drug as it binds to the nanoparticles by adsorption and is BBB impermeable. Significant anti-nociceptive effects were observed demonstrating the value of this approach to increase BBB transport of these small drugs [151]; however, in vivo investigations using capillary depletion and morphological examination have identified the capture of anti-TfR limited to the brain capillary endothelial cells [152]. The application of this technology for the transport of large and other small molecules across the BBB still remains to be proven.

Nanoparticles Modified With Synthetic Opioid Glycopeptide g7

A synthetic peptide opioid-derived [H2N-Gly-L-Phe-L-Thr-Gly-L-Phe-L-Leu-L-Ser(O-b-Glu)-CONH2] or g7 was able to cross efficiently the BBB [153]. Evidence of BBB crossing pathways was obtained after systemic administration of g7-nanoparticles (NPs) in rodents, indicating that g7-peptide, due to its peculiar amphipathic character, was able to selectively promote endocytosis at BBB level and therefore mediate BBB translocation of g7-NPs to the CNS parenchyma [154]. Preliminary experiments were conducted in order to demonstrate that this approach can be applied to the delivery of large molecular weight molecules such as enzymes. IV injection of PLGA [poly(lactide-co-glycolide)] NPs labelled with rhodamine, modified with g7, and loaded with FITC-albumin was used and showed that g7-NPs were able to cross the BBB and distributed widely in brain parenchyma [155]. This demonstrates the potential use of these g7-NPs for the delivery of high molecular weight molecules in mice [155], in addition to its use for small molecules such as loperamide [153]. The mechanism involved in the transport of g7-NPs across the BBB is not known and still under questions about involvement of a receptor or special membrane-membrane interaction resulting in pinocytosis or adsorptive transcytosis [154].

4.3.5
Adsorptive-Mediated
Transcytosis

Adsorptive mediate transcytosis (AMT) is nonspecific vesicular transport system triggered by the electrostatic interaction between the cationic peptide/protein and the anionic sites on the membrane surface [156]. As the luminal plasma membrane of brain capillary endothelial cells is overall negatively charged, cationic molecules can bind and potentially transcytose across the BBB [157]. Similar to receptor-mediated transcytosis (RMT), AMT in brain is a saturable process that is both time and concentration-dependent [157]; however, the maximum binding capacity is several thousand times greater than RMT [158]. This method of CNS delivery was first explored in the late 1980s [159] and has subsequently been applied to several peptides in vitro and in vivo [160, 161]. These peptides possess multiple positive charges and ability to interact with lipid membrane and to adopt a significant secondary structure upon binding to lipids, with some of them sharing common features such as hydrophobicity and amphipathicity.

On the other hand, limitations do exist for CNS delivery via AMT. As much as being a benefit for AMT, the high adsorptive property of cationized peptides/proteins to anionic sites on cell surface will likely favor random tissue and organ distribution. Since anionic sites are found on the surface of all living cells, direct targeting of specific cell type or organ will be a challenge. Potential toxicity and immunogenicity should also be noted given the efficient cell penetration.

Dendrimers

Dendrimers are man-made nanosized molecules consist of a central core and tree-like branches radiating symmetrically from the core [162]. With carefully crafted structure, the active surface terminal groups of the dendrimers can be utilized to modify physiochemical and biological properties [163, 164]. A number of advantages have demonstrated promising roles of dendrimers in CNS delivery, such as improving the solubility, stability, permeability, biodistribution, and efficacy of a number of therapeutics as well as being used as imaging and diagnostic molecules in animal models bearing brain tumors [165]. Dendrimer-drug interaction can be achieved by simple physical entrapment, where the micellar structure formed by the hydrophobic core and the hydrophilic shell is maintained at all concentrations [166, 167]. Due to the multivalent property of dendrimers, complexes may result through chemical bonding such as electrostatic interaction and pegylation [168]. The application of dendrimers is far-reaching, with capability of improving pharmaco-kinetics. CNS-targeting ligand modification of the dendrimer surface can potentially improve the rate and duration of drug delivery to brain cells. However, toxicity remains an important issue surrounding dendrimers. Both in vitro and in vivo studies have demonstrated various degrees of toxicity associated with dendrimers containing cationic surface groups, which tend to interact with lipid bilayer, leading to increase membrane permeability and decrease integrity [169–172]. Alternatively, anionic dendrimers and masking of primary amino groups have been shown to drastically reduce toxicity [172, 173].

Protein Transduction Domains

Protein transduction domains (PTDs) are typically amino acid sequences located on transcription factors allowing transport of larger molecules across the cell membranes; examples are TAT, homeodomain of Antennapedia, Syn-B, polyarginines, and others. These peptides are basic molecules, cations, which are positively charged and bind to negatively charged phospholipids of cell membranes and are then taken up by adsorptive-mediated endocytosis. Coupling of doxorubicin to either SynB1 (18aa) or SynB3 (10aa) vectors significantly enhances its brain uptake by about 30-fold and bypasses the efflux pump MDR1 [174]. Using this approach, brain uptake of an enkephalin analog (dalargin) was enhanced several 100-fold after vectorization [175]. Highly hydrophilic cationic polyarginines (9 mer of L-Arg, r9) have been shown to have very efficient cellular uptake rate that is $20\times$ faster than the well-known cell-penetrating peptide HIV-1 TAT_{49-57} and $100\times$ faster with D-arginine oligomer (r9) [176]. Kumar et al. [132] demonstrated the transvascular delivery of siRNA to the CNS using D-arginine oligomer (r9) joined to C-terminus of a short RVG (rabies virus glycoprotein) peptide, which allows specific binding to the acetylcholine receptor expressed on neuronal cells contributing to its CNS delivery probably by adsorptive endocytosis with the specific siRNA. Rather than the number of constituent amino acids of the peptide, C-terminal structure and the basicity of the molecule are the most important determinants of uptake by adsorptive-mediated system of cultured bovine brain capillary endothelial cells [177].

Biologically Active Core/Shell Nanoparticles Designed for Drug Delivery Across the BBB

Biologically active polymer core/shell nanoparticles enable solubilization of hydrophobic drugs through hydrophobic interaction and/or hydrogen bonding while effectively protecting against the harsh external environment by the hydrophilic shell [178]. This facilitates the durability of the drugs in the circulation. Anchoring of Tat molecules to the surface of these nanoparticles has been successfully synthesized for drug delivery across the BBB [178]. Ciprofloxacin as a model antibiotic has been tested with this technique and demonstrated the presence of Tat on the surfaces of the nanoparticles promoted their brain uptake and entered the cytoplasm of neurons [178].

Myristoylated Polyarginine Peptide (MPAP)

Utilizing a hydrophobic 14-carbon moiety of myristic acid with a polyarginine peptide, Pham et al. demonstrated delivery and in vivo distribution of fluorescent cargo in mouse brain for near-infrared (NIR) fluorescence imaging [179]. Fluorescent cargo was found primarily accumulated in the neurons. It was proposed that the myristoyl moiety guide the membrane association, while the hydrophilic polyarginine initiate electrostatic interaction with the negative charges on the BBB cell membranes facilitating adsorptive-mediated endocytosis [179]. The potential of this approach in application to targeted therapetic delivery to CNS needs to be investigated in the future.

Exosomes

Fifty to hundred nanometer specialized membranous vesicles originated by inward budding into multivesicular bodies, which either are digested by lysosomes or fuse with the plasma membrane to release internal vesicles referred to as exosomes [180]. Exosomes are secreted by a variety of cells including dendritic cells, mast cells, T cells, platelets, neurons, Schwann cells, epithelial cells, and tumor cells [181, 182]. Exosomes were initially thought to be a mechanism for eradicating unneeded membrane proteins from reticulocytes [183]. However it is commonly accepted that they are specialized in intercellular communications facilitating transfer of proteins, lipids, and RNAs [184]. Within the CNS, exosomes have been associated to a number of pathogenic proteins, such as amyloid peptide and prions, and are thought to propagate pathogenesis through interaction with recipient cells [184]. Target cell binding is thought to be accomplished by expression of adhesive proteins [185], receptor ligand interaction [186], or endocytosis [187]. Exosomes has been extensively applied to miRNA field, as majority of miRNA in serum and saliva are found within exosomes [188]. Diagnostic application of RNA profiles of exosomes has been correlated to a number of cancer models [189–191]. As therapeutic delivery system, exosomes have demonstrated their potential in targeted delivery of interference RNA [192] and tumor prevention by using specific sources of exosomes [193]. However, many aspects regarding exosomes association to diseases are not yet well understood, such that exosomes may promote the tumor-invasive activity via transfer of caveolin-1 [194] and problems surrounding its role in pathogen transmission [195, 196].

4.4 Intranasal Administration for Brain Delivery

Intranasal (IN) drug delivery to the brain was first proposed by William H. Frey II in 1989 and has since been recognized as a noninvasive method of direct delivery of a drug to the brain and the CNS [5, 197, 198]. In human body, nasal cavity is the only site where nervous system is in direct contact with the environment, hence offering great potential as an alternative route for direct drug delivery to the CNS [198]. Olfactory pathway has been attributed to the direct brain delivery [199, 200]; however more recently, the involvement of trigeminal pathway has also been demonstrated [201, 202]. Mechanism of transport may also occur via the capillaries, lymphatics, and cerebrospinal fluid present in the nasal mucosa tissue or by being excreted by nasal mucociliary movement [203]. In animals, brain delivery of a wide spectrum of therapeutics has been demonstrated ranging from small lipophilic molecules (cocaine [204], morphine [205]) to larger therapeutics (leptin [206, 207], insulin [208, 209]). Intranasal application involve siRNA [210, 211], and nose-to-brain delivery in combination with cell-penetrating peptidemodified nano-micelles have shown significant improvement [212]. While IN delivery represents a convenient noninvasive route of administration, incorporation of nanoparticles and cell-penetrating peptides may likely improve therapeutic efficacy by increasing drug stability, durability, and specificity. On the other hand, inconsistency in evidences for direct IN transport does warrant further investigation, as many of the methodologies used in animals are not translatable in humans due to the aggressiveness of the techniques [198]. Moreover, the relatively small proportion of human nasal mucosa occupied by olfactory epithelium (~3%) compare to that of rodent (up to 50%) raises question about the estimated clinical potential/translation of nose-to-brain delivery in human [198, 213].

5   Conclusions

As the population ages, increasing numbers of patients will develop brain cancer or various neurodegenerative diseases, creating a great unmet need for therapies which can treat CNS disorders. The existing market for central nervous system (CNS) diseases was $65B in 2006, dominated by antidepressant, stroke, epilepsy, and Alzheimer's medications. This market is forecasted to increase to $105B in 2015 and includes many unmet therapeutic needs: brain cancer (primary and metastatic), Alzheimer's disease, Parkinson's disease, amyotrophic lateral sclerosis (ALS), multiple sclerosis (MS), psychiatric disorders, stroke, and infections. There are many promising biopharmaceutical agents very active on brain targets that, unfortunately, cannot enter the brain in sufficient quantities to be effective. Therefore new technologies have to be developed to address this problem.

In this review the current techniques are highlighted, and new approaches in development to deliver small and large molecules such as biologics to the brain are described. The techniques used to this day involve direct injection or infusion of therapeutic compounds in the brain, but these methods present great limitations due to brain parenchyma distribution. Only the use of technologies able to cross the endothelial cells of the BBB using a physiologic approach will allow a homogenous distribution of therapeutics in the brain and provide a uniform exposure of brain cells.

Monoclonal antibodies and ligands of these receptors can be used as Trojan Horses for transcytosis of therapeutic compounds to the CNS. New vectors based on monovalent antibodies or antibodies presenting lower affinity for the transferrin receptor are being developed with some success but present some toxicity. A new peptide vector Angiopep ligand to LRP1 is being developed by Angiochem Inc. and demonstrates a high transport rate across the BBB and ability to transport large quantities of drugs to the brain parenchyma. This technology is the most advanced as it is now in Phase II for the treatment of recurrent gliomas and brain metastasis for its first product ANG1005, an Angiopep-paclitaxel conjugate. In addition, a new biotech company, BiOasis Technologies Inc., is developing $\times B^3$ as a vector for brain delivery. $\times B^3$ also known as melanotransferrin or p97 has demonstrated high transport of small and large therapeutics such as antibodies and enzymes piggy-backed on $\times B^3$ across the BBB. A new family of peptides derived from $\times B^3$ shows the same abilities and is now the second generation of $\times B^3$.

References

1. Domínguez A, Álvarez A, Hilario E et al (2013) Central nervous system diseases and the role of the blood-brain barrier in their treatment. Neurosci Discov 1:11. https://doi.org/10.7243/2052-6946-1-3

2. Ereshefshy L, Evans R, Sood R, Williamson D, English BA (2016) Venturing into a new era of CNS drug development to improve success. Waltham, MA: Parexel. https://www.parexel.com/application/files_previous/4314/4113/4032/Venturing_Into_a_New_Era_of_CNS_Drug_Development_to_Improve_Success.pdf (accessed December 2, 2016)

3. Jain KK (2014) Global drug delivery in central nervous system diseases – technologies, markets, companies. Research and Markets, Basel

4. Kesselheim AS, Hwang TJ, Franklin JM (2015) Two decades of new drug development for central nervous system disorders. Nat Rev Drug Discov 14:815–816. https://doi.org/10.1038/nrd4793

5. Thorne RG, Frey WH (2001) Delivery of neurotrophic factors to the central nervous system: pharmacokinetic considerations. Clin Pharmacokinet 40:907–946. https://doi.org/10.2165/00003088-200140120-00003

6. Abbott NJ (2013) Blood-brain barrier structure and function and the challenges for CNS drug delivery. J Inherit Metab Dis 36:437–449. https://doi.org/10.1007/s10545-013-9608-0

7. Palmer AM (2010) The role of the blood-CNS barrier in CNS disorders and their treatment. Neurobiol Dis 37:3–12. https://doi.org/10.1016/j.nbd.2009.07.029

8. Wolak DJ, Thorne RG (2013) Diffusion of macromolecules in the brain: implications for drug delivery. Mol Pharm 10:1492–1504. https://doi.org/10.1038/nature13314.A

9. Haar PJ, Broaddus WC, Chen ZJ et al (2010) Quantification of convection-enhanced delivery to the ischemic brain. Physiol Meas 31:1075–1089. https://doi.org/10.1088/0967-3334/31/9/001

10. Abbott NJ, Dolman DEM, Patabendige AK (2008) Assays to predict drug permeation across the blood-brain barrier, and distribution to brain. Curr Drug Metab 9:901–910. https://doi.org/10.2174/138920008786485182

11. de Boer AG, Gaillard PJ (2007) Drug targeting to the brain. Annu Rev Pharmacol Toxicol 47:323–355. https://doi.org/10.1146/annurev.pharmtox.47.120505.105237

12. Misra A, Ganesh S, Shahiwala A, Shah SP (2003) Drug delivery to the central nervous system: a review. J Pharm Pharm Sci

6:252–273. https://doi.org/10.1007/978-1-60761-529-3

13. Gabathuler R (2010) Approaches to transport therapeutic drugs across the blood-brain barrier to treat brain diseases. Neurobiol Dis 37:48–57. https://doi.org/10.1016/j.nbd.2009.07.028

14. Banks WA (2016) From blood-brain barrier to blood-brain interface: new opportunities for CNS drug delivery. Nat Rev Drug Discov 15:275–292. https://doi.org/10.1038/nrd.2015.21

15. Dohgu S, Fleegal-DeMotta M, Banks WA (2011) Lipopolysaccharide-enhanced transcellular transport of HIV-1 across the blood-brain barrier is mediated by luminal microvessel IL-6 and GM-CSF. J Neuroinflammation 8:167. https://doi.org/10.1186/1742-2094-8-167

16. Song DK, Lonser RR (2008) Convection-enhanced delivery for the treatment of pediatric neurologic disorders. J Child Neurol 23:1231–1237

17. Bobo RH, Laske DW, Akbasak A et al (1994) Convection-enhanced delivery of macromolecules in the brain. Proc Natl Acad Sci U S A 91:2076–2080. https://doi.org/10.1073/pnas.91.6.2076

18. Stiles DK, Zhang Z, Ge P et al (2012) Widespread suppression of huntingtin with convection-enhanced delivery of siRNA. Exp Neurol 233:463–471. https://doi.org/10.1016/j.expneurol.2011.11.020

19. Allhenn D, Shetab Boushehri MA, Lamprecht A (2012) Drug delivery strategies for the treatment of malignant gliomas. Int J Pharm 436:299–310. https://doi.org/10.1016/j.ijpharm.2012.06.025

20. Chen KS, Mitchell DA (2012) Monoclonal antibody therapy for malignant glioma. Adv Exp Med Biol 746:121–141

21. Mehta AI, Choi BD, Ajay D et al (2012) Convection enhanced delivery of macromolecules for brain tumors. Curr Drug Discov Technol 9:305–310. CDDT-EPUB-20120220-005 [pii]

22. Bidros DS, Liu JK, Vogelbaum MA (2010) Future of convection-enhanced delivery in the treatment of brain tumors. Future Oncol 6:117–125. https://doi.org/10.2217/fon.09.135

23. Saito R, Tominaga T (2012) Convection-enhanced delivery: from mechanisms to clinical drug delivery for diseases of the central nervous system. Neurol Med Chir 52:531–538. https://doi.org/10.2176/nmc.52.531

24. Barua NU, Miners JS, Bienemann AS et al (2012) Convection-enhanced delivery of neprilysin: a novel amyloid-β-degrading therapeutic strategy. J Alzheimers Dis 32:43–56. https://doi.org/10.3233/JAD-2012-120658

25. Miners JS, Barua N, Kehoe PG et al (2011) Aβ-degrading enzymes: potential for treatment of Alzheimer disease. J Neuropathol Exp Neurol 70:944–959. https://doi.org/10.1097/NEN.0b013e3182345e46

26. Lam MF, Thomas MG, Lind CRP (2011) Neurosurgical convection-enhanced delivery of treatments for Parkinson's disease. J Clin Neurosci 18:1163–1167. https://doi.org/10.1016/j.jocn.2011.01.012

27. Rogawski MA (2009) Convection-enhanced delivery in the treatment of epilepsy. Neurotherapeutics 6:344–351. https://doi.org/10.1016/j.nurt.2009.01.017

28. Yan Q, Matheson C, Sun J et al (1994) Distribution of intracerebral ventricularly administered neurotrophins in rat brain and its correlation with trk receptor expression. Exp Neurol 127:23–36. https://doi.org/10.1006/exnr.1994.1076

29. Morrison PF, Laske DW, Bobo H et al (1994) High-flow microinfusion: tissue penetration and pharmacodynamics. Am J Phys 266:R292–R305

30. Gray S, Naqabhushan Kalburgi S, McCown TJ, Samulski J (2013) Global CNS gene delivery and evasion of anti-AAV-neutralizing antibodies by intrathecal AAV administration in non-human primates. Gene Ther 20:450–459. https://doi.org/10.1038/gt.2012.101.Global

31. Calias P, Papisov M, Pan J et al (2012) CNS penetration of intrathecal-lumbar idursulfase in the monkey, dog and mouse: Implications for neurological outcomes of lysosomal storage disorder. PLoS One. https://doi.org/10.1371/journal.pone.0030341

32. La Rocca RV, Rezazadeh A (2011) Carmustine-impregnated wafers and their impact in the management of high-grade glioma. Expert Opin Pharmacother 12:1325–1332. https://doi.org/10.1517/14656566.2011.580737

33. Perry J, Chambers A, Spithoff K, Laperriere N (2007) Gliadel wafers in the treatment of malignant glioma: a systematic review. Curr Oncol 14:189–194. https://doi.org/10.3747/co.2007.147

34. Santini JT, Cima MJ, Langer R (1999) A controlled-release microchip. Nature 397:335–338. https://doi.org/10.1038/16898

35. Richards Grayson AC, Choi IS, Tyler BM et al (2003) Multi-pulse drug delivery from a resorbable polymeric microchip device. Nat Mater 2:767–772. https://doi.org/10.1038/nmat998

36. Farra R, Sheppard NF, McCabe L et al (2012) First-in-human testing of a wirelessly controlled drug delivery microchip. Sci Transl Med 4:122ra21–122ra21. https://doi.org/10.1126/scitranslmed.3003276

37. Chaichana KL, Pinheiro L, Brem H (2015) Delivery of local therapeutics to the brain: working toward advancing treatment for malignant gliomas. Ther Deliv 6:353–369. https://doi.org/10.4155/tde.14.114

38. Azad TD, Pan J, Connolly ID et al (2015) Therapeutic strategies to improve drug delivery across the blood-brain barrier. Neurosurg Focus 38:E9. https://doi.org/10.3171/2014.12.FOCUS14758

39. Etame AB, Diaz RJ, Smith CA et al (2012) Focused ultrasound disruption of the blood-brain barrier: a new frontier for therapeutic delivery in molecular neurooncology. Neurosurg Focus 32:E3. https://doi.org/10.3171/2011.10.FOCUS11252

40. Hynynen K, McDannold N, Vykhodtseva N, Jolesz FA (2001) Noninvasive MR imaging-guided focal opening of the blood-brain barrier in rabbits. Radiology 220:640–646. https://doi.org/10.1148/radiol.2202001804

41. Hynynen K (2008) Ultrasound for drug and gene delivery to the brain. Adv Drug Deliv Rev 60:1209–1217. https://doi.org/10.1016/j.addr.2008.03.010

42. Sanovich E, Bartus RT, Friden PM et al (1995) Pathway across blood-brain barrier opened by the bradykinin agonist, RMP-7. Brain Res 705:125–135. https://doi.org/10.1016/0006-8993(95)01143-9

43. Liu LB, Xue YX, Liu YH (2010) Bradykinin increases the permeability of the blood-tumor barrier by the caveolae-mediated transcellular pathway. J Neuro-Oncol 99:187–194. https://doi.org/10.1007/s11060-010-0124-x

44. Ma T, Xue Y (2016) MiRNA-200b regulates RMP7-induced increases in blood-tumor barrier permeability by targeting RhoA and ROCKII. Front Mol Neurosci 9:1–13. https://doi.org/10.3389/fnmol.2016.00009

45. Warren K, Jakacki R, Widemann B et al (2006) Phase II trial of intravenous lobradimil and carboplatin in childhood brain tumors: a report from the Children's Oncology Group. Cancer Chemother Pharmacol 58:343–347. https://doi.org/10.1007/s00280-005-0172-7

46. Lipinski CA, Lombardo F, Dominy BW, Feeney PJ (2001) Experimental and computational approaches to estimate solubility and permeability in drug discovery and development settings. Adv Drug Deliv Rev 46:3–26. https://doi.org/10.1016/S0169-409X(00)00129-0

47. Lu CT, Zhao YZ, Wong HL et al (2014) Current approaches to enhance CNS delivery of drugs across the brain barriers. Int J Nanomedicine 9:2241–2257. https://doi.org/10.2147/IJN.S61288

48. Davis SS (1997) Biomédical applications of nanotechnology – implications for drug targeting and gene therapy. Trends Biotechnol 15:217–224. https://doi.org/10.1016/S0167-7799(97)01036-6

49. Brasnjevic I, Steinbusch HWM, Schmitz C, Martinez-Martinez P (2009) Delivery of peptide and protein drugs over the blood-brain barrier. Prog Neurobiol 87:212–251. https://doi.org/10.1016/j.pneurobio.2008.12.002

50. Pardridge WM (2005) The blood-brain barrier: bottleneck in brain drug development. NeuroRx 2:3–14. https://doi.org/10.1602/neurorx.2.1.3

51. Pardridge WM (2002) Drug and gene targeting to the brain with molecular trojan horses. Nat Rev Drug Discov 1:131–139. https://doi.org/10.1038/nrd725

52. Abbott NJ, Patabendige AAK, Dolman DEM et al (2010) Structure and function of the blood-brain barrier. Neurobiol Dis 37:13–25. https://doi.org/10.1016/j.nbd.2009.07.030

53. Banks WA, Audus KL, Davis TP (1992) Permeability of the blood-brain barrier to peptides: an approach to the development of therapeutically useful analogs. Peptides 13:1289–1294. https://doi.org/10.1016/0196-9781(92)90037-4

54. Deli MA (2011) Drug transport and the blood-brain barrier. J Cereb Blood Flow Metab. https://doi.org/10.2174/978160805120511101010144

55. del Amo EM, Urtti A, Yliperttula M (2008) Pharmacokinetic role of L-type amino acid transporters LAT1 and LAT2. Eur J Pharm Sci 35:161–174. https://doi.org/10.1016/j.ejps.2008.06.015

56. Peura L, Malmioja K, Huttunen K et al (2013) Design, synthesis and brain uptake of lat1-targeted amino acid prodrugs of

dopamine. Pharm Res 30:2523–2537. https://doi.org/10.1007/s11095-012-0966-3

57. Gynther M, Ropponen J, Laine K et al (2009) Glucose promoiety enables glucose transporter mediated brain uptake of ketoprofen and indomethacin prodrugs in rats. J Med Chem 52:3348–3353. https://doi.org/10.1021/jm8015409

58. Sampaio-Maia B, Serrão MP, Soares-da-Silva P (2001) Regulatory pathways and uptake of L-DOPA by capillary cerebral endothelial cells, astrocytes, and neuronal cells. Am J Physiol Cell Physiol 280:C333–C342

59. Zuchero YJY, Chen X, Bien-Ly N et al (2016) Discovery of novel blood-brain barrier targets to enhance brain uptake of therapeutic antibodies. Neuron 89:70–82. https://doi.org/10.1016/j.neuron.2015.11.024

60. de la Ballina LR, Cano-Crespo S, González-Muñoz E et al (2016) Amino acid transport associated to cluster of differentiation 98 heavy chain (CD98hc) is at the crossroad of oxidative stress and amino acid availability. J Biol Chem 1:jbc.M115.704254. https://doi.org/10.1074/jbc.M115.704254

61. Cai Q, Wang L, Deng G et al (2016) Systemic delivery to central nervous system by engineered PLGA nanoparticles. Am J Transl Res 8:749–764

62. Gaillard PJ (2011) Case study: to-BBB's G-technology, getting the best from drug-delivery research with industry-academia partnerships. Ther Deliv 2:1391–1394. https://doi.org/10.4155/tde.11.111

63. Kannan R, Chakrabarti R, Tang D et al (2000) GSH transport in human cerebrovascular endothelial cells and human astrocytes: evidence for luminal localization of Na+-dependent GSH transport in HCEC. Brain Res 852:374–382. https://doi.org/10.1016/S0006-8993(99)02184-8

64. Jain KK (2013) Applications of biotechnology in neurology. Appl Biotechnol Neurol. https://doi.org/10.1007/978-1-62703-272-8

65. Gaillard PJ, Appeldoorn CCM, Rip J et al (2012) Enhanced brain delivery of liposomal methylprednisolone improved therapeutic efficacy in a model of neuroinflammation. J Control Release 164:364–369

66. Lindqvist A, Rip J, Gaillard PJ et al (2013) Enhanced brain delivery of the opioid peptide damgo in glutathione pegylated liposomes: a microdialysis study. Mol Pharm 10:1533–1541. https://doi.org/10.1021/mp300272a

67. Sminia P, Westerman BA (2016) Blood-brain barrier crossing and breakthroughs in glioblastoma therapy. Br J Clin Pharmacol 81:1018–1020. https://doi.org/10.1111/bcp.12881

68. Georgieva JV, Hoekstra D, Zuhorn IS (2014) Smuggling drugs into the brain: an overview of ligands targeting transcytosis for drug delivery across the blood-brain barrier. Pharmaceutics 6:557–583. https://doi.org/10.3390/pharmaceutics6040557

69. Lajoie JM, Shusta EV (2015) Targeting receptor-mediated transport for delivery of biologics across the blood-brain barrier. Annu Rev Pharmacol Toxicol 55:613–631. https://doi.org/10.1146/annurev-pharmtox-010814-124852

70. Qian ZM, Li H, Sun H, Ho K (2002) Targeted drug delivery via the transferrin receptor-mediated endocytosis pathway. Pharmacol Rev 54:561–587. https://doi.org/10.1124/pr.54.4.561

71. Zhang Y, Pardridge WM (2006) Blood-brain barrier targeting of BDNF improves motor function in rats with middle cerebral artery occlusion. Brain Res 1111:227–229. https://doi.org/10.1016/j.brainres.2006.07.005

72. Lee JH, Engler JA, Collawn JF, Moore BA (2001) Receptor mediated uptake of peptides that bind the human transferrin receptor. Eur J Biochem 268:2004–2012. https://doi.org/10.1046/j.1432-1327.2001.02073.x

73. Wang Z, Zhao Y, Jiang Y et al (2015) Enhanced anti-ischemic stroke of ZL006 by T7-conjugated PEGylated liposomes drug delivery system. Sci Rep 5:12651. https://doi.org/10.1038/srep12651

74. Jones AR, Shusta EV (2007) Blood-brain barrier transport of therapeutics via receptor-mediation. Pharm Res 24:1759–1771. https://doi.org/10.1007/s11095-007-9379-0

75. Pardridge WM (2015) Blood-brain barrier drug delivery of IgG fusion proteins with a transferrin receptor monoclonal antibody. Expert Opin Drug Deliv 12:207–222. https://doi.org/10.1517/17425247.2014.952627

76. Yu YJ, Zhang Y, Kenrick M et al (2011) Boosting brain uptake of a therapeutic antibody by reducing its affinity for a transcytosis target. Sci Transl Med 3:84ra44. https://doi.org/10.1126/scitranslmed.3002230

77. Niewoehner J, Bohrmann B, Collin L et al (2014) Increased brain penetration and potency of a therapeutic antibody using a

monovalent molecular shuttle. Neuron 81:49–60. https://doi.org/10.1016/j.neuron.2013.10.061

78. Bohrmann B, Baumann K, Benz J et al (2012) Gantenerumab: a novel human anti-Aβ antibody demonstrates sustained cerebral amyloid-β binding and elicits cell-mediated removal of human amyloid-β. J Alzheimers Dis 28:49–69. https://doi.org/10.3233/JAD-2011-110977

79. Wu C, Ying H, Grinnell C et al (2007) Simultaneous targeting of multiple disease mediators by a dual-variable-domain immunoglobulin. Nat Biotechnol 25:1290–1297. https://doi.org/10.1038/nbt1345

80. Gu J, Ghayur T (2012) Generation of dual-variable-domain immunoglobulin molecules for dual-specific targeting. Methods Enzymol 502:25–41. https://doi.org/10.1016/B978-0-12-416039-2.00002-1

81. Jakob CG, Edalji R, Judge RA et al (2013) Structure reveals function of the dual variable domain immunoglobulin (DVD-Ig™) molecule. MAbs 5:358–363. https://doi.org/10.4161/mabs.23977

82. Farid Gizatullin (AbbVie Bioresearch Center) (2014) Uptake and retention of DVD-Ig™ in mouse brain by intravenous or intraperitoneal injection no title. Biol Formul Deliv. Summit

83. Havrankova J, Brownstein M, Roth J (1981) Insulin and insulin receptors in rodent brain. Diabetologia 20(Suppl):268–273

84. Smith MW, Gumbleton M (2006) Endocytosis at the blood-brain barrier: from basic understanding to drug delivery strategies. J Drug Target 14:191–214. https://doi.org/10.1080/10611860600650086

85. Coloma MJ, Lee HJ, Kurihara A et al (2000) Transport across the primate blood-brain barrier of a genetically engineered chimeric monoclonal antibody to the human insulin receptor. Pharm Res 17:266–274. https://doi.org/10.1023/A:1007592720793

86. Boado RJ, Zhang Y, Zhang Y et al (2008) GDNF fusion protein for targeted-drug delivery across the human blood-brain barrier. Biotechnol Bioeng 100:387–396. https://doi.org/10.1002/bit.21764

87. Kingwell K (2016) Drug delivery: new targets for drug delivery across the BBB. Nat Rev Drug Discov 21:2016. https://doi.org/10.1038/nrd.2016.14

88. Willnow TE, Nykjaer A, Herz J (1999) Lipoprotein receptors: new roles for ancient proteins. Nat Cell Biol 1:E157–E162. https://doi.org/10.1038/14109

89. Herz J, Beffert U (2000) Apolipoprotein E receptors: linking brain development and Alzheimer's disease. Nat Rev Neurosci 1:51–58. https://doi.org/10.1038/35036221

90. Herz J, Strickland DK (2001) LRP: a multifunctional scavenger and signaling receptor. J Clin Invest 108:779–784. https://doi.org/10.1172/JCI200113992

91. Kounnas MZ, Moir RD, Rebeck GW et al (1995) LDL receptor-related protein, a multifunctional ApoE receptor, binds secreted β-amyloid precursor protein and mediates its degradation. Cell 82:331–340. https://doi.org/10.1016/0092-8674(95)90320-8

92. Gonias SL, Wu L, Salicioni AM (2004) Low density lipoprotein receptor-related protein: regulation of the plasma membrane proteome. Thromb Haemost 91:1056–1064. https://doi.org/10.1160/TH04-01-0023

93. Lin L, Hu K (2014) LRP-1: functions, signaling and implications in kidney and other diseases. Int J Mol Sci 15:22887–22901. https://doi.org/10.3390/ijms151222887

94. Boucher P, Herz J (2011) Signaling through LRP1: protection from atherosclerosis and beyond. Biochem Pharmacol 81:1–5. https://doi.org/10.1016/j.bcp.2010.09.018

95. Strickland DK, Ranganathan S (2003) Diverse role of LDL receptor-related protein in the clearance of proteases and in signaling. J Thromb Haemost 1:1663–1670. https://doi.org/10.1046/j.1538-7836.2003.00330.x

96. Gonias SL, Campana WM (2014) LDL receptor-related protein-1: a regulator of inflammation in atherosclerosis, cancer, and injury to the nervous system. Am J Pathol 184:18–27. https://doi.org/10.1016/j.ajpath.2013.08.029

97. Hussain MM (2001) Structural, biochemical and signaling properties of the low-density lipoprotein receptor gene family. Front Biosci 6:D417–D428

98. Rebeck GW, Reiter JS, Strickland DK, Hyman BT (1993) Apolipoprotein E in sporadic Alzheimer's disease: allelic variation and receptor interactions. Neuron 11:575–580. https://doi.org/10.1016/0896-6273(93)90070-8

99. Burgmans S, van de Haar HJ, Verhey FRJ, Backes WH (2013) Amyloid-β interacts with blood-brain barrier function in dementia: a systematic review. J Alzheimer's Dis 35:859–873. https://doi.org/10.3233/JAD-122155

100. Erickson MA, Banks WA (2013) Blood-brain barrier dysfunction as a cause and consequence of Alzheimer's disease. J Cereb

Blood Flow Metab 33:1500–1513. https://doi.org/10.1038/jcbfm.2013.135

101. Gaultier A, Wu X, Le Moan N et al (2009) Low-density lipoprotein receptor-related protein 1 is an essential receptor for myelin phagocytosis. J Cell Sci 122:1155–1162. https://doi.org/10.1242/jcs.040717

102. DAE H, Koning N, Schuurman KG et al (2013) Selective upregulation of scavenger receptors in and around demyelinating areas in multiple sclerosis. J Neuropathol Exp Neurol 72:106–118. https://doi.org/10.1097/NEN.0b013e31827fd9e8

103. Benes P, Jurajda M, Zaloudík J et al (2003) C766T low-density lipoprotein receptor-related protein 1 (LRP1) gene polymorphism and susceptibility to breast cancer. Breast Cancer Res 5:R77–R81. https://doi.org/10.1186/bcr591

104. Catasus L, Gallardo A, Llorente-Cortes V et al (2011) Low-density lipoprotein receptor-related protein 1 is associated with proliferation and invasiveness in Her-2/neu and triple-negative breast carcinomas. Hum Pathol 42:1581–1588. https://doi.org/10.1016/j.humpath.2011.01.011

105. Yamamoto M, Ikeda K, Ohshima K et al (1997) Increased expression of low density lipoprotein receptor-related protein/alpha2-macroglobulin receptor in human malignant astrocytomas. Cancer Res 57:2799–2805

106. Huang X-Y, Shi G-M, Devbhandari RP et al (2012) Low level of low-density lipoprotein receptor-related protein 1 predicts an unfavorable prognosis of hepatocellular carcinoma after curative resection. PLoS One 7:e32775. https://doi.org/10.1371/journal.pone.0032775

107. Cal R, Juan-Babot O, Brossa V et al (2012) Low density lipoprotein receptor-related protein 1 expression correlates with cholesteryl ester accumulation in the myocardium of ischemic cardiomyopathy patients. J Transl Med 10:160. https://doi.org/10.1186/1479-5876-10-160

108. Wohlfart S, Gelperina S, Kreuter J (2012) Transport of drugs across the blood-brain barrier by nanoparticles. J Control Release 161:264–273. https://doi.org/10.1016/j.jconrel.2011.08.017

109. Kreuter J, Shamenkov D, Petrov V et al (2002) Apolipoprotein-mediated transport of nanoparticle-bound drugs across the blood-brain barrier. J Drug Target 10:317–325. https://doi.org/10.1080/10611860290031877

110. Wagner S, Zensi A, Wien SL et al (2012) Uptake mechanism of ApoE-modified nanoparticles on brain capillary endothelial cells as a blood-brain barrier model. PLoS One. https://doi.org/10.1371/journal.pone.0032568

111. Lambert LA, Perri H, Meehan TJ (2005) Evolution of duplications in the transferrin family of proteins. Comp Biochem Physiol B Biochem Mol Biol 140:11–25. https://doi.org/10.1016/j.cbpc.2004.09.012

112. Jefferies WA, Food MR, Gabathuler R et al (1996) Reactive microglia specifically associated with amyloid plaques in Alzheimer's disease brain tissue express melanotransferrin. Brain Res 712:122–126. https://doi.org/10.1016/0006-8993(95)01407-1

113. Food MR, Rothenberger S, Gabathuler R et al (1994) Transport and expression in human melanomas of a transferrin-like glycosylphosphatidylinositol-anchored protein. J Biol Chem 269:3034–3040

114. Demeule M, Poirier J, Jodoin J et al (2002) High transcytosis of melanotransferrin (P97) across the blood-brain barrier. J Neurochem 83:924–933. https://doi.org/10.1046/j.1471-4159.2002.01201.x

115. Gabathuler R, Arthur G, Kennard M et al (2005) Development of a potential protein vector (NeuroTrans) to deliver drugs across the blood-brain barrier. Int Congr Ser 1277:171–184. https://doi.org/10.1016/j.ics.2005.02.021

116. Karkan D, Pfeifer C, Vitalis TZ et al (2008) A unique carrier for delivery of therapeutic compounds beyond the blood-brain barrier. PLoS One. https://doi.org/10.1371/journal.pone.0002469

117. Nounou MI, Adkins CE, Rubinchik E et al (2016) Anti-cancer antibody trastuzumab-melanotransferrin conjugate (BT2111) for the treatment of metastatic HER2+ breast cancer tumors in the brain: an in-vivo study. Pharm Res 33:2930–2942. https://doi.org/10.1007/s11095-016-2015-0

118. Pan W, Kastin AJ, Zankel TC et al (2004) Efficient transfer of receptor-associated protein (RAP) across the blood-brain barrier. J Cell Sci 117:5071–5078. https://doi.org/10.1242/jcs.01381

119. Migliorini MM, Behre EH, Brew S et al (2003) Allosteric modulation of ligand binding to low density lipoprotein receptor-related protein by the receptor-associated protein requires critical lysine residues within its carboxyl-terminal domain. J Biol Chem 278:17986–17992

120. Bu G, Rennke S (1996) Receptor-associated protein is a folding chaperone for low density lipoprotein receptor-related protein. J Biol Chem 271:22218–22224

121. Prince WS, McCormick LM, Wendt DJ et al (2004) Lipoprotein receptor binding, cellular uptake, and lysosomal delivery of fusions between the Receptor-associated Protein (RAP) and α-L-iduronidase or acid α-glucosidase. J Biol Chem 279:35037–35046. https://doi.org/10.1074/jbc.M402630200

122. Spencer BJ, Verma IM (2007) Targeted delivery of proteins across the blood-brain barrier. Proc Natl Acad Sci U S A 104:7594–7599. https://doi.org/10.1073/pnas.0702170104

123. Vodicka MA (2001) Determinants for lentiviral infection of non-dividing cells. Somat Cell Mol Genet 26:35–49. https://doi.org/10.1023/A:1021022629126

124. Parr-Brownlie LC, Bosch-Bouju C, Schoderboeck L et al (2015) Lentiviral vectors as tools to understand central nervous system biology in mammalian model organisms. Front Mol Neurosci 8:14. https://doi.org/10.3389/fnmol.2015.00014

125. Naldini L, Blömer U, Gallay P et al (1996) In vivo gene delivery and stable transduction of nondividing cells by a lentiviral vector. Science 272:263–267. https://doi.org/10.1126/science.272.5259.263

126. Zufferey R, Dull T, Mandel RJ et al (1998) Self-inactivating lentivirus vector for safe and efficient in vivo gene delivery. J Virol 72:9873–9880

127. Dull T, Zufferey R, Kelly M et al (1998) A third-generation lentivirus vector with a conditional packaging system. J Virol 72:8463–8471

128. Finkelshtein D, Werman A, Novick D et al (2013) LDL receptor and its family members serve as the cellular receptors for vesicular stomatitis virus. Proc Natl Acad Sci 110:7306–7311. https://doi.org/10.1073/pnas.1214441110

129. Bender E (2016) Gene therapy: industrial strength. Nature 537:S57–S59

130. Schwarze SR, Ho A, Vocero-Akbani A, Dowdy SF (1999) In vivo protein transduction: delivery of a biologically active protein into the mouse [see comments]. Science (80) 285:1569–1572

131. Oller-Salvia B, Sanchez-Navarro M, Giralt E, Teixido M (2016) Blood-brain barrier shuttle peptides: an emerging paradigm for brain delivery. Chem Soc Rev 45:4690–4707. https://doi.org/10.1039/C6CS00076B

132. Kumar P, Wu H, McBride JL et al (2007) Transvascular delivery of small interfering RNA to the central nervous system. Nature 448:39–43. https://doi.org/10.1038/nature05901

133. Demeule M, Régina A, Ché C et al (2008) Identification and design of peptides as a new drug delivery system for the brain. J Pharmacol Exp Ther 324:1064–1072. https://doi.org/10.1124/jpet.107.131318

134. Demeule M, Currie JC, Bertrand Y et al (2008) Involvement of the low-density lipoprotein receptor-related protein in the transcytosis of the brain delivery vector Angiopep-2. J Neurochem 106:1534–1544. https://doi.org/10.1111/j.1471-4159.2008.05492.x

135. Régina A, Demeule M, Ché C et al (2008) Antitumour activity of ANG1005, a conjugate between paclitaxel and the new brain delivery vector Angiopep-2. Br J Pharmacol 155:185–197. https://doi.org/10.1038/bjp.2008.260

136. Kurzrock R, Gabrail N, Chandhasin C et al (2012) Safety, pharmacokinetics, and activity of GRN1005, a novel conjugate of angiopep-2, a peptide facilitating brain penetration, and paclitaxel, in patients with advanced solid tumors. Mol Cancer Ther 11:308–316. https://doi.org/10.1158/1535-7163.MCT-11-0566

137. Van Rooy I, Mastrobattista E, Storm G et al (2011) Comparison of five different targeting ligands to enhance accumulation of liposomes into the brain. J Control Release 150:30–36. https://doi.org/10.1016/j.jconrel.2010.11.014

138. Malcor JD, Payrot N, David M et al (2012) Chemical optimization of new ligands of the low-density lipoprotein receptor as potential vectors for central nervous system targeting. J Med Chem 55:2227–2241. https://doi.org/10.1021/jm2014919

139. Jacquot G, Lécorché P, Malcor J-D et al (2016) Optimization and in vivo validation of peptide vectors targeting the LDL receptor. Mol Pharm. https://doi.org/10.1021/acs.molpharmaceut.6b00687

140. Hamers-Casterman C, Atarhouch T, Muyldermans S et al (1993) Naturally occurring antibodies devoid of light chains. Nature 363:446–448. https://doi.org/10.1038/363446a0

141. Harmsen MM, De Haard HJ (2007) Properties, production, and applications of camelid single-domain antibody fragments. Appl Microbiol Biotechnol 77:13–22. https://doi.org/10.1007/s00253-007-1142-2

142. Muyldermans S, Lauwereys M (1999) Unique single-domain antigen binding

fragments derived from naturally occurring camel heavy-chain antibodies. J Mol Recognit 12:131–140. https://doi.org/10.1002/(SICI)1099-1352(199903/04)12:2<131::AID-JMR454>3.0.CO;2-M

143. Hassanzadeh-Ghassabeh G, Devoogdt N, De Pauw P et al (2013) Nanobodies and their potential applications. Nanomedicine 8:1013–1026. https://doi.org/10.2217/nnm.13.86

144. Tanha J, Dubuc G, Hirama T et al (2002) Selection by phage display of llama conventional VH fragments with heavy chain antibody VHH properties. J Immunol Methods 263:97–109. https://doi.org/10.1016/S0022-1759(02)00027-3

145. Abulrob A, Sprong H, Van Bergen En Henegouwen P, Stanimirovic D (2005) The blood-brain barrier transmigrating single domain antibody: mechanisms of transport and antigenic epitopes in human brain endothelial cells. J Neurochem 95:1201–1214. https://doi.org/10.1111/j.1471-4159.2005.03463.x

146. Muruganandam A, Tanha J, Narang S, Stanimirovic D (2002) Selection of phage-displayed llama single-domain antibodies that transmigrate across human blood-brain barrier endothelium. FASEB J 16:240–242. https://doi.org/10.1096/fj.01-0343fje

147. Tanha J, ASD M (2003) Phage display technology for identifying specific antigens on brain endothelial cells. Methods Mol Med 89:435–449

148. Haqqani AS, Caram-Salas N, Ding W et al (2013) Multiplexed evaluation of serum and CSF pharmacokinetics of brain-targeting single-domain antibodies using a NanoLC-SRM-ILIS method. Mol Pharm 10:1542–1556. https://doi.org/10.1021/mp3004995

149. Pardridge WM (2007) shRNA and siRNA delivery to the brain. Adv Drug Deliv Rev 59:141–152. https://doi.org/10.1016/j.addr.2007.03.008

150. Pardridge WM (2010) Biopharmaceutical drug targeting to the brain. J Drug Target 18:157–167. https://doi.org/10.3109/10611860903548354

151. Ulbrich K, Hekmatara T, Herbert E, Kreuter J (2009) Transferrin- and transferrin-receptor-antibody-modified nanoparticles enable drug delivery across the blood-brain barrier (BBB). Eur J Pharm Biopharm 71:251–256. https://doi.org/10.1016/j.ejpb.2008.08.021

152. Paris-Robidas S, Emond V, Tremblay C et al (2011) In vivo labeling of brain capillary endothelial cells after intravenous injection of monoclonal antibodies targeting the transferrin receptor. Mol Pharmacol 80:32–39. https://doi.org/10.1124/mol.111.071027

153. Tosi G, Costantino L, Rivasi F et al (2007) Targeting the central nervous system: in vivo experiments with peptide-derivatized nanoparticles loaded with loperamide and rhodamine-123. J Control Release 122:1–9. https://doi.org/10.1016/j.jconrel.2007.05.022

154. Tosi G, Fano RA, Bondioli L et al (2011) Investigation on mechanisms of glycopeptide nanoparticles for drug delivery across the blood-brain barrier. Nanomedicine (Lond) 6:423–436. https://doi.org/10.2217/nnm.11.11

155. Salvalaio M, Rigon L, Belletti D et al (2016) Targeted polymeric nanoparticles for brain delivery of high molecular weight molecules in lysosomal storage disorders. PLoS One. https://doi.org/10.1371/journal.pone.0156452

156. Chacko A-M, Li C, Pryma DA et al (2013) Targeted delivery of antibody-based therapeutic and imaging agents to CNS tumors: crossing the blood-brain barrier divide. Expert Opin Drug Deliv 10:907–926. https://doi.org/10.1517/17425247.2013.808184

157. Bickel U, Yoshikawa T, Pardridge WM (2001) Delivery of peptides and proteins through the blood-brain barrier. Adv Drug Deliv Rev 46:247–279. https://doi.org/10.1016/S0169-409X(00)00139-3

158. Hervé F, Ghinea N, Scherrmann J-M (2008) CNS delivery via adsorptive transcytosis. AAPS J 10:455–472. https://doi.org/10.1208/s12248-008-9055-2

159. Triguero D, Buciak JB, Yang J, Pardridge WM (1989) Blood-brain barrier transport of cationized immunoglobulin G: enhanced delivery compared to native protein. Proc Natl Acad Sci U S A 86:4761–4765. https://doi.org/10.1073/pnas.86.12.4761

160. Drin G, Rousselle C, Scherrmann J-M et al (2002) Peptide delivery to the brain via adsorptive-mediated endocytosis: advances with SynB vectors. AAPS PharmSci 4:E26. https://doi.org/10.1208/ps040426

161. Drin G, Cottin S, Blanc E et al (2003) Studies on the internalization mechanism of cationic cell-penetrating peptides. J Biol Chem 278:31192–31201. https://doi.org/10.1074/jbc.M303938200

162. Dufès C, Uchegbu IF, Schätzlein AG (2005) Dendrimers in gene delivery. Adv Drug Deliv Rev 57:2177–2202. https://doi.org/10.1016/j.addr.2005.09.017

163. Bosman AW, Janssen HM, Meijer EW (1999) About dendrimers: structure, physical properties, and applications. Chem Rev 99:1665–1688. https://doi.org/10.1021/cr970069y

164. Smith DK, Diederich F (1998) Functional dendrimers: unique biological mimics. Chem Eur J 4:1353–1361. https://doi.org/10.1002/(SICI)1521-3765(19980807)4:8<1353::AID-CHEM1353>3.0.CO;2-0

165. Xu L, Zhang H, Wu Y (2014) Dendrimer advances for the central nervous system delivery of therapeutics. ACS Chem Neurosci 5:2–13. https://doi.org/10.1021/cn400182z

166. Cheng Y, Wu Q, Li Y et al (2009) New insights into the interactions between dendrimers and surfactants: 2. Design of new drug formulations based on dendrimer-surfactant aggregates. J Phys Chem B 113:8339–8346. https://doi.org/10.1021/jp9021618

167. Jansen JF, de Brabander-van den Berg EMM, Meijer EW (1994) Encapsulation of guest molecules into a dendritic box. Science (80) 266:1226–1229. https://doi.org/10.1126/science.266.5188.1226

168. Madaan K, Kumar S, Poonia N et al (2014) Dendrimers in drug delivery and targeting: drug-dendrimer interactions and toxicity issues. J Pharm Bioallied Sci 6:139–150. https://doi.org/10.4103/0975-7406.130965

169. Chen HT, Neerman MF, Parrish AR, Simanek EE (2004) Cytotoxicity, hemolysis, and acute in vivo toxicity of dendrimers based on melamine, candidate vehicles for drug delivery. J Am Chem Soc 126:10044–10048. https://doi.org/10.1021/ja048548j

170. Domański DM, Klajnert B, Bryszewska M (2004) Influence of PAMAM dendrimers on human red blood cells. Bioelectrochemistry 63:189–191. https://doi.org/10.1016/j.bioelechem.2003.09.023

171. Roberts JC, Bhalgat MK, Zera RT (1996) Preliminary biological evaluation of polyamidoamine (PAMAM) starburst dendrimers. J Biomed Mater Res 30:53–65. https://doi.org/10.1002/(SICI)1097-4636(199601)30:1<53::AID-JBM8>3.0.CO;2-Q

172. Jones CF, Campbell RA, Franks Z et al (2012) Cationic PAMAM dendrimers disrupt key platelet functions. Mol Pharm 9:1599–1611. https://doi.org/10.1021/mp2006054

173. Dutta T, Garg M, Dubey V et al (2008) Toxicological investigation of surface engineered fifth generation poly (propyleneimine) dendrimers in vivo. Nanotoxicology 2:62–70. https://doi.org/10.1080/17435390802105167

174. Rousselle C, Smirnova M, Clair P et al (2001) Enhanced delivery of doxorubicin into the brain via a peptide-vector-mediated strategy: saturation kinetics and specificity. J Pharmacol Exp Ther 296:124–131

175. Rousselle C, Clair P, Smirnova M et al (2003) Improved brain uptake and pharmacological activity of dalargin using a peptide-vector-mediated strategy. J Pharmacol Exp Ther 306:371–376. https://doi.org/10.1124/jpet.102.048520

176. Wender PA, Mitchell DJ, Pattabiraman K et al (2000) The design, synthesis, and evaluation of molecules that enable or enhance cellular uptake: peptoid molecular transporters. Proc Natl Acad Sci U S A 97:13003–13008. https://doi.org/10.1073/pnas.97.24.13003

177. Temsamani J, Rousselle C, Rees AR, Scherrmann JM (2001) Vector-mediated drug delivery to the brain. Expert Opin Biol Ther 1:773–782. https://doi.org/10.1517/14712598.1.5.773

178. Liu L, Guo K, Lu J et al (2008) Biologically active core/shell nanoparticles self-assembled from cholesterol-terminated PEG-TAT for drug delivery across the blood-brain barrier. Biomaterials 29:1509–1517. https://doi.org/10.1016/j.biomaterials.2007.11.014

179. Pham W, Zhao BQ, Lo EH et al (2005) Crossing the blood-brain barrier: a potential application of myristoylated polyarginine for in vivo neuroimaging. NeuroImage 28:287–292. https://doi.org/10.1016/j.neuroimage.2005.06.007

180. van der Pol E, Böing AN, Harrison P et al (2012) Classification, functions, and clinical relevance of extracellular vesicles. Pharmacol Rev 64:676–705. https://doi.org/10.1124/pr.112.005983

181. Théry C, Ostrowski M, Segura E (2009) Membrane vesicles as conveyors of immune responses. Nat Rev Immunol 9:581–593. https://doi.org/10.1038/nri2567

182. Muralidharan-Chari V, Clancy J, Plou C et al (2009) ARF6-regulated shedding of tumor cell-derived plasma membrane microvesicles. Curr Biol 19:1875–1885. https://doi.org/10.1016/j.cub.2009.09.059

183. Harding CV, Heuser JE, Stahl PD (2013) Exosomes: looking back three decades and into the future. J Cell Biol 200:367–371. https://doi.org/10.1083/jcb.201212113

184. Raposo G, Stoorvogel W (2013) Extracellular vesicles: exosomes, microvesicles, and friends. J Cell Biol 200:373–383. https://doi.org/10.1083/jcb.201211138

185. Clayton A, Turkes A, Dewitt S et al (2004) Adhesion and signaling by B cell-derived exosomes: the role of integrins. FASEB J 18:977–979. https://doi.org/10.1096/fj.03-1094fje

186. Köppler B, Cohen C, Schlöndorff D, Mack M (2006) Differential mechanisms of microparticle transfer toB cells and monocytes: anti-inflammatory propertiesof microparticles. Eur J Immunol 36:648–660. https://doi.org/10.1002/eji.200535435

187. Parolini I, Federici C, Raggi C et al (2009) Microenvironmental pH is a key factor for exosome traffic in tumor cells. J Biol Chem 284:34211–34222. https://doi.org/10.1074/jbc.M109.041152

188. Gallo A, Tandon M, Alevizos I, Illei GG (2012) The majority of microRNAs detectable in serum and saliva is concentrated in exosomes. PLoS One. https://doi.org/10.1371/journal.pone.0030679

189. Mitchell PS, Parkin RK, Kroh EM et al (2008) Circulating microRNAs as stable blood-based markers for cancer detection. Proc Natl Acad Sci U S A 105:10513–10518. https://doi.org/10.1073/pnas.0804549105

190. Taylor DD, Gercel-Taylor C (2008) MicroRNA signatures of tumor-derived exosomes as diagnostic biomarkers of ovarian cancer. Gynecol Oncol 110:13–21. https://doi.org/10.1016/j.ygyno.2008.04.033

191. Peinado H, Alečković M, Lavotshkin S et al (2012) Melanoma exosomes educate bone marrow progenitor cells toward a pro-metastatic phenotype through MET. Nat Med 18:883–891. https://doi.org/10.1038/nm.2753

192. Alvarez-Erviti L, Seow Y, Yin H et al (2011) Delivery of siRNA to the mouse brain by systemic injection of targeted exosomes. Nat Biotechnol 29:3–4. https://doi.org/10.1038/nbt.1807

193. Van Niel G, Porto-Carreiro I, Simoes S, Raposo G (2006) Exosomes: a common pathway for a specialized function. J Biochem 140:13–21. https://doi.org/10.1093/jb/mvj128

194. Felicetti F, Parolini I, Bottero L et al (2009) Caveolin-1 tumor-promoting role in human melanoma. Int J Cancer 125:1514–1522. https://doi.org/10.1002/ijc.24451

195. Silverman JM, Reiner NE (2011) Exosomes and other microvesicles in infection biology: organelles with unanticipated phenotypes. Cell Microbiol 13:1–9. https://doi.org/10.1111/j.1462-5822.2010.01537.x

196. Fevrier B, Vilette D, Archer F et al (2004) Cells release prions in association with exosomes. Proc Natl Acad Sci U S A 101:9683–9688. https://doi.org/10.1073/pnas.0308413101

197. Dhuria SV, Hanson LR, Frey WH (2010) Intranasal delivery to the central nervous system: mechanisms and experimental considerations. J Pharm Sci 99:1654–1673. https://doi.org/10.1002/jps.21924

198. Miyake MM, Bleier BS (2015) The blood-brain barrier and nasal drug delivery to the central nervous system. Am J Rhinol Allergy 29:124–127. https://doi.org/10.2500/ajra.2015.29.4149

199. Chen X-Q, Fawcett JR, Rahman Y-E et al (1998) Delivery of nerve growth factor to the brain via the olfactory pathway. J Alzheimers Dis 1:35–44

200. Frey WH, Liu J, Chen X et al (1997) Delivery of 125 I-NGF to the brain via the olfactory route. Drug Deliv 4:87–92. https://doi.org/10.3109/10717549709051878

201. Thorne RG, Pronk GJ, Padmanabhan V, Frey WH (2004) Delivery of insulin-like growth factor-I to the rat brain and spinal cord along olfactory and trigeminal pathways following intranasal administration. Neuroscience 127:481–496. https://doi.org/10.1016/j.neuroscience.2004.05.029

202. Ross TM, Martinez PM, Renner JC et al (2004) Intranasal administration of interferon beta bypasses the blood-brain barrier to target the central nervous system and cervical lymph nodes: a non-invasive treatment strategy for multiple sclerosis. J Neuroimmunol 151:66–77. https://doi.org/10.1016/j.jneuroim.2004.02.011

203. Kanazawa T (2015) Brain delivery of small interfering ribonucleic acid and drugs through intranasal administration with nanosized polymer micelles. Med Devices 8:57–64. https://doi.org/10.2147/MDER.S70856

204. Chow HHS, Chen Z, Matsuura GT (1999) Direct transport of cocaine from the nasal cavity to the brain following intranasal cocaine administration in rats. J Pharm Sci 88:754–758. https://doi.org/10.1021/js9900295

205. Westin U, Piras E, Jansson B et al (2005) Transfer of morphine along the olfactory pathway to the central nervous system after nasal administration to rodents. Eur J Pharm

Sci 24:565–573. https://doi.org/10.1016/j.ejps.2005.01.009

206. Schulz C, Paulus K, Lehnert H (2004) Central nervous and metabolic effects of intranasally applied leptin. Endocrinology 145:2696–2701. https://doi.org/10.1210/en.2003-1431

207. Shimizu H, Oh-I S, Okada S, Mori M (2005) Inhibition of appetite by nasal leptin administration in rats. Int J Obes 29:858–863. https://doi.org/10.1038/sj.ijo.0802951

208. Benedict C, Hallschmid M, Hatke A et al (2004) Intranasal insulin improves memory in humans. Psychoneuroendocrinology 29:1326–1334. https://doi.org/10.1016/j.psyneuen.2004.04.003

209. Reger MA, Watson GS, Green PS et al (2008) Intranasal insulin improves cognition and modulates beta-amyloid in early AD. Neurology 70:440–448. https://doi.org/10.1212/01.WNL.0000265401.62434.36

210. Renner DB, Frey WH, Hanson LR (2012) Intranasal delivery of siRNA to the olfactory bulbs of mice via the olfactory nerve pathway. Neurosci Lett 513:193–197. https://doi.org/10.1016/j.neulet.2012.02.037

211. Nishina K, Mizusawa H, Yokota T (2013) Short interfering RNA and the central nervous system: development of nonviral delivery systems. Expert Opin Drug Deliv 10:289–292. https://doi.org/10.1517/17425247.2013.748746

212. Kanazawa T, Akiyama F, Kakizaki S et al (2013) Delivery of siRNA to the brain using a combination of nose-to-brain delivery and cell-penetrating peptide-modified nanomicelles. Biomaterials 34:9220–9226. https://doi.org/10.1016/j.biomaterials.2013.08.036

213. Morrison EE, Costanzo RM (1992) Morphology of olfactory epithelium in humans and other vertebrates. Microsc Res Tech 23:49–61. https://doi.org/10.1002/jemt.1070230105

Chapter 6

Inorganic Nanoparticles and Their Strategies to Enhance Brain Drug Delivery

Eduardo Gallardo-Toledo, Carolina Velasco-Aguirre, and Marcelo Javier Kogan

Abstract

The main obstacle for brain drug delivery after systemic administration is the presence of the blood-brain barrier (BBB). The use of drug nanocarriers to overcome this selective barrier that isolates the central nervous system (CNS) has been widely studied in recent years. Among the different nanoparticles described in literature, inorganic nanoparticles such as gold, iron oxide, carbon, silver, and others have been studied for brain drug delivery. Here, we describe the strategies employed with different inorganic nanoparticles to reach the CNS, which in general used targeting molecules that can facilitate the transport across the BBB through transport mechanisms such as adsorptive-mediated transcytosis and receptor-mediated transport or the use of peptide vectors. Throughout this chapter, examples of diverse nanosystems will be given, highlighting the main research objectives, their characteristics, and the results obtained.

Key words Inorganic nanoparticle, Brain drug delivery, Central nervous system, Blood-brain barrier, Gold nanoparticles, Iron oxide nanoparticles, Carbon nanotubes

1 Introduction

In the last decades, there has been an increase in the drug delivery development of the central nervous system (CNS) in order to treat diverse cerebral diseases, which incidence is greater every day due to the worldwide aging. The main limitation for systemic administration has been the presence of the blood-brain barrier (BBB), which is both a defensive and highly selective barrier which isolates the CNS, generating a stable environment for neuronal function. Besides, for the presence of this barrier, 98% of small molecules and 100% of big molecules (molecular weight >1 kDa) are not capable to reach the brain [1]. In order to overcome this natural barrier, it has developed strategies such as intracerebroventricular and intracerebral administration or methods that produce a transient BBB disruption induced by chemical, physical, or biological

Javier O. Morales and Pieter J. Gaillard (eds.), *Nanomedicines for Brain Drug Delivery*, Neuromethods, vol. 157,
https://doi.org/10.1007/978-1-0716-0838-8_6, © Springer Science+Business Media, LLC, part of Springer Nature 2021

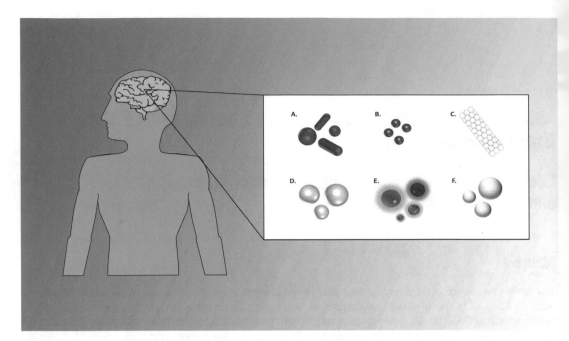

Fig. 1 Representative illustration of main inorganic nanoparticles addressed during this chapter that are studied for brain drug delivery. (A) Gold nanoparticles, (B) Iron oxide nanoparticles, (C) carbon nanotubes, (D) silver nanoparticles, (E) quantum dots, and (F) silica nanoparticles

agents (e.g., magnetic heating and ultrasound) [2, 3]. However, these brain drug delivery strategies have not been widely used because they are considered as highly invasive and risky methods [4]. For this reason, improving the BBB permeability of drug nanocarriers appears as the main strategy to be used without side effects.

Among the different nanocarriers, inorganic nanoparticles including gold, iron oxide, carbon, silver, silica, quantum dots, and others have been widely used for brain drug delivery (Fig. 1). Characteristics such as shape, size, or porosity of inorganic nanoparticles can be precisely tuned. The surface of these nanoparticles can be easily modified with ligands, polymers, drugs, or biological molecules, which improve their stability and facilitate their BBB penetration [5]. Additionally, the application of external stimuli as magnetic field or near-infrared radiation can be used to produce a temporary space release of drug in the CNS. Furthermore, inorganic nanoparticles can be used not only as vehicles but also as imaging agents or both (theranostic agents) [6]. All these unique properties make inorganic nanoparticles take advantage over their polymeric and lipid-based counterparts for brain drug delivery [7].

In this chapter, a general description will be made of how inorganic nanoparticles of different materials are used as vehicles for brain drug delivery or as theranostic agents for CNS disorders.

In each section, several examples will be named, highlighting the main achievements of each study and the advantages of each nanosystem used.

2 Gold Nanoparticles

An impressive and emerging field is the use of gold nanoparticles (AuNPs) for the delivery of drugs to treat brain-related diseases such as neurodegeneration and brain tumors. For example, there has been described the use of gold nanospheres (GNSs) to destroy the B-amyloid aggregates involved in Alzheimer's disease [8–10], which make them promising in biomedical applications.

These nanomaterials have different characteristic shapes, such as spheres, rods, nanoshells, among others, with tunable sizes [11]. The spherical gold can be obtained using simple established laboratory procedures, such as sodium borohydride, citrate, and carbohydrate reduction methods [12].

However, as reviewed by Krol in 2012, the delivery of nanoparticles to the brain remains unsolved [13]. This point is mainly due to (1) the interaction of nanoparticles with biological components after administration; (2) physical parameters such as residence time on target molecules; and (3) the lack of unique target biomarkers on the diseased cells and the presence of highly selective barriers. Consequently, the treatment of many brain-related disorders using different drugs or nanoparticles is limited because of the presence of the BBB, which thoroughly regulates the crossing of drugs. In this context, peptide conjugation could represent an attractive strategy to enhance the delivery of AuNPs. Moreover, AuNPs can be capped with molecules as peptides for tagging them to selective targets which is called active targeting.

Peptides can cross the BBB via different mechanisms. Passive diffusion of peptides is insufficient, while the use of carrier systems (i.e., carrier-mediated transcytosis (CMT)) expressed at the BBB is a useful strategy to deliver peptides to the brain. However, for CMT, it is necessary to bind specific chemical groups to render them substrates for endogen carriers, e.g., by the glycosylation of the peptide to enable transport through the GLT-1 receptors [14, 15]. Adsorptive-mediated transcytosis (AMT) and receptor-mediated transcytosis (RMT) are mechanisms by which different peptides have been transported across the BBB. Binding the peptide with affinity for the membrane or to a specific membrane receptor on the cell surface can trigger endocytosis. For AMT cell-penetrating peptides (CPPs) with the sequence TAT, penetratin, D-penetratin, Syn-B, pegelin, and the heptapeptide Gly-L-Phe-D-Thr-Gly-L-Phe-L-Leu-L-Ser (O-Beta-D-Glucose) (g7) are used as vectors for different cargos. The CPPs share a standard feature, which is the ability to interact with the lipid membranes

[16, 17]. For example, the sequence TAT, which is an arginine-rich CPP originating from the immunodeficiency virus type-I, contains arginines that enable interaction with negatively charged membranes. The guanidinium group of arginine is required for peptide uptake and is more potent than other cations [18]. This mechanism is independent of cell receptors and temperature.

In another hand, the Giralt group has described another example of a CPP-rich arginine-peptide that they called the sweet arrow peptide (VRLPPP)3 [19]. As they discussed for the cell penetration process of CPPs, the essential factors for cell penetration are the interaction between the arginines with the negatively charged groups in the membranes, secondary structure of the peptide and its aggregation state [20].

There are limited reports related to the use of peptides or proteins to improve the crossing of AuNPs through the BBB. Our group improved the brain delivery by conjugation with the amphipathic peptide LPFFD [21]. This last peptide is a β-breaker peptide designed by Soto et al. (which is based on the original sequence of β-amyloid) [22], where a Cys was added at the N-terminal peptide extreme for chemisorption to the gold surface for microwave destruction of β-amyloid toxic aggregates. An enhancement from approximately 0.01% to approximately 0.05% of the ID/g in the brain tissue was observed for AuNP-CLPFFD in comparison with AuNPs. The improvement in the delivery to the brain could be attributed to an interaction of the conjugate with the receptor for advanced glycation end products (RAGE), which play an essential role in the influx of β-amyloid into the brain. RAGE is a multiligand receptor in the immunoglobulin (IgG) superfamily that binds β-amyloid and mediates transcytosis across the BBB [23]. However, in that paper, the mechanism of penetration was not investigated.

In other reports to increase the crossing through the BBB, we conjugated AuNPs with a peptide that recognizes the transferrin receptor (THRPPMWSPVWP) to AuNP-CLPFFD conjugate [8]. This peptide sequence interacts with the transferrin receptor present in the microvascular endothelial cells of the BBB, thus causing an increment in the permeability of the conjugate in the brain, as demonstrated by experiments in vitro and in vivo (Fig. 2).

Another study related to the functionalization of AuNPs was reported by Sousa et al. where AuNPs where functionalized with cationized human serum albumin to induce AMT through the BBB [24]. These researchers followed the biodistribution with near-infrared time-domain imaging in mice up to 7 days after the intravenous injection of the nanoparticles. The peak concentration in the head of the mice was detected between 19 and 24 h. The precise particle distribution in the brain was studied ex vivo using X-ray microtomography, confocal laser microscopy, and fluorescence microscopy. Also, they observed that the particles mainly

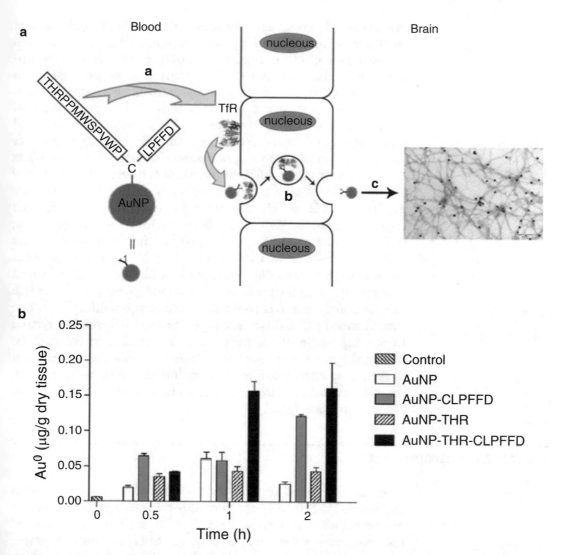

Fig. 2 (**A**) Scheme of conjugate AuNP-THR-CLPFFD. (a) The peptide THR-CLPFFD anchored to the AuNP. This peptide contains the THR sequence that recognizes the transferrin receptor and LPFFD that recognizes Ab aggregates; (b) transport of the endocytic vesicle through the endothelial cells of the blood-brain barrier; and (c) recognition and binding of the conjugate to Aβ aggregates inside the CNS. (**B**) Gold content in rat brains (*n* = 4) after administration of AuNP, AuNP-CLPFFD, AuNP-THR, AuNP-THR-CLPFFD, and PBS as control. Before organ extraction, the animals were perfused with PBS. The gold content was determined by neutron activation. *TfR* transferrin receptor (Reproduced from Prades et al. (2012) with permission from Biomaterials [8])

accumulate in the hippocampus, thalamus, hypothalamus, and the cerebral cortex. Additionally, Schäffler and colleagues capped AuNPs with albumin, observing a higher accumulation in the brain [25].

Shilo et al. reported the use of AuNPs conjugated with insulin to promote their translocation through RMT [26]. Consequently, insulin-coated gold nanoparticles (INS-AuNPs) and control AuNPs were injected into the tail vein of male BALB/c mice, and the gold content in the organs was analyzed using atomic absorption. The amount of INS-AuNPs observed in mouse brains was over five times greater than that of the control, untargeted AuNPs.

An important aspect to consider is that the structure of the peptide anchored to the nanoparticle should not be changed on the surface of the nanoparticles if the delivery strategy involves the recognition of a receptor, as discussed in previous works [5]. It is thus relevant to use different techniques to determine the structure of the ligand on the AuNPs surface as the number of peptides per nanoparticle, which is relevant for the interaction with the target. Different techniques have been employed for such purposes, such as circular dichroism, which provides information about the secondary structure of the peptide [27]; NMR, which provides information related to the structure of the molecules [28]; infrared spectroscopy, with the presence of functional groups on the surface [29]; and surface-enhanced Raman spectroscopy (SERS) [30]. The data obtained using these techniques provide information related to the disposition of the peptides on the surface of the AuNPs. Finally, the structure of the peptide and the surface that is exposed to the biological media determine the interaction of the entity with the cell membranes, triggering the entrance to the cell as was discussed previously [31].

3   Iron Oxide Nanoparticles

Iron oxide nanoparticles (IONPs) were one of the first nanocrystals extensively studied [32] due their intrinsic characteristics, such as biocompatibility and superparamagnetic properties that can be used for magnetic resonance imaging (MRI) as contrast probe, magnetic hyperthermia, targeted drug delivery, or a combination of these applications for theranostic [6, 33–37]. These nanoparticles are approved by the US Food and Drug Administration (FDA) for some imageology uses in medicine [38]. For example, Feridex particles (AMAG Pharmaceuticals) were approved for the detection of spleen and liver lesions [39, 40], whereas their analog Combidex and other dextran-IONPs have entered or passed clinical trials to their use as MRI contrast agents [41].

IONPs are synthetic particles with sizes ranging between 10 and 100 nm, which composition can be Fe_3O_4, α-Fe_2O_3, and γ-Fe_2O_3, known as magnetite, hematite, and maghemite, respectively.

One of the most relevant characteristics of the IONPs is that they are superparamagnetic when their size is less than 20 nm (superparamagnetic iron oxide nanoparticles or SPIONs), which means that they present zero magnetism in the absence of an external magnetic field but can be magnetized when there is one. The reason is because at such a small scale, the behavior of the nanoparticles when the thermal energy overcomes their anisotropy energy differs of bulk material, leading to random fluctuations of the magnetizations that result in zero magnetic moment and net coercivity at macroscopic scale [42, 43]. This property makes them effective in reducing T2 relaxation time, which leads to signal attenuation on a T2- or T2*-weighted map.

One interesting in vitro/ex vivo article has reported the use of near-infrared fluorescent maghemite nanoparticles (γ-Fe_2O_3) modified with specific anti-Aβ monoclonal antibody clone BAM10 [44], as potential targeting agents for Aβ aggregates, the main fibrillar component of plaque deposits found in the brains affected by Alzheimer's disease. Cy7-γ-Fe_2O_3-PEG nanoparticles (15.0 ± 1.3 nm) were covalently binding to BAM10 to study the kinetics of the Aβ_{40} fibril formation in the presence of BAM10-conjugated nanoparticles and the ex vivo detection of Aβ_{40} by the dual-modal MRI and fluorescence. These results showed that the conjugation of the BAM10 to the NIR fluorescent iron oxide nanoparticles significantly inhibits the Aβ_{40} fibrillation kinetics and specifically marks these fibrils. In another hand, the selective labeling of the Aβ_{40} fibrils with the BAM10-conjugated nanoparticles enabled specific detection of Aβ_{40} fibrils ex vivo by MRI and fluorescence imaging [44]. These dual-modal nanoparticles may, therefore, be an effective tool for anti-Aβ monoclonal antibody delivery and may be used for the development of both therapeutic and diagnostic agents for Alzheimer's disease.

As it was mentioned before, the main advantage of SPIONs is their capacity of being magnetized in the presence of a magnetic field, a characteristic that can be used to direct the delivery of the SPIONs to a specific organ. Thomsen et al. tried to find whether SPIONs were able to cross a monolayer of brain capillary endothelial cells cultured in an in vitro BBB model and the effect of external magnetic force over SPIONs penetration rate and efficiency [45]. This article shows SPIONs are able to pass into and through a brain capillary endothelial cell monolayer and enter astrocytes cultured at the bottom of lower chambers in a manner that is significantly enhanced by the use of an external magnetic force. Besides, data suggest external magnetic force did not affect the integrity of the endothelial monolayer, nor is the cell viability affected by the fluorescent SPIONs or by the magnetic force transporting the nanoparticles inside the cells.

In addition to the examples mentioned above, different groups have evaluated in vivo if IONPs can reach the CNS and their therapeutic effect after systemic administration. Shin and Yim et al. developed a facile approach for the delivery of magnetic metal ferrite nanoparticles coated with cross-linked serum albumin (SA-MNPs), in order to use the adsorptive-transcytosis mechanism present in the BBB [46]. The nanoparticles obtained were stable with a hydrodynamic diameter of 30 nm and a metal core of 15 nm. SA-MNPs were incubated 1 h at 37 °C with three different types of cells (primary cultured brain microvascular endothelial cells, neurons, and astrocytes) to evaluate their in vitro penetration. Besides, the nanoparticles were intravenously administered to BALB/c mice to evaluate their delivery to the CNS through MRI. Both in vitro and in vivo tests demonstrated their capability to cross the BBB without any breakdown.

In order to improve the brain delivery of PEG-coated Fe_3O_4 nanoparticles by lactoferring-receptor-mediated transcytosis of cerebral endothelial cells, lactoferrin (Lf) was covalently conjugated over the nanosystem [47]. The authors compared both in vitro and in vivo the efficacy of PEG-Fe_3O_4 and Lf-PEG-Fe_3O_4 nanoparticles. Experiments carried out using an in vitro porcine BBB model and SD rats for in vivo assays showed that Lf-conjugated PEG--Fe_3O_4 nanoparticles had a better capacity to cross the BBB in comparison to PEG-Fe_3O_4 nanoparticles.

Zhao et al. developed magnetic paclitaxel nanoparticles and evaluated, both in vitro and in vivo, their cytotoxicity against glioma [48]. They showed that the nanosystem was readily internalized into glioma cells and presented a therapeutic effect similar to free paclitaxel in vitro. As in vivo model, the authors used glioma-bearing rats that were treated with an intravenous injection of paclitaxel-loaded superparamagnetic nanoparticles. To target the nanosystem to the brain, a 0.5 T magnet was used, showing an increment in drug content of 6- to 14-fold in implanted glioma and 4.6- to 12.1-fold in the normal brain (in comparison to free paclitaxel, Table 1). Besides, the survival rate of animals was significantly extended after SPIONs magnetic targeting (Fig. 3). Other approaches for the treatment of glioblastoma were described by Agemy et al. through antiangiogenic therapy [49]. The authors developed a novel nanosystem that is based on nanoworm vectors

Table 1
Paclitaxel concentration in rat normal brain and glioma after intravenous administration of commercial paclitaxel and magnetic paclitaxel nanoparticles

	MPNPs + magnetic targeting		p-Value	Paclitaxel (μg/g)		p-Value
	Glioma	Brain		Glioma	Brain	
1 h	1.138 ± 0.199	0.854 ± 0.067	0.048	0.190 ± 0.045	0.186 ± 0.047	0.769
8 h	0.920 ± 0.284	0.765 ± 0.144	0.040	0.064 ± 0.034	0.063 ± 0.029	0.722
16 h	0.592 ± 0.157	0.382 ± 0.053	0.022	0.000 ± 0.000	0.000 ± 0.000	–

The paclitaxel concentration was analyzed at 1, 8, and 16 h by HPLC. Reproduced from Zhao et al. (2010) with permission from Anticancer Res [48]
Each value represents the mean ± SD of five independent determinations

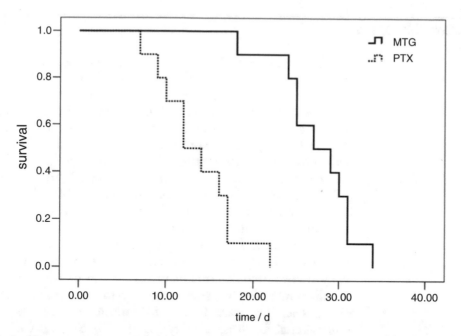

Fig. 3 Kaplan-Meier survival curves for C6 glioma-bearing rats. Magnetic paclitaxel nanoparticles (magnetic targeting group, MTG) or commercial paclitaxel (PTX) was intravenously administered. MTG was combined with 0.5 T magnetic field across the head for targeting (Reproduced from Zhao et al. 2010 with permission from Anticancer Res [48])

as theranostic platform, which incorporates three characteristics for targeting antitumoral activity and imaging (CGKRK peptide, (D [KLAKLAK]2) peptide, iron oxide component, respectively). The nanosystem showed both diagnostic and effective therapeutic capacity in glioblastoma-bearing mice, removing most tumors and delaying tumor development.

Several therapies against diseases produced by viruses have the problem that they are unable to eradicate the virus of the CNS. One example is HIV, where viral replication occurs despite a complete viral suppression in peripheral blood. Fiandra et al. studied the crossing of the antiretroviral drug enfuvirtide (Enf) through the BBB in both in vitro and in vivo [50]. Because this drug is normally unable to penetrate the cerebrospinal fluid, the authors used iron oxide nanoparticles coated with a suitable amphiphilic polymer, poly(isobutylene-alt-1-tetradecene-maleic anhydride), to increase Enf translocation across the BBB. Results showed that the nanoformulation enhanced the antiretroviral penetration across the BBB in mice model. The experiments carried out revealed that the mechanism involved corresponds to the uptake of the nanosystem by endothelial cells, nanocomplex dissociation, and the release of the peptide, which finally reaches the brain parenchyma.

4   Carbon Nanotubes

Carbon nanotubes (CNTs) are allotropes of carbon with a cylindrical structure formed by a graphene sheet. This graphene sheet can be wrapped in multiple ways, leading to different imaginary cut lines or chiral vectors [51, 52]. Mainly, there are three chiral vectors that denote three types of CNT structures with distinct properties: chiral, armchair, and zigzag [53]. Besides, depending in the number of graphene sheets wrapped that conforms the CNT, this can be categorized into two main groups: single-walled carbon nanotubes (SWCNTs) or multi-walled carbon nanotubes (MWCNTs), which have only one graphene sheet wrapped or more than one wrapped in a concentric fashion, respectively [54].

Due to their interesting properties, CNTs are valuable for nanomedicine as a diagnostic and/or therapeutic platform. As many studies have demonstrated, CNTs can be taken up by cells via different pathways, such as endocytosis or passive diffusion, which is an important feature to consider them as a potential nanomaterial for drug delivery [6, 55, 56]. Shityakov et al. have studied the capability of CNTs to cross the BBB using an in vitro model [57]. They used a Transwell® device (Corning Inc., NY, USA) to evaluate the transport of fluorescein isothiocyanate-labeled MWCNTs (MWCNTs-FITC) across microvascular cerebral endothelial cells (cEND cells). The results indicated that MWCNTs-FITC are able to penetrate the monolayer and did not have effects on cell viability. Besides, the chemical and physical properties make them a suitable nanocarrier. For example, the hollow interior of CNTs can be loaded with hydrophobic drugs through non-covalent π–π stacking, whereas the outer surface can be functionalized with molecules that recognize a specific receptor or target where the therapeutic effect is desired [54]. Besides, CNTs can be used to perform thermal ablation, to detect specific antigens, to enhance drug cytotoxicity, or as an adjuvant contrast agent [58–61]. All these features allow to create a multifunctional nanosystem that could be used for CNS pathologies.

Kafa et al. evaluated the penetration properties (in vitro) and brain uptake and the in vivo biodistribution of amino-functionalized MWCNTs [62]. For an in vitro study, the authors used a co-culture of primary porcine brain endothelial cells and primary rat astrocytes as BBB model. Cellular uptake was observed by TEM, which showed that MWCNTs accumulate in endocytic vesicles. Radiolabeled MWCNTs were used to quantify the transcytosis process by γ-scintigraphy, observing a maximum transport after 72 h incubation (13% of MWCNTs). For in vivo experiments, the radiolabeled MWCNTs were detected using γ-scintigraphy too. The results demonstrated the presence of MWCNTs in brain capillaries and brain parenchyma.

Some reports have shown that CNTs can be used as neuroprotective agents. Lee et al. investigated the potential of use SWCNTs functionalized with amine (a-SWCNTs) to protect neurons after ischemic injury [63]. The authors used a rat stroke model (MCAO), which was administered with a-SWCNTs via lateral ventricle before the induction of stroke. Animals treated with a-SWCNTs had smaller infarct regions in the brain, compared to control group (PBS administration), and showed an enhancement of motor function assessed by rota-rod treadmill test, despite that SWCNTs were administered without any therapeutic molecule. Another work used amino-functionalized MWCNTs to deliver therapeutic caspase-3 siRNA (siCas-3) into the brain of endothelin-1 stroke murine model [64]. The cortical injection of the nanosystem successfully improved the antiapoptotic effect of siCas-3 and promoted functional motor recovery. Besides, MWCNTs were observed in neuron cells at 48 h post-administration by TEM.

In the case of neurodegenerative diseases, SWCNTs were successfully used to deliver acetylcholine into the brain of a mouse model of Alzheimer's disease after multiple gastric gavage administrations [65]. The study revealed that SWCNTs were taken up the intestinal wall by macrophages and adsorptive cells but were also observed in other organs as the brain. TEM analysis showed that SWCNTs localized in the lysosomes of neurons and neurites. Due to the delivery of acetylcholine, therapeutic effects, as improvement in the learning and memory capability, were observed on induced Alzheimer's disease mouse model.

CNTs have also been used in the treatment of brain tumors. VanHandel et al. evaluated the uptake and toxicity of MWCNTs coated with Pluronic F108 and tagged with PKH26, a hydrophobic red fluorescent dye, after intratumoral injection [66]. They used a GL261 glioma model and determined that most MWCNTs (75%) were phagocyted by the tumor macrophages. Microglia showed to be MWCNTs-positive 24 h after injection, and the presence of the carbon nanotubes in glioma was confirmed 48 h post-administration by immunohistochemistry. Despite that a transient and self-limiting local inflammation was observed, no significant toxicity was noted in mice. Zhao et al. have demonstrated that SWCNTs chemically functionalized with PL-PEG which enhanced the uptake of CpG oligonucleotide (immunoadjuvant) in tumor-associated inflammatory cells after intratumoral injections [67]. The authors used an in vivo model mice bearing intracranial GL261 gliomas and measured the tumor growth by bioluminescent imaging, histology, and animal survival. CNTs potentiated CpG immunotherapy, resulting in elevated levels of proinflammatory cytokines and reduced tumor growth. Oxidized MWCNTs (O-MWCNTs) were conjugated with PEG and Angiopep-2 (ANG), a brain-targeting ligand, and loaded with the anticancer drug doxorubicin (DOX) [68]. The nanosystem was evaluated in vitro (intracellular tracking) and in vivo (fluorescence imaging).

Male Balb/c mice were intravenously injected with O-MWCNTs-ANG-DOX, which demonstrated an improvement in the brain uptake of DOX when it was compared with non-ANG and DOX alone counterparts. Besides increased glioma targeting, O-MWCNTs-ANG-DOX increased the median survival time the most.

5 Silver Nanoparticles

Silver nanoparticles (AgNP) have dimensions between 1 and 100 nm in size and represent another type of metallic colloidal nanoparticles. This nanomaterial shows stable physicochemical performance functions better in electricity, optics, and catalysis and exhibits excellent antibacterial activity [69], attracting attention in scientific research due to their versatility of applications in different areas of engineering, medicine, chemistry, and physics [70–72].

AgNP in biomedicine has shown an increase in its application, such as molecular imaging, drug delivery, diagnosis in the treatment of wound healing, and development of materials and medical devices with antimicrobial properties. Other therapeutic uses of AgNP have been studied, using them as carriers or as active compounds for therapy and diagnostic in cancer, as well as an antibacterial, antifungal, and antiviral agents. However, it has been agreed that AgNP could induce cell apoptosis owing to their strong cytotoxicity [73, 74]. Studies have shown that silver nanoparticles introduced into the systemic blood supply can induce BBB dysfunction and astrocyte swelling, in addition to causing neuronal degeneration in vivo [75–77]. For this reason, it is necessary to look forward to surface modifications in order to reduce their intrinsic toxic effect and therefore allow the use of these nanoparticles in biomedicine's applications.

Some studies have revealed that AgNP can reach the blood circulation system via the blood-pulmonary barrier, resulting in a systemic distribution, mainly in the brain [78]. In this context, the exact region where AgNPs are distributed in the brain ultrastructure and the BBB was investigated by Tang et al. They observed accumulation of silver nanoparticles in the brain, where AgNPs have been shown to cross the BBB using two different mechanisms: transcytosis through endothelial cells of the brain-blood capillary and by reducing the close connection between endothelial cells or dissolving the endothelial cell membranes [79]. They concluded that AgNPs have toxic effects on neurons, resulting in neuronal degeneration and cell membrane thrombolysis. Besides, the nanoparticles could move from the pyknotic neuron to influence other neurons. Because it has been demonstrated that AgNP can be accumulated in the brain over a long period of time, they may induce negative effects as neuronal necrosis.

Tang et al. in 2010 studied in an in vitro BBB model the spatial distribution of AgNP that crossed through the BBB model. They demonstrated that AgNPs were able to cross the BBB in a significant larger quantity, compared to silver microparticles. This result confirmed that silver nanoparticles could easily cross the blood-brain barrier as was reported in animal models [80]. Another in vitro report examines the interactions of AgNP with the endothelial microvascular cells to identify the connection of proinflammatory mediators that can increase BBB permeability [81]. The potential proinflammatory mediators such as interleukin IL-1b, IL-2, tumor necrosis factor TNFα, and prostaglandin E2 were determined. The investigators reported a size-dependent increase in BBB permeability correlated with the severity of immunotoxicity. Further, this study suggests that AgNP may interact with the cerebral microvasculature producing a proinflammatory cascade which may further induce brain inflammation and neurotoxicity.

However, recently a study by Gonzalez-Carter et al. has been published where the effect of AgNP on the resident immune cells of the brain was examined [82]. It is well-understood that microglia are implicated in neurodegenerative disorders such as Parkinson's disease, a reason why it is relevant to examine how AgNPs affect microglial inflammation to fully assess AgNP neurotoxicity. In the reported study, in vitro uptake and intracellular transformation of citrate-capped AgNPs by microglia, as well as their effects on microglial inflammation and related neurotoxicity, were examined. The results demonstrated an internalization and dissolution of AgNP within microglia and formation of nonreactive silver sulfide (Ag_2S) on the surface of AgNP. Besides, AgNP treatment upregulated microglial expression of the hydrogen sulfide (H_2S)-synthesizing enzyme cystathionine-γ-lyase (CSE), which might be a general mechanism of silver toxicity limitation. Concomitantly, this diminishes microglial inflammation, reducing lipopolysaccharide (LPS)-stimulated ROS, nitric oxide, and TNFα production.

In another hand, an in vivo study was reported using a targeted drug delivery against glioblastoma [83]. In this case, polymeric nanoparticles containing the drug alisertib, a selective Aurora A kinase (AAK) inhibitor and AgNP were conjugated with a chlorotoxin, an active targeting 36-amino acid-long peptide that specifically binds to MMP-2, a receptor overexpressed in brain cancer cells. The results showed a tumor reduction in vivo when using silver/alisertib-polymer-chlorotoxin nanoparticles. This can be attributed to a synergistic effect based on the coexistence of the drug with Ag + ions formed from slow AgNP dissolution once they are released from the protective polymeric shell.

6 Silica Nanoparticles

Silica-based nanoparticles (SiNPs) have roused a significant interest in the drug delivery field, owing to their intrinsic properties, such as biocompatibility, synthesis tunability, and low cost of production. In the biomedical field, SiNPs can be classified as mesoporous or nonporous nanoparticles, which exhibit amorphous structure silica. The former is mainly characterized by the mesopores (2–50 nm pore size) that have been broadly used for drug delivery based on physical or chemical adsorption of active compounds onto nanoparticle surface [84]. In another hand, nonporous silica nanoparticles may transport active compounds through conjugation using different chemical linkers or by encapsulation inside the silica matrix [85–87].

The drug delivery application of mesoporous silica nanoparticles to overcome the BBB has been studied by diverse investigation groups. Recently, Song et al. have studied the effect of conjugation in vitro, by attaching lactoferrin on the nanoparticles surface and evaluating their transport efficiency across the BBB model [88]. Silica nanoparticles were synthesized by using water-in-oil microemulsion method. Later, to modify its surface with primary amines, (3-Aminopropyl)triethoxysilane (APTES) was poured into the SiNPs solution and then stirred for a specific period of time. The surface-modified SiNPs were next resuspended in 2-(N-morpholino)ethanesulfonic acid (MES) solution after rinsing. Afterward, the obtained suspension was mixed with EDC and NHS, following by addition of bis-carboxy-polyethylene glycol (PEG). The resulted PEG-labeled SiNPs were washed with phosphate-buffered saline by centrifugation and finally dispersed in PBS. In this case, the nanoparticle surface was modified with polyethylene glycol to reduce nonspecific interactions. PEGylated silica nanoparticles have been reported to exhibit relatively longer blood circulation times and lower uptake by the liver than non-PEGylated silica nanoparticles. To increase the BBB penetration efficiency, previously obtained nanoparticles were modified with lactoferrin by attaching this probe to the PEG-labeled SiNPs surface. The lactoferrin conjugation procedure comprises first an EDC and NHS addition in PEG-labeled SiNPs solution, with the aim of activating terminal carboxylic groups present in PEG chain. Subsequent, different amounts of lactoferrin were introduced to trigger the reaction. The transport efficiency of obtained PEG-labeled SiNPs conjugated with lactoferrin was studied using an in vitro BBB model which consists of three distinct types of cells: endocytes, pericytes, and astrocytes. Transfer of NPs from the apical side to the basolateral side was observed, and the results indicated that lactoferrin-attached SiNPs improve transport efficiency across the BBB compared to uncovered SiNPs. This receptor-mediated transcytosis of SiNPs over the cerebral endothelial cells represents an interesting alternative to deliver drugs or imaging probes to the brain.

Another study that has shed light on the potentiality of this nanosystem corresponds to the investigation done by Ku et al., where in vivo assay was performed [89]. In this study, fluorescein-doped magnetic silica nanoparticles (FMSNs) were combined with magnetic materials (e.g., magnetic iron oxide) and a fluorescent probe (e.g., FITC) into mesoporous silica material. FMSNs were firstly modified through PEGylation, to increase the surface functional groups and protect dye molecules trapped in FMSNs from photobleaching. Subsequently, the second generation (G2) PAMAM-NH$_2$ dendrimers, which have been reported to be a nontoxic and nonimmunogenic drug/gene delivery agent, were covalently conjugated onto FMSNs through 3-(triethoxysilyl) propyl isocyanate (ICP) to yield PFMSNs. Finally, the amino groups of PFMSNs were reacted with tresylated MPEG-5000 to yield PEGylated PFMSNs. They demonstrated that PEGylated PFMSNs could penetrate the BBB through transcytosis of vascular endothelial cells, subsequently diffuse into cerebral parenchyma, and distribute in the neurons. In contrast, non-PEGylated FMSNs were not found to cross the BBB.

In another in vivo study, the authors proposed the use of nanoshells, an optically tunable nanoparticle consisting of a dielectric core (silica) surrounded by a thin metallic layer (gold), as a new contrast-enhancing agent for photoacoustic tomography [90]. The size of the nanoparticle core relative to the thickness of the gold shell can be adjusting, whence the optical resonance of nanoshells can be precisely and systematically varied over a broad spectrum including the near-infrared (NIR) region where the optical transmission through biological tissues is optimal. This study was abled to show for the first time the rat brain vasculature in vivo through photoacustic tomography, using nanoshells to create an enhanced optical contrast with a high spatial resolution. This nanosystem allowed to monitor the dynamics of the nanoshells in the circulatory system of the rat and its reach to the brain.

Another investigation group reports a study where transport of PEGylated silica nanoparticles with diameters of 100, 50, and 25 nm across BBB was evaluated [91]. For this purpose, an in vitro BBB model based on mouse cerebral endothelial cells (bEnd.3) cultured on Transwell inserts within a chamber was used. Moreover, in vivo animal experiments were further performed by noninvasive in vivo imaging and ex vivo optical imaging after injection via carotid artery. The results showed that PEGylated silica nanoparticles can traverse the BBB in vitro and in vivo. The transport efficiency of PEGylated SiNPs across BBB was found to be size-dependent, with increased particle size resulting in decreased efficiency.

7   Quantum Dots

Quantum dots (QDs) are semiconductor nanocrystals that typically have a heavy metal core composed of materials such as CdTe, CdSe, and PbS (first generation) or CdTe/CdSe, Cd3P2, and InAs/ZnSe (second generation), which is surrounded by an unreactive metallic shell (such as ZnS). QDs have unique optical and electronic properties such as high brightness and long-term photostability and possess a narrow emission spectrum. These properties can be accurately adjusted by tuning their size and composition, which make QDs a revolutionary platform for diagnostic purposes. Moreover, the option to incorporate a therapeutic agent over the outer coating of QDs makes them a suitable candidate for therapy and diagnosis (theranosis) of CNS disorders.

However, there are few reports about QD-based drug delivery to the CNS, with innate toxicity, relative poor stability, and low BBB permeability of these nanomaterials being the main issue. Gao et al. used QDs encapsulated into the core of PEG-PLA nanoparticles and functionalized with wheat germ agglutinin, observing an improvement in their stability and safe brain targeting after intranasal administration, as an alternative route to bypass the BBB [92]. One of the most used strategies to increase the translocation of nanoparticles from blood to CNS through BBB is the transferrin receptor. In this line, lysine-coated CdSe/CdS/ZnS QDs were synthesized and conjugated with transferrin [93]. The authors used an in vitro BBB model to demonstrate the receptor-mediated transport of their bioconjugated QDs. The migration rate of these nanoparticles was concentration- and time-dependent, illustrating a QD-based platform that could be used for theranostic purposes. Other research groups demonstrated that CdS:Mn/ZnS quantum dots conjugated with TAT (a cell-penetrating peptide) can rapidly reach the brain parenchyma (within few minutes) without manipulating the BBB, after intra-arterial administration [94]. Paris-Robidas et al. used QDs conjugated with Ri7 (monoclonal antibody) to target the murine transferrin receptor and obtain direct subcellular evidence of vectorized transport of their nanosystem [95]. First, they demonstrated that QDs-Ri7 internalization is due to specific transferrin receptor-mediated endocytosis using N2A and bEnd5 cell lines. A fourfold increase in the volume of distribution in brain tissues was found after intravenous administration of QDs-Ri7 in mice when they compared it with control animals. These nanoparticles remain in the cerebral vasculature 0.5, 1, and 4 h post-administration, with a decline in signal intensity after 24 h (Immunofluorescence analysis). However, authors described that parenchymal penetration of QDs-Ri7 was tremendously low and comparable with control (IgG), but the systematic administration of QDs-Ri7 complexes undergoes extensive internalization by brain capillary endothelial cells, thus showing a novel therapeutic approach for brain endothelial cell drug delivery [95].

8 Other Nanoparticles

Besides all kind of nanoparticles mentioned before, there are few examples of nanosystems developed for the treatment of CNS disorders, which are fabricated with inorganic materials, such as cerium oxide, yttrium oxide, zinc oxide, and selenium. Schubert et al. have shown that nanoparticles composed of cerium oxide or yttrium oxide can be used to diminish neuronal death by the reduction of the oxidative stress [96]. They used the HT22 hippocampal nerve cell line to demonstrate that ceria and yttria nanoparticles act as direct antioxidants to control the amount of ROS necessary to kill cells after γ-irradiation. Another group uses these potent-free radical-scavenging properties of ceria nanoparticles to use them as a potential therapeutic agent for stroke [97]. Using a mouse hippocampal brain slice model of cerebral ischemia, they demonstrated the neuroprotective effect of ceria nanoparticles on this in vitro model and their potential to prevent cell death by reduced ROS production, suggesting that cerium oxide nanoparticles mitigate ischemic brain injury through the scavenging of peroxynitrite. However, both studies just evaluated the use of these nanoparticles by in vitro assays, being necessary in more studies.

Xie et al. investigated if zinc oxide nanoparticles had effects on animal model of depression [98]. They gave lipopolysaccharides to male Swiss mice to induce depressive-like behaviors in one group and co-administered zinc oxide nanoparticles in another group (i.p.). Results showed that lipopolysaccharide injections elicited a cognitive and behavioral impairment in model mice that could be improved after zinc oxide nanoparticle co-administration. However, more studies are necessary to understand if these nanoparticles reach the CNS, in what quantity, and what mechanisms are involved.

In the last decade, there has been great interest in the use of selenium (Se) for the treatment of neurodegenerative diseases, such as Alzheimer's disease. There are accumulative evidences which indicate that the presence of hyperphosphorylated tau and Aβ plaques elicits apoptotic neuronal loss by the increment of oxidative stress [99]. As Se is an antioxidant bioactive compound, there is a correlation between the loss of cognitive function and low plasma selenium levels; it has seemed very interesting to explore its use. It has been reported that levels of oxidative stress and Aβ formation can be decreased using selenium as treatment for animal models of Alzheimer's disease [100]. Yin et al. synthetized sialic acid-modified selenium nanoparticles conjugated with B6 peptide (B6-Sa-SeNPs), as a substitute for transferrin to mediate brain drug delivery [101]. The authors used bEnd.3 cell line to evaluate the cellular uptake and an in vitro BBB model (Transwell experiment) to determine the capacity of their nanosystem to cross this barrier. The different assays performed, such as flow cytometry, laser-scanning confocal microscopy, and inductively coupled plasma atomic emission spectroscopy, showed high cellular uptake of B6-Sa-SeNPs and provided evidence that nanoparticles crossed the BBB model. Besides, the authors demonstrated that B6-Sa-SeNPs could effectively inhibit Aβ aggregation and disaggregate preformed Aβ fibrils. Although these results showed a promising platform for the treatment of Alzheimer's disease and other CNS pathologies related to damage produced by oxidative stress, in vivo studies are necessary to determinate the brain delivery of this nanoparticles.

9   Conclusions

It is possible to conclude that there are different kinds of nanomaterials with potential applications for drug delivery, therapy, and diagnostic for CNS pathologies. Remarkably, the development of new nanomaterials for theranostic applications, or their use to reduce the oxidative stress in CNS, appear as interesting new fields to be explored.In other hand, although the application of nanomaterials for pathologies is a promising field, it is necessary to achieve higher local concentrations into the tissues to determine the potential toxic effects of them on the CNS.

For the mentioned applications, it is necessary to explore the conjugation with molecules as peptides that allow the improvement of delivery to the CNS. In relation with the improvement of delivery of nanomaterials, it is relevant to explore other routes for administration, as the intranasal, evaluating also the quantity of nanoparticles delivered to the nervous tissue.

Acknowledgments

Fondap 15130011 and Fondecyt 1170929.

References

1. Jiang X (2013) Brain drug delivery systems. Pharm Res 30(10):2427–2428. https://doi.org/10.1007/s11095-013-1148-7

2. Hwang SR, Kim K (2014) Nano-enabled delivery systems across the blood-brain barrier. Arch Pharm Res 37(1):24–30. https://doi.org/10.1007/s12272-013-0272-6

3. Leyva-Gomez G, Cortes H, Magana JJ, Leyva-Garcia N, Quintanar-Guerrero D, Floran B (2015) Nanoparticle technology for treatment of Parkinson's disease: the role of surface phenomena in reaching the brain. Drug Discov Today 20(7):824–837. https://doi.org/10.1016/j.drudis.2015.02.009

4. Li X, Tsibouklis J, Weng T, Zhang B, Yin G, Feng G, Cui Y, Savina IN, Mikhalovska LI, Sandeman SR, Howel CA, Mikhalovsky SV (2017) Nano carriers for drug transport across the blood-brain barrier. J Drug Target 25(1):17–28. https://doi.org/10.1080/1061186X.2016.1184272

5. Velasco-Aguirre C, Morales F, Gallardo-Toledo E, Guerrero S, Giralt E, Araya E, Kogan MJ (2015) Peptides and proteins used to enhance gold nanoparticle delivery to the brain: preclinical approaches. Int J Nanomedicine 10:4919–4936. https://doi.org/10.2147/IJN.S82310

6. Xie J, Lee S, Chen XY (2010) Nanoparticle-based theranostic agents. Adv Drug Deliver Rev 62(11):1064–1079. https://doi.org/10.1016/j.addr.2010.07.009

7. Tsou YH, Zhang XQ, Zhu H, Syed S, Xu X (2017) Drug delivery to the brain across the blood-brain barrier using nanomaterials. Small 13(43). https://doi.org/10.1002/smll.201701921

8. Prades R, Guerrero S, Araya E, Molina C, Salas E, Zurita E, Selva J, Egea G, Lopez-Iglesias C, Teixido M, Kogan MJ, Giralt E (2012) Delivery of gold nanoparticles to the brain by conjugation with a peptide that recognizes the transferrin receptor. Biomaterials 33(29):7194–7205. https://doi.org/10.1016/j.biomaterials.2012.06.063

9. Araya E, Olmedo I, Bastus NG, Guerrero S, Puntes VF, Giralt E, Kogan MJ (2008) Gold nanoparticles and microwave irradiation inhibit beta-amyloid amyloidogenesis. Nanoscale Res Lett 3(11):435–443. https://doi.org/10.1007/s11671-008-9178-5

10. Triulzi RC, Dai Q, Zou J, Leblanc RM, Gu Q, Orbulescu J, Huo Q (2008) Photothermal ablation of amyloid aggregates by gold nanoparticles. Colloids Surf B Biointerfaces 63 (2):200–208. https://doi.org/10.1016/j.colsurfb.2007.12.006

11. Huang X, El-Sayed MA (2010) Gold nanoparticles: optical properties and implementations in cancer diagnosis and photothermal therapy. J Adv Res 1(1):13–28. https://doi.org/10.1016/j.jare.2010.02.002

12. Perrault SD, Chan WC (2009) Synthesis and surface modification of highly monodispersed, spherical gold nanoparticles of 50–200 nm. J Am Chem Soc 131 (47):17042–17043. https://doi.org/10.1021/ja907069u

13. Krol S (2012) Challenges in drug delivery to the brain: nature is against us. J Control Release 164(2):145–155. https://doi.org/10.1016/j.jconrel.2012.04.044

14. Egleton RD, Mitchell SA, Huber JD, Janders J, Stropova D, Polt R, Yamamura HI, Hruby VJ, Davis TP (2000) Improved bioavailability to the brain of glycosylated Met-enkephalin analogs. Brain Res 881 (1):37–46

15. Bilsky EJ, Egleton RD, Mitchell SA, Palian MM, Davis P, Huber JD, Jones H, Yamamura HI, Janders J, Davis TP, Porreca F, Hruby VJ, Polt R (2000) Enkephalin glycopeptide analogues produce analgesia with reduced dependence liability. J Med Chem 43 (13):2586–2590

16. Morris MC, Deshayes S, Heitz F, Divita G (2008) Cell-penetrating peptides: from molecular mechanisms to therapeutics. Biol Cell 100(4):201–217. https://doi.org/10.1042/BC20070116

17. Mager I, Eiriksdottir E, Langel K, El Andaloussi S, Langel U (2010) Assessing the uptake kinetics and internalization mechanisms of cell-penetrating peptides using a quenched fluorescence assay. Biochim

Biophys Acta 1798(3):338–343. https://doi.org/10.1016/j.bbamem.2009.11.001

18. Fernandez-Carneado J, Kogan MJ, Castel S, Giralt E (2004) Potential peptide carriers: amphipathic proline-rich peptides derived from the N-terminal domain of gamma-zein. Angew Chem 43(14):1811–1814. https://doi.org/10.1002/anie.200352540

19. Pujals S, Sabido E, Tarrago T, Giralt E (2007) all-D proline-rich cell-penetrating peptides: a preliminary in vivo internalization study. Biochem Soc Trans 35(Pt 4):794–796. https://doi.org/10.1042/BST0350794

20. Madani F, Lindberg S, Langel U, Futaki S, Graslund A (2011) Mechanisms of cellular uptake of cell-penetrating peptides. J Biophys 2011:414729. https://doi.org/10.1155/2011/414729

21. Guerrero S, Araya E, Fiedler JL, Arias JI, Adura C, Albericio F, Giralt E, Arias JL, Fernandez MS, Kogan MJ (2010) Improving the brain delivery of gold nanoparticles by conjugation with an amphipathic peptide. Nanomedicine 5(6):897–913. https://doi.org/10.2217/nnm.10.74

22. Soto C, Sigurdsson EM, Morelli L, Kumar RA, Castano EM, Frangione B (1998) Beta-sheet breaker peptides inhibit fibrillogenesis in a rat brain model of amyloidosis: implications for Alzheimer's therapy. Nat Med 4 (7):822–826

23. Deane R, Du Yan S, Submamaryan RK, LaRue B, Jovanovic S, Hogg E, Welch D, Manness L, Lin C, Yu J, Zhu H, Ghiso J, Frangione B, Stern A, Schmidt AM, Armstrong DL, Arnold B, Liliensiek B, Nawroth P, Hofman F, Kindy M, Stern D, Zlokovic B (2003) RAGE mediates amyloid-beta peptide transport across the blood-brain barrier and accumulation in brain. Nat Med 9 (7):907–913. https://doi.org/10.1038/nm890

24. Sousa F, Mandal S, Garrovo C, Astolfo A, Bonifacio A, Latawiec D, Menk RH, Arfelli F, Huewel S, Legname G, Galla HJ, Krol S (2010) Functionalized gold nanoparticles: a detailed in vivo multimodal microscopic brain distribution study. Nanoscale 2 (12):2826–2834. https://doi.org/10.1039/c0nr00345j

25. Schaffler M, Sousa F, Wenk A, Sitia L, Hirn S, Schleh C, Haberl N, Violatto M, Canovi M, Andreozzi P, Salmona M, Bigini P, Kreyling WG, Krol S (2014) Blood protein coating of gold nanoparticles as potential tool for organ targeting. Biomaterials 35(10):3455–3466. https://doi.org/10.1016/j.biomaterials.2013.12.100

26. Shilo M, Berenstein P, Dreifuss T, Nash Y, Goldsmith G, Kazimirsky G, Motiei M, Frenkel D, Brodie C, Popovtzer R (2015) Insulin-coated gold nanoparticles as a new concept for personalized and adjustable glucose regulation. Nanoscale 7 (48):20489–20496. https://doi.org/10.1039/c5nr04881h

27. Olmedo I, Araya E, Sanz F, Medina E, Arbiol J, Toledo P, Alvarez-Lueje A, Giralt E, Kogan MJ (2008) How changes in the sequence of the peptide CLPFFD-NH2 can modify the conjugation and stability of gold nanoparticles and their affinity for beta-amyloid fibrils. Bioconjug Chem 19 (6):1154–1163. https://doi.org/10.1021/bc800016y

28. Bower PV, Louie EA, Long JR, Stayton PS, Drobny GP (2005) Solid-state NMR structural studies of peptides immobilized on gold nanoparticles. Langmuir 21 (7):3002–3007. https://doi.org/10.1021/la040092w

29. Barth A (2007) Infrared spectroscopy of proteins. Biochim Biophys Acta 1767 (9):1073–1101. https://doi.org/10.1016/j.bbabio.2007.06.004

30. Arif M, Karthigeyan D, Siddhanta S, Kumar GV, Narayana C, Kundu TK (2013) Analysis of protein acetyltransferase structure-function relation by surface-enhanced raman scattering (SERS): a tool to screen and characterize small molecule modulators. Methods Mol Biol 981:239–261. https://doi.org/10.1007/978-1-62703-305-3_19

31. Riveros A, Dadlani K, Salas-Huenuleo E, Caballero L, Melo F, Kogan M (2013) Gold nanoparticle-membrane interactions: implications in biomedicine. J Biomater Tissue Eng 3. https://doi.org/10.1166/jbt.2013.1067

32. Enochs WS, Harsh G, Hochberg F, Weissleder R (1999) Improved delineation of human brain tumors on MR images using a long-circulating, superparamagnetic iron oxide agent. J Magn Reson Imaging 9(2):228–232

33. Wu W, He Q, Jiang C (2008) Magnetic iron oxide nanoparticles: synthesis and surface functionalization strategies. Nanoscale Res Lett 3(11):397–415. https://doi.org/10.1007/s11671-008-9174-9

34. Laurent S, Forge D, Port M, Roch A, Robic C, Vander Elst L, Muller RN (2008) Magnetic iron oxide nanoparticles: synthesis, stabilization, vectorization, physicochemical characterizations, and biological applications. Chem Rev 108(6):2064–2110. https://doi.org/10.1021/cr068445e

35. Gupta AK, Gupta M (2005) Synthesis and surface engineering of iron oxide nanoparticles for biomedical applications. Biomaterials 26(18):3995–4021. https://doi.org/10.1016/j.biomaterials.2004.10.012

36. Corot C, Violas X, Robert P, Gagneur G, Port M (2003) Comparison of different types of blood pool agents (P792, MS325, USPIO) in a rabbit MR angiography-like protocol. Investig Radiol 38(6):311–319

37. Duguet E, Vasseur S, Mornet S, Devoisselle JM (2006) Magnetic nanoparticles and their applications in medicine. Nanomedicine 1 (2):157–168. https://doi.org/10.2217/17435889.1.2.157

38. Wang YX, Hussain SM, Krestin GP (2001) Superparamagnetic iron oxide contrast agents: physicochemical characteristics and applications in MR imaging. Eur Radiol 11 (11):2319–2331. https://doi.org/10.1007/s003300100908

39. Lanza GM, Winter PM, Caruthers SD, Morawski AM, Schmieder AH, Crowder KC, Wickline SA (2004) Magnetic resonance molecular imaging with nanoparticles. J Nucl Cardiol 11(6):733–743

40. Mornet S, Vasseur S, Grasset F, Duguet E (2004) Magnetic nanoparticle design for medical diagnosis and therapy. J Mater Chem 14(14):2161–2175. https://doi.org/10.1039/b402025a

41. Harisinghani MG, Barentsz J, Hahn PF, Deserno WM, Tabatabaei S, van de Kaa CH, de la Rosette J, Weissleder R (2003) Noninvasive detection of clinically occult lymph-node metastases in prostate cancer. New Engl J Med 348(25):2491–U2495. https://doi.org/10.1056/Nejmoa022749

42. Xu CJ, Sun SH (2007) Monodisperse magnetic nanoparticles for biomedical applications. Polym Int 56(7):821–826. https://doi.org/10.1002/pi.2251

43. Jun YW, Seo JW, Cheon J (2008) Nanoscaling laws of magnetic nanoparticles and their applicabilities in biomedical sciences. Acc Chem Res 41(2):179–189. https://doi.org/10.1021/ar700121f

44. Skaat H, Corem-Slakmon E, Grinberg I, Last D, Goez D, Mardor Y, Margel S (2013) Antibody-conjugated, dual-modal, near-infrared fluorescent iron oxide nanoparticles for antiamyloidogenic activity and specific detection of amyloid-beta fibrils. Int J Nanomedicine 8:4063–4076. https://doi.org/10.2147/IJN.S52833

45. Thomsen LB, Linemann T, Pondman KM, Lichota J, Kim KS, Pieters RJ, Visser GM,

Moos T (2013) Uptake and transport of superparamagnetic iron oxide nanoparticles through human brain capillary endothelial cells. ACS Chem Neurosci 4 (10):1352–1360. https://doi.org/10.1021/cn400093z

46. Yim YS, Choi JS, Kim GT, Kim CH, Shin TH, Kim DG, Cheon J (2012) A facile approach for the delivery of inorganic nanoparticles into the brain by passing through the blood-brain barrier (BBB). Chem Commun 48 (1):61–63. https://doi.org/10.1039/c1cc15113d

47. Qiao RR, Jia QJ, Huwel S, Xia R, Liu T, Gao FB, Galla HJ, Gao MY (2012) Receptor-mediated delivery of magnetic nanoparticles across the blood-brain barrier. ACS Nano 6 (4):3304–3310. https://doi.org/10.1021/nn300240p

48. Zhao M, Liang C, Li A, Chang J, Wang H, Yan R, Zhang J, Tai J (2010) Magnetic paclitaxel nanoparticles inhibit glioma growth and improve the survival of rats bearing glioma xenografts. Anticancer Res 30(6):2217–2223

49. Agemy L, Friedmann-Morvinski D, Kotamraju VR, Roth L, Sugahara KN, Girard OM, Mattrey RF, Verma IM, Ruoslahti E (2011) Targeted nanoparticle enhanced proapoptotic peptide as potential therapy for glioblastoma. Proc Natl Acad Sci U S A 108 (42):17450–17455. https://doi.org/10.1073/pnas.1114518108

50. Fiandra L, Colombo M, Mazzucchelli S, Truffi M, Santini B, Allevi R, Nebuloni M, Capetti A, Rizzardini G, Prosperi D, Corsi F (2015) Nanoformulation of antiretroviral drugs enhances their penetration across the blood brain barrier in mice. Nanomedicine 11(6):1387–1397. https://doi.org/10.1016/j.nano.2015.03.009

51. Dresselhaus MS, Dresselhaus G, Saito R (1995) Physics of carbon nanotubes. Carbon 33(7):883–891. https://doi.org/10.1016/0008-6223(95)00017-8

52. Wilder JWG, Venema LC, Rinzler AG, Smalley RE, Dekker C (1998) Electronic structure of atomically resolved carbon nanotubes. Nature 391:59. https://doi.org/10.1038/34139

53. Mahar B, Laslau C, Yip R, Sun Y (2007) Development of carbon nanotube-based sensors—a review. IEEE Sensors J 7(2):266–284. https://doi.org/10.1109/JSEN.2006.886863

54. Sanginario A, Miccoli B, Demarchi D (2017) Carbon nanotubes as an effective opportunity for cancer diagnosis and treatment. Biosensors 7(1). https://doi.org/10.3390/bios7010009

55. Kostarelos K, Lacerda L, Pastorin G, Wu W, Wieckowski S, Luangsivilay J, Godefroy S, Pantarotto D, Briand JP, Muller S, Prato M, Bianco A (2007) Cellular uptake of functionalized carbon nanotubes is independent of functional group and cell type. Nat Nanotechnol 2(2):108–113. https://doi.org/10.1038/nnano.2006.209

56. Jin H, Heller DA, Strano MS (2008) Single-particle tracking of endocytosis and exocytosis of single-walled carbon nanotubes in NIH-3T3 cells. Nano Lett 8(6):1577–1585. https://doi.org/10.1021/nl072969s

57. Shityakov S, Salvador E, Pastorin G, Forster C (2015) Blood-brain barrier transport studies, aggregation, and molecular dynamics simulation of multiwalled carbon nanotube functionalized with fluorescein isothiocyanate. Int J Nanomedicine 10:1703–1713. https://doi.org/10.2147/IJN.S68429

58. Hernandez-Rivera M, Zaibaq NG, Wilson LJ (2016) Toward carbon nanotube-based imaging agents for the clinic. Biomaterials 101:229–240. https://doi.org/10.1016/j.biomaterials.2016.05.045

59. Wong BS, Yoong SL, Jagusiak A, Panczyk T, Ho HK, Ang WH, Pastorin G (2013) Carbon nanotubes for delivery of small molecule drugs. Adv Drug Deliv Rev 65 (15):1964–2015. https://doi.org/10.1016/j.addr.2013.08.005

60. Singh R, Torti SV (2013) Carbon nanotubes in hyperthermia therapy. Adv Drug Deliv Rev 65(15):2045–2060. https://doi.org/10.1016/j.addr.2013.08.001

61. Zhang L, Lv D, Su W, Liu Y, Chen Y, Xiang R (2013) Detection of cancer biomarkers with nanotechnology. Am J Biochem Biotechnol 9 (1):71

62. Kafa H, Wang JT, Rubio N, Venner K, Anderson G, Pach E, Ballesteros B, Preston JE, Abbott NJ, Al-Jamal KT (2015) The interaction of carbon nanotubes with an in vitro blood-brain barrier model and mouse brain in vivo. Biomaterials 53:437–452. https://doi.org/10.1016/j.biomaterials.2015.02.083

63. Lee HJ, Park J, Yoon OJ, Kim HW, Lee DY, Kim DH, Lee WB, Lee NE, Bonventre JV, Kim SS (2011) Amine-modified single-walled carbon nanotubes protect neurons from injury in a rat stroke model. Nat Nanotechnol 6(2):121–125. https://doi.org/10.1038/nnano.2010.281

64. Al-Jamal KT, Gherardini L, Bardi G, Nunes A, Guo C, Bussy C, Herrero MA, Bianco A, Prato M, Kostarelos K, Pizzorusso T (2011) Functional motor recovery from brain ischemic insult by carbon nanotube-mediated siRNA silencing. Proc Natl Acad Sci U S A 108(27):10952–10957. https://doi.org/10.1073/pnas.1100930108

65. Yang Z, Zhang Y, Yang Y, Sun L, Han D, Li H, Wang C (2010) Pharmacological and toxicological target organelles and safe use of single-walled carbon nanotubes as drug carriers in treating Alzheimer disease. Nanomedicine 6(3):427–441. https://doi.org/10.1016/j.nano.2009.11.007

66. VanHandel M, Alizadeh D, Zhang L, Kateb B, Bronikowski M, Manohara H, Badie B (2009) Selective uptake of multiwalled carbon nanotubes by tumor macrophages in a murine glioma model. J Neuroimmunol 208(1–2):3–9. https://doi.org/10.1016/j.jneuroim.2008.12.006

67. Zhao D, Alizadeh D, Zhang L, Liu W, Farrukh O, Manuel E, Diamond DJ, Badie B (2011) Carbon nanotubes enhance CpG uptake and potentiate antiglioma immunity. Clin Cancer Res 17(4):771–782. https://doi.org/10.1158/1078-0432.CCR-10-2444

68. Ren J, Shen S, Wang D, Xi Z, Guo L, Pang Z, Qian Y, Sun X, Jiang X (2012) The targeted delivery of anticancer drugs to brain glioma by PEGylated oxidized multi-walled carbon nanotubes modified with angiopep-2. Biomaterials 33(11):3324–3333. https://doi.org/10.1016/j.biomaterials.2012.01.025

69. Raffi M, Hussain F, Bhatti T, Akhter J, Hameed A, Hassan M (2007) Antibacterial characterization of silver nanoparticles against E. coli ATCC-15224. J Mater Sci Technol 24:192–196

70. Chen J, Han CM, Lin XW, Tang ZJ, Su SJ (2006) Effect of silver nanoparticle dressing on second degree burn wound. Chin J Surg 44(1):50–52

71. Lu S, Gao W, Gu HY (2008) Construction, application and biosafety of silver nanocrystalline chitosan wound dressing. Burns 34(5):623–628. https://doi.org/10.1016/j.burns.2007.08.020

72. Sun H, Choy TS, Zhu DR, Yam WC, Fung YS (2009) Nano-silver-modified PQC/DNA biosensor for detecting E. coli in environmental water. Biosens Bioelectron 24(5):1405–1410. https://doi.org/10.1016/j.bios.2008.08.008

73. Gurunathan S, Lee KJ, Kalishwaralal K, Sheikpranbabu S, Vaidyanathan R, Eom SH (2009) Antiangiogenic properties of silver nanoparticles. Biomaterials 30(31):6341–6350. https://doi.org/10.1016/j.biomaterials.2009.08.008

74. Kalishwaralal K, Banumathi E, Ram Kumar Pandian S, Deepak V, Muniyandi J, Eom SH, Gurunathan S (2009) Silver nanoparticles inhibit VEGF induced cell proliferation and migration in bovine retinal endothelial cells. Colloids Surf B Biointerfaces 73(1):51–57. https://doi.org/10.1016/j.colsurfb.2009.04.025

75. Sharma HS, Ali SF, Hussain SM, Schlager JJ, Sharma A (2009) Influence of engineered nanoparticles from metals on the blood-brain barrier permeability, cerebral blood flow, brain edema and neurotoxicity. An experimental study in the rat and mice using biochemical and morphological approaches. J Nanosci Nanotechnol 9(8):5055–5072

76. Sharma HS, Ali SF, Tian ZR, Hussain SM, Schlager JJ, Sjoquist PO, Sharma A, Muresanu DF (2009) Chronic treatment with nanoparticles exacerbate hyperthermia induced blood-brain barrier breakdown, cognitive dysfunction and brain pathology in the rat. Neuroprotective effects of nanowired-antioxidant compound H-290/51. J Nanosci Nanotechnol 9(8):5073–5090

77. Tang J, Xiong L, Wang S, Wang J, Liu L, Li J, Yuan F, Xi T (2009) Distribution, translocation and accumulation of silver nanoparticles in rats. J Nanosci Nanotechnol 9(8):4924–4932

78. Takenaka S, Karg E, Roth C, Schulz H, Ziesenis A, Heinzmann U, Schramel P, Heyder J (2001) Pulmonary and systemic distribution of inhaled ultrafine silver particles in rats. Environ Health Perspect 109(Suppl 4):547–551

79. Tang J, Xiong L, Wang S, Wang J, Liu L, Li J, Wan Z, Xi T (2008) Influence of silver nanoparticles on neurons and blood-brain barrier via subcutaneous injection in rats. Appl Surf Sci 255(2):502–504. https://doi.org/10.1016/j.apsusc.2008.06.058

80. Tang J, Xiong L, Zhou G, Wang S, Wang J, Liu L, Li J, Yuan F, Lu S, Wan Z, Chou L, Xi T (2010) Silver nanoparticles crossing through and distribution in the blood-brain barrier in vitro. J Nanosci Nanotechnol 10(10):6313–6317

81. Trickler WJ, Lantz SM, Murdock RC, Schrand AM, Robinson BL, Newport GD, Schlager JJ, Oldenburg SJ, Paule MG, Slikker W Jr, Hussain SM, Ali SF (2010) Silver nanoparticle induced blood-brain barrier inflammation and increased permeability in primary

rat brain microvessel endothelial cells. Toxicol Sci 118(1):160–170. https://doi.org/10.1093/toxsci/kfq244

82. Gonzalez-Carter DA, Leo BF, Ruenraroengsak P, Chen S, Goode AE, Theodorou IG, Chung KF, Carzaniga R, Shaffer MS, Dexter DT, Ryan MP, Porter AE (2017) Silver nanoparticles reduce brain inflammation and related neurotoxicity through induction of H2S-synthesizing enzymes. Sci Rep 7:42871. https://doi.org/10.1038/srep42871

83. Locatelli E, Naddaka M, Uboldi C, Loudos G, Fragogeorgi E, Molinari V, Pucci A, Tsotakos T, Psimadas D, Ponti J, Franchini MC (2014) Targeted delivery of silver nanoparticles and alisertib: in vitro and in vivo synergistic effect against glioblastoma. Nanomedicine 9(6):839–849. https://doi.org/10.2217/nnm.14.1

84. de la Torre C, Domínguez-Berrocal L, Murguía JR, Marcos MD, Martínez-Máñez R, Bravo J, Sancenón F (2018) ε-Polylysine-capped mesoporous silica nanoparticles as carrier of the C9h peptide to induce apoptosis in cancer cells. Chem Eur J 24(8):1890–1897. https://doi.org/10.1002/chem.201704161

85. Li ZZ, Xu SA, Wen LX, Liu F, Liu AQ, Wang Q, Sun HY, Yu W, Chen JF (2006) Controlled release of avermectin from porous hollow silica nanoparticles: influence of shell thickness on loading efficiency, UV-shielding property and release. J Control Release 111 (1–2):81–88. https://doi.org/10.1016/j.jconrel.2005.10.020

86. Stevens EV, Carpenter AW, Shin JH, Liu J, Der CJ, Schoenfisch MH (2010) Nitric oxide-releasing silica nanoparticle inhibition of ovarian cancer cell growth. Mol Pharm 7 (3):775–785. https://doi.org/10.1021/mp9002865

87. Tang L, Gabrielson NP, Uckun FM, Fan TM, Cheng J (2013) Size-dependent tumor penetration and in vivo efficacy of monodisperse drug-silica nanoconjugates. Mol Pharm 10 (3):883–892. https://doi.org/10.1021/mp300684a

88. Song Y, Du D, Li L, Xu J, Dutta P, Lin Y (2017) In vitro study of receptor-mediated silica nanoparticles delivery across blood-brain barrier. ACS Appl Mater Interfaces 9 (24):20410–20416. https://doi.org/10.1021/acsami.7b03504

89. Ku S, Yan F, Wang Y, Sun Y, Yang N, Ye L (2010) The blood-brain barrier penetration and distribution of PEGylated fluorescein-doped magnetic silica nanoparticles in rat brain. Biochem Biophys Res Commun 394

(4):871–876. https://doi.org/10.1016/j.bbrc.2010.03.006

90. Wang Y, Xie X, Wang X, Ku G, Gill KL, O'Neal DP, Stoica G, Wang LV (2004) Photoacoustic tomography of a nanoshell contrast agent in the in vivo rat brain. Nano Lett 4(9):1689–1692. https://doi.org/10.1021/nl049126a

91. Liu D, Lin B, Shao W, Zhu Z, Ji T, Yang C (2014) In vitro and in vivo studies on the transport of PEGylated silica nanoparticles across the blood-brain barrier. ACS Appl Mater Interfaces 6(3):2131–2136. https://doi.org/10.1021/am405219u

92. Gao X, Chen J, Chen J, Wu B, Chen H, Jiang X (2008) Quantum dots bearing lectin-functionalized nanoparticles as a platform for in vivo brain imaging. Bioconjug Chem 19 (11):2189–2195. https://doi.org/10.1021/bc8002698

93. Xu G, Yong KT, Roy I, Mahajan SD, Ding H, Schwartz SA, Prasad PN (2008) Bioconjugated quantum rods as targeted probes for efficient transmigration across an in vitro blood-brain barrier. Bioconjug Chem 19 (6):1179–1185. https://doi.org/10.1021/bc700477u

94. Santra S, Yang H, Stanley JT, Holloway PH, Moudgil BM, Walter G, Mericle RA (2005) Rapid and effective labeling of brain tissue using TAT-conjugated CdS:Mn/ZnS quantum dots. Chem Commun 25:3144–3146. https://doi.org/10.1039/b503234b

95. Paris-Robidas S, Brouard D, Emond V, Parent M, Calon F (2016) Internalization of targeted quantum dots by brain capillary endothelial cells in vivo. J Cerebr Blood Flow Metab 36(4):731–742. https://doi.org/10.1177/0271678X15608201

96. Schubert D, Dargusch R, Raitano J, Chan SW (2006) Cerium and yttrium oxide nanoparticles are neuroprotective. Biochem Biophys Res Commun 342(1):86–91. https://doi.org/10.1016/j.bbrc.2006.01.129

97. Estevez AY, Pritchard S, Harper K, Aston JW, Lynch A, Lucky JJ, Ludington JS, Chatani P, Mosenthal WP, Leiter JC, Andreescu S, Erlichman JS (2011) Neuroprotective mechanisms of cerium oxide nanoparticles in a mouse hippocampal brain slice model of ischemia. Free Radic Biol Med 51 (6):1155–1163. https://doi.org/10.1016/j.freeradbiomed.2011.06.006

98. Xie Y, Wang Y, Zhang T, Ren G, Yang Z (2012) Effects of nanoparticle zinc oxide on spatial cognition and synaptic plasticity in mice with depressive-like behaviors. J Biomed

Sci 19:14. https://doi.org/10.1186/1423-0127-19-14

99. Naziroglu M, Muhamad S, Pecze L (2017) Nanoparticles as potential clinical therapeutic agents in Alzheimer's disease: focus on selenium nanoparticles. Expert Rev Clin Pharmacol 10(7):773–782. https://doi.org/10.1080/17512433.2017.1324781

100. Ishrat T, Parveen K, Khan MM, Khuwaja G, Khan MB, Yousuf S, Ahmad A, Shrivastav P, Islam F (2009) Selenium prevents cognitive decline and oxidative damage in rat model of streptozotocin-induced experimental dementia of Alzheimer's type. Brain Res 1281:117–127. https://doi.org/10.1016/j.brainres.2009.04.010

101. Yin T, Yang L, Liu Y, Zhou X, Sun J, Liu J (2015) Sialic acid (SA)-modified selenium nanoparticles coated with a high blood-brain barrier permeability peptide-B6 peptide for potential use in Alzheimer's disease. Acta Biomater 25:172–183. https://doi.org/10.1016/j.actbio.2015.06.035

Chapter 7

Magnetic Nanoparticles as Delivery Systems to Penetrate the Blood-Brain Barrier

Joan Estelrich and Maria Antònia Busquets

Abstract

Crossing the blood-brain barrier (BBB) is essential for effective treatment of brain disorders. Due to their physical properties, biocompatibility, and biodegradability, superparamagnetic iron oxide nanoparticles (SPIONs) show promise as carriers of therapeutics or as a therapeutic system in themselves. The use of SPIONs as a therapeutic system with the capacity to penetrate the BBB is based on three different strategies: (a) SPIONs are encapsulated in a therapeutic nanoscale system thought to cross the BBB. The presence of SPIONs inside the brain is easily detectable by techniques such as magnetic resonance imaging (MRI), confirming entry of the nanoscale system into the brain; (b) movement of the SPION-encapsulated load toward the site of action is assisted by the action of an external magnetic field; and (c) SPIONs generate moderate or high heat after applying radiofrequency or microwave radiation; this heat can either locally open the BBB or kill the cancerous cells.

This review summarizes the advances that have been made when SPIONs have been applied in some diseases involving the brain, such as neurodegenerative diseases, gliomas, and neuro-acquired immunodeficiency syndrome.

Key words Acquired immunodeficiency syndrome, Brain, Drug delivery, Glioma, Magnetic hyperthermia, Magnetic nanoparticles, Magnetic targeting, Nanotechnology, Neurodegenerative diseases, Superparamagnetism

1 Introduction

One of the most commonly used nanoscale materials are magnetic nanoparticles (MNPs), a type of core/shell nanoparticle structure that consists of a magnetic core encapsulated in an organic or polymeric coating. The magnetic core is composed of elements such as iron, nickel, and cobalt, and their chemical compounds. Without a coating, MNPs have hydrophobic surfaces with large surface-to-volume ratios and a propensity to agglomerate [1]. MNPs exhibit a variety of unique magnetic phenomena that are drastically different from those of their bulk counterparts. Such phenomena endow the magnetic nanoparticles with attractive properties that can be used in a large number of applications in

Javier O. Morales and Pieter J. Gaillard (eds.), *Nanomedicines for Brain Drug Delivery*, Neuromethods, vol. 157,
https://doi.org/10.1007/978-1-0716-0838-8_7, © Springer Science+Business Media, LLC, part of Springer Nature 2021

different scientific fields. When MNPs possess a special magnetic property, superparamagnetism, they can be utilized in biomedical applications [2]. Properly synthesized and functionalized, MNPs formed by iron oxides with superparamagnetic properties are non-toxic and well-tolerated by living organisms. In addition, they are biocompatible and enable coating with affinity biomolecules for highly specific binding to a target biomaterial [3]. As with other nanoparticles, MNPs have the capacity to cross the blood-brain barrier (BBB), either by modulating BBB integrity or exploiting transport systems present on the endothelium.

In this chapter, we review the use of MNPs, and more specifically, of iron oxide nanoparticles (IONs), as a promising strategy for drug delivery in diseases affecting the least accessible organ, the brain.

2   Magnetic Materials

All materials are magnetic to some extent, with their response depending on their atomic structure and temperature. The existence of electrons circulating around atomic nuclei and spinning on their axes, and the rotation of positively charged atomic nuclei, produce magnetic dipoles, also called magnetons. In combination, these effects usually cancel out so that a given type of atom is not a magnetic dipole. However, materials with unpaired electrons (for instance, iron, cobalt, nickel, and some of the rare earths) form permanent magnetic dipoles. In this case, when an external magnetic field (H) is applied, the atomic dipoles align themselves with it, generating a magnetic moment within the material. The quantity of magnetic moment per unit volume is defined as magnetization (M). The relationship between magnetization and magnetic field is given by.

$$M = \chi H \tag{1}$$

where χ is the volumetric magnetic susceptibility, which in SI units is dimensionless, and both M and H are expressed in A/m.

Macroscopic regions in which billions of dipoles are aligned and coupled in a preferential direction form a magnetic domain. A bulk material spontaneously subdivides into a multidomain structure to reduce the magnetostatic energy associated with a large stray field [4]. Magnetization does not vary within each domain, but between domains there are relatively thin domain walls in which the direction of magnetization rotates from the direction of one domain to another. Formation of the domain walls is driven by the balance between magnetostatic energy and domain wall energy. Within the domain, the magnetic field and force are intense in one direction, but the alignment directions of the separate regions are random with respect to one another throughout the material. Hence, the bulk sample of the material is usually unmagnetized or weakly magnetized in absence of an external magnetic field (Fig. 1a [left]). This type of material is called a ferromagnet, and its inherent property, ferromagnetism. A ferromagnet is a material that exhibits large positive susceptibility (χ in the range of 10^3–10^5). If a ferromagnetic material is kept in a strong magnetic field, the magnetic domains are forced to align with the external magnetic field (EMF) (Fig. 1b [right]). When the EMF is removed, the dipoles maintain this alignment and some magnetizations are retained. Other materials, called paramagnets, show small positive susceptibility (χ in the range of 10^{-6}–10^{-1}) and after removing the EMF, the magnetic properties do not persist.

All ferromagnets have a maximum temperature, called the Curie temperature (T_C), at which the thermal motion of the dipoles becomes violent, and the permanent magnetic properties are lost. At a temperature above the T_C, a ferromagnet becomes a paramagnet.

Fundamental changes in the magnetic structure of macroscopic, magnetically ordered materials occur when physical size is reduced [1]. Below a critical value, called the critical diameter D_{CR}, it requires more energy to create a domain wall than to support the external magnetostatic energy of the single domain state (Fig. 2), and the material becomes a single domain. A single domain particle is uniformly magnetized with all the spins aligned in the same direction. The magnetization will be reversed by spin rotation since there are no domain walls to move. A further reduction in size, below the superparamagnetic diameter D_{SPM}, causes the thermal energy to exceed the energy barrier which separates the two energetically equivalent easy directions of magnetization (the easy axis is the preferred direction of the total magnetization of the dipoles of a given material), and the direction of the magnetization fluctuates randomly. This system is termed a superparamagnet. The magnetic moments of individual crystallites compensate for each other and the overall magnetic moment becomes null. When an EMF is applied, the behavior is similar to paramagnetism except that instead of each individual atom being independently influenced by an EMF, the magnetic moment of the crystallite aligns with the magnetic field. In consequence, superparamagnetic nanoparticles become magnetic in the presence of an external magnet but revert to a nonmagnetic state when the external magnet is removed. This fact is of paramount importance when these particles are introduced into living systems (e.g., in drug delivery), because once the EMF is removed, magnetization disappears, and thus agglomeration (and the possible embolization of capillary vessels) is avoided [5].

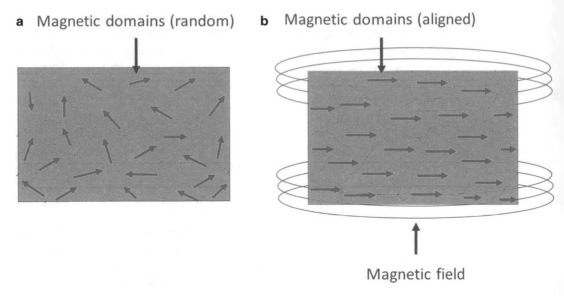

a Magnetic domains (random)   **b** Magnetic domains (aligned)

Magnetic field

Fig. 1 Multidomain structure of a ferromagnetic material. (**a**) The directions of the alignment of magnetic dipoles of the domains are randomized in absence of a magnetic field. (**b**) Alignment of the magnetic domains with an external magnetic field

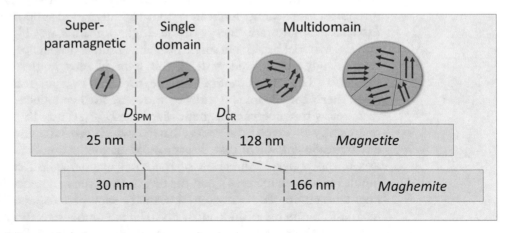

Fig. 2 Changes in the magnetic structure of a ferromagnetic material as a function of the size of the crystallite. Below a size, the critical diameter, D_{CR}, the multidomain structure becomes a single domain. A single domain with a size below the superparamagnetic diameter, D_{SPM}, implies that the material presents superparamagnetic properties, that is, the material becomes magnetic only in the presence of an external magnet. The critical sizes for magnetite and magnetite have been displayed

3   Iron Oxide Nanoparticles

Among the MNPs, almost isotropic nanometer-sized iron oxide remains the most studied system since it is considered relatively safe at reasonable doses. Iron is one of the most abundant transition metals in the human body (the average adult possesses ca. 5 g) [6]. IONs are highly magnetizable and have a core of iron oxide particles composed of magnetite (Fe_3O_4) or maghemite (γ-Fe_2O_3), or most commonly, a nonstoichiometric combination of the two. Other iron oxide are ferrites, whose general formula is MFe_2O_4 (M = Mn, Zn, Co, Ni). As noted previously, the iron core exerts a low level of toxicity as it gradually degrades to Fe^{3+} and is integrated into the iron stores of the body, which are used for metabolic processes and eventually eliminated [7]. Because the crystalline structures of magnetite and maghemite are nearly identical, the two oxides have very similar properties. Both are considered ferromagnetic, although magnetite has a larger bulk saturation magnetization (92–100 emu/g) than maghemite (60–80 emu/g) and a lower Curie temperature.

To date, a variety of synthetic methods, such as coprecipitation, thermal decomposition, hydrothermal and solvothermal syntheses, sol–gel synthesis, microemulsion, ultrasound irradiation, and biological synthesis have been used to produce IONs. The methods of synthesis can be divided into aqueous and nonaqueous routes. Aqueous approaches are characterized by their low cost and sustainability; however, directly obtaining water-soluble monodisperse superparamagnetic IONs (SPIONs) with no size selection poses a generic challenge. Nonaqueous routes generally yield SPIONs that only dissolve in nonpolar solvents [8]. The most usual and straightforward method of synthesizing SPIONs is to coprecipitate Fe^{2+} and Fe^{3+} aqueous salt solutions by adding a base. Control of the size, shape, and composition of SPIONs depends on the type of salts used, the Fe^{2+} to Fe^{3+} ratio, and the pH and ionic strength of the media [9]. A major advantage of SPIONs is that they are only magnetized in the presence of a magnetic field, under the influence of which they can be moved toward the magnet to concentrate near the target location, so that delivery can be influenced and directed to a specific region (i.e., the brain). Thus, application of an external magnet can control the spatial and temporal delivery of SPIONs at the desired site [10].

The European Medicines Agency (EMEA) and the US Food and Drug Administration (FDA) have approved several SPION formulations for intravenous or oral administration in clinical use as contrast agents in magnetic resonance imaging (MRI) [11]. Furthermore, SPIONs are candidates for use as therapeutic carriers, because they have several desirable properties. (1) Drugs can be incorporated into SPIONs by encapsulation, adsorption, or covalent linkage. A hydrophilic coating can provide reduced monocyte-phagocyte system (MPS) uptake and improved plasma circulation time. (2) A high payload can be achieved by functional linkers on SPIONs. (3) Drug release can be controlled to avoid the "burst effect" by conjugating environmentally sensitive moieties (e.g., pH and temperature) onto SPIONs. (4) Attachment of tumor-specific targeting moieties renders delivery more efficient. (5) SPIONs prevent drugs from degrading and overcome drug resistance by masking the drugs entrapped within the nanoparticles. (6) Co-delivery of multiple therapeutic agents may generate synergistic effects and produce a better therapeutic outcome. However, their therapeutic use in treating central nervous system (CNS) pathologies in vivo is limited by insufficient local accumulation and retention as a result of their incapacity to traverse biological barriers, mainly the BBB. The combined use of approaches to temporarily open the BBB (by means of pharmacological, osmotic, or ultrasound mechanisms) and magnetic targeting synergistically delivers therapeutic magnetic nanoparticles across the BBB to enter the brain both passively and actively [12]. In this chapter, we examine the possibilities of using MNPs as therapeutic systems that can penetrate the BBB. Thus, we explain how application of an EMF facilitates guidance of MNPs across the BBB, discuss the enormous therapeutic potential of combining an EMF with molecular targeting at the BBB and describe how the unique magnetic properties of MNPs can also be exploited to generate heat when the MNPs are under the influence of radiofrequency or microwave radiation [7]. We also explore the use of MNPs to treat brain cancers in humans, although in this case the MNPs do not need to cross the BBB since they are injected directly into the tumor.

4   SPIONs as Systems with the Capacity to Penetrate the Brain

The use of SPIONs as a delivery system with the capacity to penetrate the BBB is generally based on three strategies. In the first, SPIONs, and occasionally drugs, are encapsulated in nanoscale systems grafted with ligands that bind to specific receptors on endothelial cells, triggering uptake at the blood interface, transport across the barrier, and release at the brain interface (transcytosis), e.g., specific antibodies targeting the transferrin (Tf) receptor [13]. In the second strategy, an external magnetic field (EMF) is applied to direct the movement of a SPION-encapsulated load to the brain and enhance SPION penetration [14]. Finally, the third strategy consists of applying a regulated RF to the SPIONs to generate moderate or high heat and thereby transiently and locally open the BBB or kill the cancerous cells, respectively.

In the first strategy, SPIONs are used in a nanoscale system because they make it possible to confirm penetration of such systems into the brain as they can be easily detected by MRI, since they act as T_2 contrast agents. However, the targeting approach of the first strategy can present a problem: the nanoparticles may be restricted to entering brain endothelial cells without further passage into the brain, unless specific antibodies are designed to limit their binding affinity for the receptor. Most studies have been carried out using tumor-bearing rodents, but there is the possibility that the

BBB óf these rats has been changed, and thus the status of the BBB integrity is unclear [15].

We indicated earlier that in the first strategy, SPIONs are located inside a nanoscale system, yielding a hybrid nanoparticle. Hybrid nanoparticles have been developed to combine the capabilities of different nanomaterials in a single platform. For instance, SPIONs can be hybridized with liposomes to obtain magnetoliposomes (MLPs) [16]. The encapsulation of any drug in liposomes is usually advantageous since it protects the loaded drug from biological degradation and increases circulation time, resulting in increased bioavailability. A potential approach to cross the BBB is to mimic the active transport mechanisms by which natural body substances (e.g., nutrients) and natural body carriers (e.g., lipoproteins, exosomes) travel from the blood into the brain. This "Trojan horse" approach leverages the presence of specific channels in blood-brain barrier cells. Thus, when non-sterically stabilized, MLPs are taken up by MPS cells, which naturally transmigrate across the BBB. In consequence, MLPs behave as "Trojan nanocarriers" that can be delivered into the brain. However, if one wants to avoid rapid clearance of MLPs from circulation, polyethylene glycol (PEG) can be grafted onto the liposome surface, which reduces protein adsorption and increases the circulation time of MLPs in the organism. Figure 3 depicts a sterically stabilized MLP decorated with antibodies for targeting purposes; in addition, the MLP can contain a hydrophobic and/or hydrophilic drug.

The main advantage of MLPs over other counterparts such as naked liposomes, micelles and polymeric nanoparticles is that this unique superparamagnetic property can be utilized for simultaneous monitoring and quantitation of their tissue-specific or

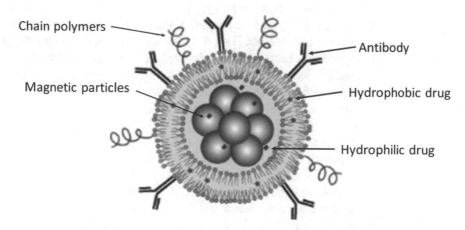

Chain polymers

Magnetic particles

Antibody

Hydrophobic drug

Hydrophilic drug

Fig. 3 Picture showing a hypothetical liposome encapsulating magnetic nanoparticles as well as hydrophilic drugs, embedding hydrophobic drugs into the lipid bilayer. Furthermore, the liposomes is protected from the biological environment by chains of hydrophilic polymers attached to the bilayer, and it is decorated with specific antibodies to increase the cellular targeting

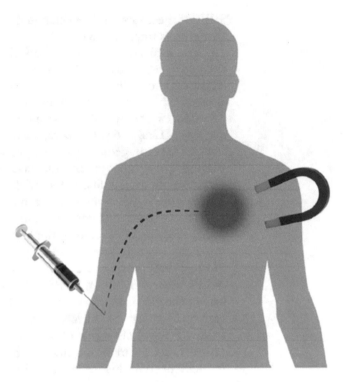

Fig. 4 Magnetic particles are administered by intravenous route, and they are concentrated in the desired zone by the effect of a magnet located on the surface of the body

nonspecific distribution. Thus, techniques such as MRI and magnetometry can be applied to indirectly measure or localize MLP-associated drugs, which may help in determining optimal or suboptimal site-specific dosing [17].

In the second strategy, the force of the external magnet placed near the body surface drags magnetic nanoparticles in one direction toward the magnet (Fig. 4). The strength of this magnetic force is dependent on the distance between the magnet and the particle and decreases with increasing distance. The working distance in humans is around 30–50 cm and a stronger magnet is therefore necessary compared with animals [18]. Hence, for human applications, the magnetic susceptibility of SPIONs must be very high to ensure responsiveness over long distances. According to FDA guidelines, exposure to magnetic field devices of up to 8 T for adults and 4 T for children does not represent a safety concern. From the few studies that have been reported, it can be postulated that the size of the magnetic nanoparticle dictates the extent to which it can affect the BBB, and thus SPIONs are the preferred magnetic particles as they exert a low impact on BBB integrity [19]. It has been speculated that the "pulling" forces of the magnetic field may harm the cells targeted for uptake of the magnetic nanoparticles. Very large, 800 nm magnetic nanoparticles led to BBB leakage after systemic administration of silicone-coated magnetic nanoparticles to rats followed by application of a 0.4 T magnetic field [20]. These large magnetic nanoparticles may have broken down the BBB when being dragged through the barrier, thereby compromising its integrity. In cellular culture, SPIONs with a size of 117 nm added to endothelial cells combined with a magnetic field strength of 0.39 T revealed that this magnetic force did not cause damage to brain endothelial cells [21].

There are several examples of studies based on this strategy. For instance, Kong et al. applied an EMF to drag magnetic polystyrene nanospheres (~100 nm) through the intact BBB of mice [18]. After intravenous injection of MNPs into the mice, no preferential accumulation of these in the brain was observed in absence of a magnetic field. However, when magnets were implanted intracranially in the right hemisphere of mouse cerebral cortex in vivo, the MNPs were enriched in the ipsilateral hemisphere where the magnet was implanted compared with the contralateral hemisphere. These results indicate that the spatial distribution of MNPs can be controlled by application of an EMF. To avoid tissue damage caused by invasive magnet implantation in the brain, an assay was conducted to test whether the distribution of MNPs could be altered by a noninvasive EMF, applied by placing a Nd-Fe-B magnet outside the skull of mice. In comparison with the no-magnet control, application of the external magnetic field increased the passage of MNPs across the BBB into the perivascular space and parenchyma. No histological changes were observed, suggesting that the blood vessels where the particles were found were intact.

A modification of this type of strategy has recently been reported by Sun et al. [12], who injected mice with lysophosphatidic acid (LPA) together with SPIONs and placed a magnet outside the mouse skull on the skin surface. LPA induces rapid disruption of tight junction complexes in brain capillary endothelial cells (BCEC), resulting in increased BBB permeability to both small and large molecules. Moreover, the short circulation half-life of LPA ensures restoration of BBB integrity. This method enhanced accumulation of SPIONs in the brain (approximately fourfold in comparison with the control).

The first and second strategies can be combined when the vectorization achieved with the ligand is reinforced with the effect of an external magnet. Thus, external magnetic force-mediated movement helps prevent nanocarrier's uptake by the MPS and accelerates active targeting.

This combination is the most commonly used strategy for delivering molecules with therapeutic potential to the brain. As an example of use of the combined strategy, Ding et al. encapsulated SPIONs in PEGylated liposomes with Tf attached to their surface [22]. An in vitro transmigration study in the presence or absence of

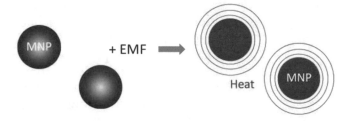

Fig. 5 Magnetic nanoparticles (MNP) can absorb the radiation afforded by a high-frequency alternating magnetic field (AMF) and transform the radiant energy into heat

an external magnetic force revealed that compared with magnetic force or transferrin receptor-mediated transportation alone, their synergy resulted in a 50–100% increase in transmigration without affecting BBB integrity. In another study, Huang et al. [23] explored a strategy based on the use of Tween 80, which as with all surfactants, binds to low-density lipoprotein (LDL) receptors present on the surface of the brain endothelium and facilitates transcytosis [24]. In this study, SPIONs were modified with PEG, poly(ethylene imine) (PEI), and Tween 80 (polysorbate 80) (Tween-SPIONs). The results demonstrated the effective passage of tail vein injected Tween-SPIONs across normal BBB in rats under an EMF, and MNPs are accumulated in the cortex near the magnet. Moreover, the study confirmed that both the Tween 80 modification and the EMF play crucial roles in the effective passage of MNPs across the intact BBB.

As regards the third strategy, the conversion of energy into heat afforded by a high-frequency alternating magnetic field (AMF), after absorption by MNPs, is the basis of one of the most promising applications of MNPs: magnetic hyperthermia (Fig. 5). Some of the main heat-generating mechanisms include hysteresis loss, Néel relaxation, and Brown relaxation. For multidomain materials, the production of heat is due to hysteresis losses, but in single domain particles (i.e., SPIONs), the heat is due to Néel relaxation (the energy disappears when the particle moment relaxes to its equilibrium orientation) and to Brown relaxation (the rotational movement of particles in a fluid dissipates energy by friction). Theoretically, Néel relaxation predominates over Brown relaxation in SPIONs. One promising example of the third strategy is wireless magnetothermal stimulation of the brain in vivo. Thus, Chen et al. developed a method to stimulate regions inside the brain without the need to implant wires or optical fibers [25]. Heat-sensitive capsaicin receptors were introduced into nerve cell of mice and then SPIONs were injected into specific brain regions. After applying an external RF, SPIONs dissipated heat, which activated the ion channel-expressing neurons.

Next, we describe the most notable applications of SPIONs as delivery systems in diseases involving the brain, including neurodegenerative diseases, cancer, and neuro-acquired immunodeficiency syndrome (neuroAIDS).

5   Neurodegenerative Diseases

Neurodegenerative diseases are characterized by neuron loss, accumulation of aggregated and misfolded proteins, cognitive decline, and locomotive dysfunction with Alzheimer's disease (AD), Parkinson's disease (PD), and Huntington's disease (HD) being the most common neurodegenerative diseases in this century.

5.1   Alzheimer's Disease

AD is a devastating neurodegenerative disorder that predominantly affects people over 65 years old. The two main pathophysiological hallmarks of AD are the presence in the brain of extracellular amyloid plaques and intracellular neurofibrillary tangles composed of hyperphosphorylated tau proteins. Although a causal relationship between amyloid deposits and the development of AD has not been conclusively demonstrated, a considerable body of experimental data suggests that amyloid aggregates are important in the etiology of AD [26]. Unfortunately, clinical development of drugs for symptomatic and disease-modifying treatment of AD has brought both promise and disappointment. Consequently, researchers are seeking new ways to treat AD, including inhibition of Aβ assembly, considered the primary therapeutic strategy for this neurodegenerative disease. Thus, passive anti-Aβ immunotherapy has emerged as a promising therapeutic approach for AD, based on the administration of specific anti-Aβ monoclonal antibodies to delay Aβ aggregation in the brain. However, any system for treating AD must possess the capacity to cross the BBB, but drugs have limited access to the brain, hence the development of nanoparticles as drug carriers [27]. A large number of studies have described the effect of nanoparticulate systems on Aβ fibrillation, although the majority of these studies were in vitro tests [28, 29]. Fibril formation occurs by means of nucleation-dependent kinetics where the formation of a critical nucleus is the key determining step, after which fibrillation proceeds rapidly. An exogenous material capable of reducing peptide toxicity may act by two opposite, postulated mechanisms: (1) by decreasing monomer nucleation and hence blocking aggregation, which would result in a reduction in the formation of oligomers, fibrils, and plaques or (2) by disaggregating amyloid plaques or fibrils. To date, it is the former mechanism that has received most research attention [30].

Although some reports in the literature have discussed the use of SPIONs in AD, the vast majority has focused on diagnostic applications; since as indicated previously, MNPs serve as a contrast agent in MRI. Besides the capacity of MNPs to detect the fibrils, some of these studies have reported an influence on Aβ fibrillation. For instance, Skaat et al. immobilized the anti-Aβ monoclonal antibody clone BAM10 using near-infrared fluorescent maghemite nanoparticles in order to inhibit Aβ fibrillation kinetics and specifically detect Aβ fibrils [31]. Although they reported observing significant inhibition of Aβ fibrillation kinetics, such inhibition has also been observed with other nanoparticulate systems. The main advantage was that selective labeling of the Aβ fibrils with these nanoparticles enabled specific detection of Aβ fibrils ex vivo by both MRI and fluorescence imaging. Most interestingly, another report [32] has discussed the effect of MNPs on Aβ fibrillation, and more specifically, the effect of dextran polymer coating charge and thickness (i.e., a single polymer layer or a double polymer layer) on fibrillation kinetics. Depending on the surface coating charge, a dual surface area-dependent effect was observed, with lower concentrations of SPIONs inhibiting fibrillation, whereas higher concentrations enhanced the rate of Aβ fibrillation. Coating charge influenced the concentration at which acceleratory effects were observed, with positive SPIONs promoting fibrillation at significantly lower particle concentrations than either negatively charged or essentially uncharged SPIONs. The authors concluded that SPIONs designed for in vivo medical imaging applications should preferentially have a negatively charged or uncharged surface coating.

Using a mouse model, Do et al. demonstrated that a drug conjugated with a magnetic nanocarrier can cross the normal BBB when subjected to a strong electromagnetic force [33], in this case external magnetic fields of 28 mT (0.43 T/m) and 79.8 mT (1.39 T/m). Moreover, the study assessed the differential effects of pulsed (0.25, 0.5, and 1 Hz) and constant magnetic fields on particle transport across the BBB in mice injected with MNPs via a tail vein and found that the rate of MNP uptake by the brain was significantly enhanced by a pulsed magnetic field. This strategy might be a valuable targeting system for AD diagnosis and therapy.

5.2   Parkinson's Disease

Parkinson's disease is the second most common neurodegenerative disorder, characterized by aggregation of α-synuclein in Lewy bodies and Lewy neurites as well as loss of dopaminergic neurons in the *substantia nigra*. As a result, PD patients exhibit distinctive symptoms including resting tremors, bradykinesia, stooped posture, and in some cases, dementia [34]. The most common treatment for PD is to enhance the level of dopamine in the brain. More recently, stem cell-based therapies have focused on in vivo cell tracking in cell replacement therapies against PD. Of all the current methods for tracking stem cells in vivo, the use of magnetic nanoparticles in MRI is the most promising. An MRI scan also gives a high-resolution 3D image, rendering stem cell tracking relatively straightforward regardless of where the stem cells are located in the body [35]. As with AD, MNPs have mainly been used as a biomarker component [36]; however, in vivo tracking of stem cells labeled with magnetic contrast agents is of particular interest as this makes it possible to monitor the cell's distribution, viability, and physiological response. For instance, human neural precursor cells (hNPCs), which are candidates for cell therapies in PD, have been labeled with MNPs [37]. Grafting MNP-labeled hNPCs allows cell tracking by MRI without impairment of cell survival, proliferation, self-renewal, or multipotency. The MNPs continued to be detected by MRI for a long time after transplantation in a rat model of PD (up to 5 months). However, in long-term grafting, activated microglia and macrophages could contribute to the MRI signal by engulfing dead labeled cells or MNPs freely dispersed in the brain parenchyma over time [38].

One way of reducing or eliminating tremors associated with PD is to stimulate the brain with electric pulses. This treatment has proven effective but has remained a last resort because it requires highly invasive implanted wires connected to a power source outside the brain. Chen et al. [25] have developed a method to stimulate brain tissue using external magnetic fields and injected MNPs – a technique allowing direct stimulation of neurons, which could be an effective treatment for a variety of neurological diseases, and more specifically for PD, without the need for implants or external connections. In this study, 22 nm SPIONs were injected into the brain and rapidly heated by exposure to an EMF. The resulting local increase in temperature induced neural activation by triggering heat-sensitive capsaicin receptors. In addition, viral gene delivery was used to induce sensitivity to heat in selected neurons in the brain.

Although oxygen is imperative for life, metabolic imbalances and excessive generation of reactive oxygen species (ROS) can provoke a range of disorders such as AD, PD, aging, and many other neural disorders. H_2O_2, superoxide ($O_2{}^{-}$), and $\cdot OH$ are the major forms of cellular ROS. Increased levels of oxidative damage to DNA, lipids, and proteins have been detected in a range of assays on postmortem tissues from patients with AD and PD. In relation to oxidative stress, SPIONs possess pH-dependent enzyme-like activities [39]. In a solution with acidic pH, SPIONs show peroxidase-like activity, degrading H_2O_2 to generate the hydroxyl radical ($\cdot OH$), whereas in solution with neutral pH, they display catalase-like activity, decomposing H_2O_2 to H_2O and O_2. Zhang et al. have explored the effects of SPIONs on intracellular ROS levels [40]. In studies on cell models and *Drosophila*, they found that SPIONs possess catalase-like activity, which can reduce intracellular oxidative stress, delay animal aging, and protect against neurodegeneration.

5.3 Huntington's Disease

While both familial and sporadic forms of AD and PD are possible, HD is an autosomal dominant neurodegenerative disease that is caused by mutations in the *huntingtin* gene. The resulting clinical manifestations of HD include chorea and cognitive and behavioral decline. Although some drugs can alleviate symptoms, there are no effective pharmaceutical treatments. As with other neurodegenerative diseases, an alternative treatment is cell-based therapy, which replenishes the lost population of striatal neurons by transplanting neural cells intracranially into the striatum. In general, cell sources that have been reported in HD cell-based therapy include fetal tissue cells, stem cells, and neural progenitor cells. Mesenchymal stem cells (MSCs) are known to play a therapeutic role in brain lesions and can be safely cultured in vitro without the risk of malignant transformation [41]. Here, MNPs were again the tool of choice for tracking stem cells in vivo and subsequent detection by MRI. For the first time, Moreas et al. investigated the therapeutic potential of MNP-labeled MSCs in a rat model of HD [42]. They observed that MSC transplantation significantly decreased the number of degenerating neurons in the damaged striatum. The MSC-labeled MNPs produced an MRI signal even after 60 days posttransplantation, which established the use of MNPs as a valuable potential tool for cell tracking therapy.

6   Use of Magnetic Nanoparticles to Treat Brain Cancer

Brain tumors, a heterogeneous group of primary and metastatic neoplasms in the CNS, are life-threatening diseases characterized by a low survival rate. Malignant gliomas are primary tumors of glial origin. Glioblastoma multiforme (GBM) is the commonest and most aggressive intracranial tumor, and despite over four decades of technological advances in diagnosis and treatment, less than 5% of GBM patients remains alive 5 years after diagnosis [43]. At present, median overall survival after first-line therapy does not exceed 12–15 months, because the invasive nature of this tumor renders a complete resection almost impossible.

Some genetic alterations characterize GBM [44], of which the commonest is overexpression of the epidermal growth factor receptor (EGFR). Consequently, this receptor is a candidate for targeting GBM cells for imaging and therapeutic purposes. A characteristic of wild-type EGFR is that it must be activated: EGFR dimerizes upon binding with its ligands. The majority of GBM tumors (54%) overexpress the wild-type EGFR protein and 31% overexpress both wild-type EGFR and the EGFRvIII mutant [45]. In contrast to other EGFRs, the EGFRvIII mutant form does not require ligand activation and is tumor-specific, since it is not expressed in normal tissues.

Diagnosis of the disease is usually performed by medical examination together with computed tomography and MRI. The treatment of these tumors is based on surgery, radiotherapy, and chemotherapy. The surgical technique is highly invasive and challenging, since the excision of tumor cells may affect normal brain cells. With the development of chemotherapeutic drugs, chemotherapy has been considered effective in inhibiting brain tumor growth. However, the efficacy of chemotherapy is greatly reduced by the existence of the BBB, which prevents drug penetration of the CNS and the blood-brain tumor barrier (BBTB), thus hampering drug accumulation and uptake in the tumor (the high intratumoral interstitial pressure created by the leaky tumor vasculature limits drug penetration from the bloodstream into the tumor). Moreover, because of this poor transport efficacy, a high drug dosage is often necessary in order to reach therapeutic concentrations in the tumor, and this is associated with severe toxicity to normal tissues. Several drug delivery strategies have been developed to overcome these challenges and improve the treatment of brain tumors, one of the most promising of which is the use of nanoparticulate systems [46, 47], although most of these systems lack the capacity to specifically target tumor cells or deliver adequate local concentrations of chemotherapeutic agents. Among the nanoparticulate systems, MNPs appear to be the most promising, since the uptake of SPIONs by malignant brain tumor cells was demonstrated in cellular cultures and in vivo some time ago [48, 49].

In an increasing number of studies in vitro, it has been shown that MNPs can reach the brain cells [18, 21, 22, 50–57]. BCECs in a cellular model (Transwell™, Corning, MA, USA) have usually been employed to assess the passage of drugs and nanoparticles [7].

Here we provide an overview of the therapeutic applications (drug delivery and magnetic hyperthermia) of SPIONs in brain cancer.

6.1 Ligand Targeting of Magnetic Particles to Brain Cells

Penetration into a tumor area can be improved by simple targeting receptors that are normally expressed in BCECs. Insulin, Tf, apolipoprotein E, and α2-macroglobulin are some examples of proteins that reach the brain by receptor-mediated endocytosis. Once nanoparticles have crossed the BBB, they must be able to specifically target glioma cells.

As indicated, the Tf receptor is present in the brain endothelial cell membrane. Cui et al. [58] formed a core-shell-structured composite consisting of a magnetic core and a silica shell. Poly (D, L-lactic-co-glycolic acid) (PLGA) was added to increase biocompatibility and biodegradability, and Tf was conjugated on the surface of the resulting nanoparticles, which were loaded with doxorubicin (DOX) and paclitaxel (PTX). Therapeutic efficacy was evaluated in intracranial U-87 MG-luc2 xenograft of BalB/c nude mice, and strongest inhibition of tumor growth was observed after administering nanoparticles containing both DOX and PTX. Tf, as ligand, has also been used by Yan et al. [59] and Jiang et al. [60].

Another ligand belonging to the same family as Tf is lactoferrin (Lf). Mammalian Lf is a cationic iron-binding glycoprotein (Mw = 80 kDa). The Lf receptor has been demonstrated to exist in the endothelial cells of the BBB and has been shown to be involved in receptor-mediated transcytosis in vitro and in vivo [61]. In a study by Qiao [54], Lf was covalently conjugated onto PEG-coated SPIONs and the efficacy of these conjugates at crossing the BBB was evaluated in an in vitro model as well as in vivo in rats. The study suggested that besides acting as a brain MRI contrast agent, the conjugates could potentially be used as a brain delivery vehicle for molecules of interest for brain diseases. Recently, a new dual-targeting magnetic polydiacetylene nanocarrier (PDNC) delivery system modified with Lf has been developed [62]. The system can control drug release and cross the BBB. When Lf-curcumin(Cur)-PDNCs were administered to rats, the retention time of the encapsulated Cur was improved, and the amount of Cur in the brain was fourfold higher compared with free Cur. Animal studies also confirmed that Lf targeting and controlled release act synergistically to significantly suppress tumors in orthotopic brain tumor-bearing rats.

Managing to cross the BBB by receptor-mediated transcytosis is the first step toward treating gliomas, but the most important challenge is to reach and penetrate the tumor membrane. Generally, GBM vasculature permeability changes at different stages of tumor development. When the cluster of tumor cells reaches a volume of more than 0.2 mm^3, the BBB is damaged (the presence of GBM causes a local disruption), and the BBTB is formed. Once both the BBB and BBTB are impaired, endothelial gaps in tumor microvessels and an enhanced permeability and retention (EPR) effect appear. This results in the accumulation of suitably sized nanoparticles in tumor tissue, even without active-targeting modifications. However, brain tumors have an extremely low permeability compared with peripheral tumors because they have fewer caveolae, transendothelial cell fenestrations, and vesiculo-vacuolar organelles [63].

Given the unique process of GBM with continuously changing vascular characteristics, various targeting strategies aimed at different growth and development stages have been explored to achieve a more effective therapy. These strategies include: (1) BBB targeting for early stage GBM with still intact BBB and no neovascularization; (2) BBTB targeting for GBM with neovascularization and gradual BBB impairment; and (3) passive targeting based on the EPR effect of GBM with impaired BBB and BBTB.

As indicated previously, one of the main approaches to generate targeted SPIONs is to incorporate a myriad of GBM-specific ligands such as peptides, peptidomimetics, proteins, antibodies, aptamers, and small molecules. These moieties target the corresponding receptors that are highly expressed in tumor cells [64]. Moreover, drugs can be attached or encapsulated in such nanoparticles, concentrating the drug in tumor tissue and thereby improving efficacy and reducing the exposure of normal tissues [65].

Hadjipanayis et al. [66] targeted glioma cells with SPIONs conjugated to an antibody that selectively binds to EGFRvIII. Nanoparticles were administered by convection-enhanced delivery (CED), in which one or more infusion catheters are stereotactically placed directly within the brain tumor [67]. After administration, nanoparticles accumulated inside the tumor, inducing tumor cell apoptosis and yielding a significant increase in overall animal survival. However, this approach has limited clinical relevance due to the invasive application of nanoparticles. Furthermore, the specific antibodies only targeted a subpopulation of EGFRvIII-positive cells, but did not affect subsets of tumor cells with wild-type EGFR or other mutant receptor forms. More recently, Shevtsov [68] demonstrated in a C6 glioma model the possibility of targeting EGFR-overexpressing brain tumors via SPIONs conjugated to epidermal growth factor (EGF). The same group studied the targeted effects of SPIONs conjugated to cetuximab, an EGFR inhibitor. They observed a significant increase in survival after treatment of three intracranial rodent GBM models employing human EGFR-expressing GBM xenografts [69].

SPIONs can be also conjugated with peptides that target receptors on the tumor cell surface. In this case, internalization of the conjugate is via receptor-mediated endocytosis. Examples of peptides to target SPIONs to GBM cells are chlorotoxin (CTX) and F3. CTX is a 30-amino acid peptide derived from scorpion venom and specifically binds to matrix metalloproteinase-2, which is overexpressed on the surface of GBM cells and other cancer cells [70]. Sun et al. conjugated CTX to PEG-SPIONs and showed that internalization of the conjugated SPION by 9L glioma tumor cells in a xenograft mouse model was tenfold higher than that of nontargeted SPIONs after 2 h incubation [71]. Kievit et al. developed a similar nanovector for targeted gene delivery in the same animal model [72]. The nanovector consisted of a SPION core coated with a copolymer of chitosan, PEG, and polyethylenimine (PEI). DNA-encoding green fluorescent protein (GFP) was bound to the resulting nanoparticle, and CTX was then attached using a short PEG linker. The same conjugate, but without CTX, was also prepared as a control. Mice bearing C6 xenograft tumors were injected via an intravenous (IV) route with the DNA-bound MNPs, and the accumulation of these at the tumor site was monitored using MRI and analyzed by histology, while GFP gene expression was monitored by fluorescence imaging and confocal fluorescence microscopy. Interestingly, CTX did not affect MNP accumulation at the tumor site, but it specifically enhanced MNP uptake into cancer cells as evidenced by higher gene expression. These results indicate that this targeted gene delivery system could potentially improve treatment outcomes of gene therapy for GBM and other deadly cancers.

F3 is a 31 amino acid peptide that specifically binds to nucleolin overexpressed in proliferating tumor endothelial cells and the associated vasculature [73]. Multifunctional nanoparticles conjugated with F3 peptides have been used to deliver encapsulated MRI contrast enhancers and photosensitizers to malignant brain tumors implanted in rats. When administered intravenously, these F3-coated SPIONs can provide significant MRI contrast enhancement of intracranial rat-implanted tumors, compared with non-coated F3 nanoparticles [74].

Tumor necrosis factor (TNF)-related apoptosis inducing ligand (TRAIL), a transmembrane protein that belongs to the TNF gene superfamily, is a promising candidate for cancer therapy because of its selective apoptotic effect on cancer cells without affecting normal cells. However, TRAIL lacks clinical applicability because of its short half-life, inefficient delivery, and unfavorable pharmacokinetic profile. Moreover, some glioma cells exhibit resistance to the apoptotic effect of TRAIL. However, when TRAIL was conjugated to SPIONs, it showed antitumor activity in glioma cells and glioma stem cells in vitro and in xenografted rats [75].

As a tumor grows, the blood vessels grow with it, and this growth primarily takes place through angiogenesis. Therefore, inhibiting angiogenesis has become a mainstream therapeutic strategy. The $\alpha_V\beta_3$ integrin is a marker of angiogenesis and its expression correlates with tumor grade. Therefore, $\alpha_V\beta_3$ integrin is an ideal target for in vivo tumor imaging since the target is present on the surface of the vessels and can be directly accessed from the blood. The Arg-Gly-Asp (RGD) peptide is the specific ligand for this receptor. Based on the properties of vascular markers, Agemy et al. assembled a multifunctional theranostic nanoparticle that incorporated the following elements: SPIONs (of elongated shape), the Cys-Gly-Lys-Arg-Lys (CGKRK) peptide, an amphipathic peptide ($_D$[KLAKLAK]), and the iRGD peptide (sequence: CRGDKGPDC) [76]. CGKRK is a vascular marker, KLAKLAK disrupts the mitochondrial membrane and initiates apoptotic cell death, and iRGD is a tumor-penetrating peptide that enhances nanoparticle penetration of extravascular tumor tissue. Two GBM models were used: mice bearing either 005 tumors or xenograft tumors generated with human GBM spheres or U87 cells. Interestingly, although SPIONs were introduced in the formulation as MRI contrast agents, the results showed that the global particle was more effective in inducing cell death than the soluble conjugate of CGKRK and KLAKLAK. In general, the nanoparticles eradicated the tumors in the majority of the mice and delayed tumor development in other animals.

6.2 Magnetic Targeting of Magnetic Particles to Brain Cells

The accumulation and retention of drug-loaded magnetic nanoparticles in tumors can be enhanced by using an externally applied magnetic field to attract the nanoparticles to the tumor location [77]. However, application of this strategy to treat brain tumors is complicated by the deep intracranial location of the tumors. External magnets exert most force on MNPs within the body at the point of contact between the body and the magnet, and therefore noninvasively applied fields cannot confer magnetic force-induced selectivity between the tumor lesion and normal brain tissue. When magnetic particles are administered systemically into the blood stream and external magnets are used to attract them to a desired target, the first question is whether the magnetic field is sufficiently strong to counteract the effect of blood flow on particles at the target region. Since directed transport of SPIONs in the bloodstream is dependent on a dynamic equilibrium between the magnetic and hydrodynamic forces acting on the SPIONs [78], a large number of factors (mainly the physical properties of nanoparticles and hydrodynamic parameters of the organism) can influence targeting. To determine the optimal conditions for targeting, Chertok et al. [79] firstly assessed MNP capture in vitro using a simple flow system under theoretically estimated glioma and normal brain flow conditions. Then, the accumulation of nanoparticles via magnetic targeting was evaluated in vivo using 9L-glioma-bearing rats. Their results showed that a higher magnetic field strength offered better magnetic targeting (Fig. 6). Another interesting finding was that the predicted in vitro capture of nanoparticles in the glioma at a pathophysiologically relevant flow rate was consistent with the in vivo results.

As the results reported by Chertok et al. were obtained from a custom-made magnetic setup and intra-arterial administration of SPIONs, Al-Jamal et al. tried to elucidate the parameters necessary for in vivo magnetic tumor targeting [80]. They utilized polymeric magnetic nanocarriers encapsulating increasing amounts of SPIONs and studied the effects of SPION loading and applied magnetic field strength on magnetic tumor targeting in tumor-bearing mice. Under controlled conditions, in vivo magnetic targeting was quantified and found to be directly proportional to SPION loading and magnetic field strength. However, with the highest SPION loading, blood circulation time is reduced, and magnetic targeting is plateaued. In addition, the results obtained from mouse were computed and extrapolated by means of a mathematical model to obtain the expected behavior in humans.

Studies based on this strategy have used different experimental conditions. The animal models employed have been rats or mice in which a brain tumor has been induced by implanting tumor cells. After a period ranging from 7 to 21 days after tumor induction, SPIONs have been injected at 0.25–25 mg Fe/kg of body weight, through the tail vein, carotid artery, or jugular vein. The magnetic

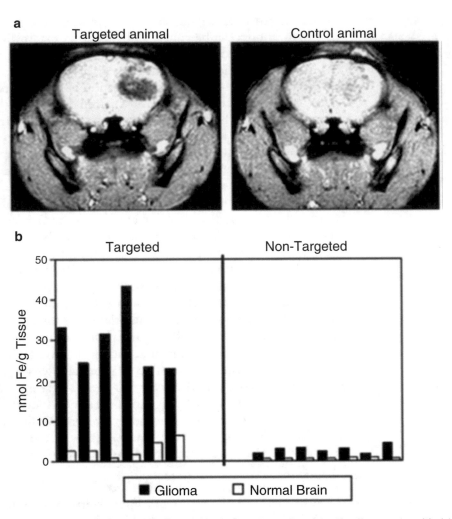

Fig. 6 Comparison between the magnetic nanoparticles accumulated in 9L-glioma rats with (**a**) or without (**b**) the effect of external magnetic targeting (Extracted from ref. 79)

field has ranged from 0.2 to 1.2 T. As is evident, no common pattern has been used in research.

Chertok et al. observed that intravascularly administered nanoparticles were passively delivered to a brain tumor even in the absence of a magnetic field. A magnetic force of 0.4 T increased the concentration of starch-coated SPIONs (which accumulated approximately fivefold more in a rat brain tumor than the concentration found in nontargeted (no magnetic force applied) brain tumors) [81]. Those authors observed the presence of SPIONs injected intravenously into the brain parenchyma of normal rat brain tissue. Hence, a possible application of SPIONs for drug delivery seems valid not only for brain capillary endothelial BCECs, but also for neurons and glial cells located deeper inside the brain, as SPIONs passing through BCECs are likely to be taken up by these cells too. This approach was improved by administering the nanoparticles via a non-occluded carotid artery, since this route increased the passive exposure of tumor vasculature to the nanoparticles [82, 83]. The main potential advantage of intra-arterial (IA) administration over IV is that the vasculature of the tissue perfused by the injected artery receives a higher plasma concentration during the first passage through the circulation. However, IA administration in conjunction with magnetic targeting can also have a serious drawback: the artery exposed to a magnetic field can become mechanically occluded due to nanoparticle aggregation, since the extent of nanoparticle aggregation at arterial flow rates depends on the magnetic field topography and strength [84]. Therefore, the magnetic field was reduced at the carotid injection site to achieve desirable tumor retention of nanoparticles while avoiding undesirable arterial aggregation. Besides IV and IA, MNPs can also be administered via the intracarotid (IC) route. When nanoparticles are coated with cationic polymers such as PEI, they present several drug/gene advantages (i.e., a positive surface charge enhances nanoparticle cellular uptake). However, the cationic surface also exerts a negative effect on nanoparticle pharmacokinetics, as the positively charged nanoparticles are cleared from plasma extremely rapidly. IC administration overcomes this pharmacokinetic drawback. For example, PEI-SPIONs were administered to rats with induced intracerebral 9L tumors [82]. After subjecting them to a magnetic field of 150 mT for 30 min, rats were sacrificed and dissected. An ex vivo analysis of the brains showed a 5.2-fold higher tumor accumulation of nanoparticles in animals receiving PEI-SPIONs than in those to which SPIONs without PEI were administered. Furthermore, the accumulation was found to be tumor-specific and was not accompanied by a corresponding rise in nanoparticle concentration in the normal brain.

As indicated previously, Kong et al. [18] demonstrated that polystyrene nanospheres encapsulating SPIONs can cross the normal BBB when subjected to an EMF. In this study, the nanoparticles containing a fluorophore were injected into mice. To assess the responsiveness of the MNPs to a locally applied EMF, a Nd-Fe-B magnet was implanted in the right hemisphere of the cerebral cortex of the mice. One week postimplantation, MNPs were administered systemically by IV injection. After obtaining brain sections, these were laser imaged using scanning confocal microscopy to track MNP distribution in the brain. An ex vivo examination showed more SPIONs in the hemisphere where the magnet had been implanted. Within this hemisphere, a higher SPION accumulation was observed in the cortex near the magnet, whereas areas farther from the magnet displayed a lower accumulation. Since implanting a magnet in the brain could provoke tissue damage, the effect of applying a noninvasive EMF (~1 T) near the head of the mice was assessed. The number of MNPs accumulated after systemic delivery was increased ~25-fold compared with the no-magnet control.

MLPs have been used to target gliomas in vivo. Thus, Marie et al. [85] subjected these nanoparticles to selective magnetic sorting to target GBM. GBM targeting was assessed in vivo on GMBs orthotopically implanted in mice. A magnetic field gradient of 190 T/m produced by a permanent external 0.4 T magnet placed on the head of the animals effectively concentrated the MLPs in the malignant neoplasm next to the healthy brain. In vivo tracking of the MLPs in the brain was performed by quantitative MRI. Ex vivo electron spin resonance (ESR) spectroscopy was used to calculate the iron oxide delivered to healthy parenchyma and tumor tissue. Histological analysis using confocal fluorescence microscopy confirmed a significant increase in the accumulation of MLPs in the malignant tissue, up to the intracellular level. Electron transmission microscopy revealed effective internalization of the magnetically conveyed MLPs by endothelial and glioblastoma cells as preserved vesicle structures.

PCT has fared poorly in clinical trials against brain glioma. Zhao et al. hypothesized that SPIONs might enhance its bioactivity by selectively delivering this drug into the brain [86] and found that PCT-SPIONs were cytotoxic both in vitro and in glioma-bearing rats. In the in vivo test, the nanoparticles were directed by a magnetic field of 0.5 T. The results showed that drug content increased by 6- to 14-fold in implanted glioma compared to free PCT. Moreover, the survival of glioma-bearing rats was significantly prolonged after magnetic targeting.

Another drug used to treat malignant brain tumors is carmustine (also known as BCNU). Although it improves patient survival, its efficiency is limited by side effects such as myelosuppression, hepatic toxicity, and pulmonary fibrosis. Furthermore, BCNU hydrolyzes with increasing temperature, resulting in a short half-life in the human body [87]. Such drawbacks can be overcome in part by developing a carrier to enhance the concentration of the drug specifically at the tumor site. Hua et al. immobilized this drug on SPIONs by covalent binding [88], which increased the half-life of BCNU from 12 to 30 h and used magnetic field guidance to concentrate the drug-enhanced local delivery of bound-BCNU to target sites. Thus, magnetic delivery of BCNU in vivo made it possible to use a lower concentration of the drug while simultaneously providing more efficient tumor suppression and reducing the likelihood of adverse systemic effects. BCNU has been also immobilized in other types of magnetic nanocarriers that reduce hydrolysis of the drug and significantly improve tumor progression control and prolong animal survival [89]. Recently, a new formulation of this drug in a magnetic nanoparticulate system for administration via the olfactory route has been developed [90].

The integration of magnetic targeting and focused ultrasound (FUS) as a synergistic delivery system could enhance the delivery of therapeutic agents to a desired region. In the presence of micro-bubbles and using a low-energy burst tone, FUS provides transient and reversible disruption of the local BBB without damaging the surrounding neural tissues. Consequently, the EPR effect is increased locally within the brain. The FUS-induced BBB opening is reversible within several hours [91–93]. Subsequent application of an external magnetic field actively enhances localization of a chemotherapeutic agent immobilized on a SPION. Using this approach, several studies have administered anticancer drugs to brain tumors in rats: Liu et al., epirubicin [94], Fan et al., DOX [95], and Chen et al., BCNU [96].

It is known that MPS cells engulf and destroy bacteria, viruses, and foreign substances. When these cells uptake MNPs, they become magnetized and therefore respond to a magnetic field [97]. Thus, diclofenac-encapsulating RGD-coated SPIONs were engulfed by monocytes and/or neutrophils. This capture conferred the magnetic properties of these to the MPS cells, and because of the tendency of such cells to migrate exclusively toward inflammatory sites, the drug could be actively targeted via EMF guidance to any poorly accessible site, e.g., the brain [98].

6.3 Use of Magnetic Nanoparticles to Generate Heat in the Brain

Under the effect of externally applied electromagnetic waves (e.g., RF or microwaves), SPIONs generate heat by Brownian and Néel processes. At high AMF frequencies, they generate sufficient heat to produce temperatures in tissues of between 40 and 45 °C. This form of thermotherapy is known as magnetic hyperthermia (if the temperature of the tissue is above 45 °C, the technique is called magnetic thermoablation). At low frequencies, the heat should open up the BBB junctions. In order for this to become a clinically viable technique, it is first essential to show that the BBB opening is local and entirely reversible.

6.3.1 Magnetic Hyperthermia

Generally, a temperature increase of between 40 and 45 °C generated via AMF can induce tumor cell death by triggering apoptosis, necrosis, and an antitumor and immune response, as tumor cells are less resistant to sudden heat stress compared with healthy surrounding cells. Due to the excellent heat transfer efficiency of SPIONs and uniform temperature distribution, SPION-based magnetic hyperthermia is a promising method for glioma treatment because further therapeutic options are limited after conventional therapies have been applied [63].

From a clinical standpoint, one of the main challenges of magnetic hyperthermia is to achieve therapeutic heating of the tumor with as small a dosage of SPIONs as possible. Minimizing the dosage of SPIONs is desirable in order to minimize possible side effects and facilitate their elimination [99]. This goal can be achieved by using the correct amplitude and frequency of the applied magnetic field and an optimal SPION size. In most studies, the SPIONs employed had diameters of less than 50 nm, and the oscillation frequencies and intensities of AMF ranged from 88.9 to 150 kHz and from 11.0 to 30.6 kA/m, respectively [100]. The empirical threshold of 4.85×10 A/m/s obtained from the product of the frequency, and the amplitude is usually considered to ensure treatment safety. However, in 2010, Bellizzi et al. suggested an optimization criterion for determining the optimal operative conditions in magnetic hyperthermia [99], and more recently, the same authors have carried out a study to numerically validate this criterion in the more challenging and clinically relevant case of brain tumors, by using a realistic 3D model of the human body [101]. The numerical estimates show that acceptable values for the product between the magnetic field amplitude and frequency may be two or four times larger than 4.85×10^9 A/m/s.

Magnetic hyperthermia has been evaluated for feasibility in animal models and in human patients with malignant brain tumors. Dextran- or aminosilane-coated SPIONs have been used for thermotherapy in a rodent GBM model [102]). Aminosilane-coated SPIONs prolonged survival by up to 4.5-fold, whereas dextran-coated particles did not prolong the life span. Only the aminosilane-coated SPIONs formed stable deposits, thus allowing for repeated magnetic field treatments without repeated administration of the particles.

Postmortem studies in humans have been performed with three GBM patients treated with thermotherapy using aminosilane-coated SPIONs [103]. In brain autopsies, SPIONs were found to be dispersed and distributed as aggregates at sites of intracranial injection. Moreover, SPIONs were mainly driven by macrophages and minimally by tumor cells. This indicates the necessity for multiple sites of administration to achieve complete glioma coverage. Interestingly, magnetotherapy allows for an increase in perfusion at the glioma region, leading to much more powerful γ radiation for killing glioma cells.

From March 2003 to January 2005, a phase I trial on 14 patients suffering from GBM was performed in Germany [104]. The patients received an intratumoral injection of aminosilane-coated SPIONs and were subjected to AMF (100 kHz and variable magnetic field strengths of 2.5–18/kA/m). The patients received six treatments (two per week). The median maximal intratumoral temperature was 44.6 °C. Although patient survival was not reported, the trial demonstrated the safety of thermotherapy and the potential to controllably apply hyperthermic temperatures in human patients.

A phase II study on 59 patients, combining the treatment with radiotherapy, began in April 2005 and concluded in September

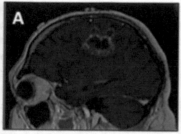
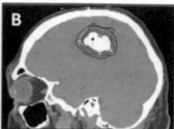
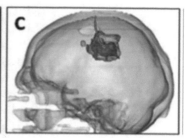

Fig. 7 Glioblastoma recurrence. (**a**) Brain with a glioblastoma visualized by magnetic resonance imaging (MRI); (**b**) post-instillation computer tomography (CT) showing magnetic nanoparticle deposits as hyperdense areas. Isothermal lines indicate calculated treatment temperatures between 40 °C (blue) and 50 °C (red). The brown line represents the tumor area; (**c**) 3-D reconstruction of fused MRI and CT showing the tumor (brown), magnetic fluid (blue), and thermometry catheter (green) (Extracted from ref. 105)

2009 [105]. A high concentration of SPIONs (112 mg/mL) was required to achieve effective thermotherapy, with a median peak temperature within the tumor of 51.2 °C (Fig. 7). This clinical trial successfully demonstrated the safety and efficacy of thermotherapy using SPIONs to treat malignant brain tumors in humans [106]. In March 2014, a new clinical trial was launched in Germany with up to 269 patients with recurrent GBM. The objective of this study was to determine whether this therapy alone or in combination with an optimized radiation therapy was equally or more effective than the optimized radiation therapy alone.

As indicated in a study by Hadjipanayis et al. [66], CED can be used to introduce nanoparticles into the tumor. CED involves continuous delivery of an agent with a specific infusion rate and volume through one or more infusion catheters placed directly within and around brain tumors [67]. This results in a greater volume of distribution compared with diffusion-controlled delivery methods such as general intratumoral injection or drug-loaded wafers [107]. One major drawback of CED is that it is difficult to visualize the distribution of infused agents and tumor targeting efficacy. Owing to the small and uniform size of SPIONs and the fact that they can be imaged directly, MRI-guided CED is an optimal choice for effective glioma treatment [63].

6.3.2 Use of Magnetic Nanoparticles to Open the BBB

Tabatabaei et al. [108, 109] examined BBB permeability in rats in the presence of moderate heat dissipated via magnetic heating of MNPs by a low RF field. To verify BBB integrity before MNP administration via the external carotid artery, fluorescent Evans blue dye was injected into the tail vein. Observation of brain samples using a 7 T magnetic resonance scanner and an epifluorescent microscope indicated a substantial but reversible opening of the BBB where hyperthermia was induced.

Dan et al. [110] used two types of SPIONs, namely, cross-linked nanoassemblies loaded with IONs and conventional

SPIONs, to study the effects of AMF-induced hyperthermia on SPION permeability and flux across the BBB. They used two in vitro models in Transwells™ under normothermic, conventional hyperthermic, and AMF-induced hyperthermic conditions. Their results showed that the flux across BBB models was low under normothermic condition, whereas AMF-induced hyperthermia for 0.5 h enhanced cross-linked SPION cell association/uptake and flux in the absence of cell death. In contrast, SPIONs agglomerated in a cell culture medium and were taken up by, but did not flux through, the BBB model.

AMF-induced hyperthermia enhanced the BBB association/ uptake and permeability of cross-linked SPIONs more effectively than conventional hyperthermia via other mechanisms in addition to the elevated temperature around the IONs. Cross-linked SPIONs activated by an AMF produced quantifiable, controllable hyperthermia in a defined area, as required for clinical applications. In conclusion, AMF-induced hyperthermia is a low toxicity approach that could potentially deliver SPIONs across the BBB for therapeutic and diagnostic CNS applications.

7   Use of Magnetic Nanoparticles to Treat NeuroAIDS

Nervous system alterations due to direct or indirect effects of human immunodeficiency virus (HIV-1) infection, collectively known as neuroAIDS, are frequently associated with AIDS patients [17]. HIV in the periphery has been significantly reduced with antiretroviral therapy (ART) composed of multiple small-molecule therapeutics. Since the introduction of antiretroviral therapy (ART), HIV infection-related morbidity and mortality have dramatically decreased. However, none of the currently available anti-retroviral agents effects a cure, because ART drugs are substrates for BBB-active efflux transporters and present minimal penetration into brain parenchyma. Consequently, the brain remains a sanctuary for HIV in AIDS even with ART [111, 112], and HIV infection has become a chronic disease requiring a lifelong commitment to daily oral treatment. However, most antiretroviral drugs have a short half-life and must be in constant circulation to control virus replication. Moreover, most antiretroviral drugs possess inadequate or zero delivery across the brain barriers. Thus, it is necessary to combine the antiretroviral drug with a carrier that facilitates its entry into the brain. Once in the brain, the HIV drug must be released from the carrier at a constant rate for a longer period to increase therapeutic efficacy without threatening the integrity of the BBB. Recent improvements in synthesis techniques have made it possible to obtain super paramagnetic MNPs as small as 10 nm, which can cross the BBB without affecting its integrity [18, 113].

Several studies have reported monocyte/macrophage-based nanocarrier drug delivery systems for delivering ART to the brain [114, 115], an approach that leverages the capacity of phagocytes to cross the BBB. Thus, an antiretroviral drug can be encapsulated in a nanovehicle (e.g., a non-sterically stabilized liposome), which is captured by the MPS following administration. MNPs can be incorporated in liposomes, and the resulting magnetoliposomes act as a new Trojan horse, penetrating the brain after their capture by the MPS.

In one of the first studies to employ MLPs specifically designed to be taken up by blood phagocytes, Jain et al. [98] covalently coupled RGD-peptide to negatively charged liposomes. Their results showed the presence of magnetic particles inside monocytes and neutrophils in the brain. Later, Saiyed et al. explored the potential of MNPs to deliver antiretroviral drugs in the brain [53, 116]. This group immobilized 3-azido-3-deoxythymidine-5-triphosphate (AZTTP) (a nucleotide analog reverse transcriptase inhibitor) on the surface of MNPs. The inhibition efficiency of the MNP-bound drug, as determined by suppression of HIV replication, remained comparable with the free drug. Then, the MNPs were incorporated into MLPs. To evaluate the potential of AZPPT-MLPs in the treatment of neuroAIDS, migration of the AZTTP-MNPs and AZTTP-MLPs across an in vitro model of BBB was measured under the influence of an EMF. Both AZTTP-magnetic particles and AZTTP-magnetoliposomes yielded an almost three-fold increase in in vitro BBB transmigration when compared with free AZTTP, and did not affect BBB integrity. However, a limitation of this delivery system is the lack of certainty about drug release from the carrier when this reaches the target. To obtain a relatively small physiological change and ensure drug release, the binding force between the carrier and the drug must remain relatively weak. Consequently, more than 99% of the drug/carriers are deposited in the liver, lungs, and other lymphoid organs before they reach the target [117]. To achieve controlled, on-demand release of AZTPP, the same group replaced the MLPs with magnetoelectric nanoparticles (MENP), a kind of nanoparticle in which magnetic and electric fields can be strongly coupled at body temperature. The greatest advantage of MENPs is that there is no heat dissipation when an alternating current field is applied, thus yielding an unprecedentedly high efficacy of drug release without any side effects or toxic effects to target or neighboring cells [117].

The same group [118] explored the use of MNPs to transport HIV-negative factor (Nef) across an in vitro BBB. Nef is an accessory protein that is thought to be integral to HIV-associated immune and neuroimmune pathogenesis.

Wen et al. prepared magnetic lipid nanoparticles formed by PLGA, phospholipids, and PEG-phospholipid. Then, the transactivation transcription peptide (Tat) was conjugated and hesperidin, naringin, and glutathione were encapsulated. The therapeutic efficiency of the drug-loaded Tat-nanoparticles versus drug-loaded nanoparticles was compared in bEnd.3 cells (endothelial cell line). The results showed a dose- and time-dependent higher accumulation in cells of Tat-nanoparticles compared with magnetic lipid nanoparticles and demonstrated that Tat-nanoparticles could serve as an effective drug delivery approach that crosses the BBB [119].

Antiretroviral drugs help in suppressing HIV replication, but they do not eliminate the virus from infected individuals. Thus, there is a need to target viral reservoirs in the body such as the brain, which serves as one of the sources of virus production and migration to the periphery. Some drugs (latency-breaking agents) can be used as HIV-reactivating agents to break latency and thus completely eradicate HIV from the body. The general scientific consensus is that a cure will likely require a combined approach (antiretroviral drug + latency-breaking agent). For this multifunctional purpose, Jayant et al. developed a layer-by-layer assembly [120] of an anti-HIV drug (tenofovir) and a latency-breaking agent (vorinostat) on SPIONs as carrier, whose penetration across the BBB was facilitated by applying an external magnetic force. The in vitro study showed a sustained long-term release (5 days) across the BBB. Moreover, this nanotechnological approach demonstrated the capacity to activate and kill latent viruses [121].

Fiandra et al. demonstrated that the antiretroviral drug enfuvirtide, which is normally unable to penetrate the cerebrospinal fluid, could cross the BBB in mice when conjugated with amphiphilic-coated SPIONs [122].

Epidemiological data demonstrate that opioid abuse is a risk factor for HIV-1 infection and progression to AIDS and other neurodegenerative changes [123]. Therefore, the use of neuroprotective agents to protect neuronal cells against the toxic effect of drugs of abuse is of therapeutic importance. Brain-derived neurotrophic factor (BDNF) is one of the most powerful neuroprotective agents for those neurons that degenerate in HIV-associated dementia (HAD) [124–126], since it has the capacity to regulate neuronal development and survival. However, the major problem with approaches using BDNF is its incapacity to effectively cross the BBB into the CNS [127]. Thus, Pilakka-Kanthikeel et al. bound BDNF with magnetic nanoparticles and reported the efficacy and capacity of such nanoparticles to cross an in vitro cellular model of BBB [128]. As the tight junctions of the BBB cause a much higher transendothelial electrical resistance (TEER) than that of other tissues, one way to confirm the passage of molecules through the BBB is to measure changes in the TEER [7]. This study showed that under the influence of an external magnetic field, a MNP-BDNF formulation could pass through the BBB. MNP-BDNF permeability was 3.5-fold higher than free BDNF

(~73% of total BDNF administered on the upper side of the BBB crossed it). Hence, the association with MNP increases the transmigration of BDNF across the BBB. Moreover, transport of the MNP-BDNF formulation did not affect BBB integrity, as indicated by the TEER reading. In sum, these results indicate that even after binding with MNP, BDNF retains its activity and is as effective as the free form of BDNF.

8 Conclusions and Outlook

This chapter presents a detailed review of the most recent approaches based on the use of magnetic nanoparticles designed to treat and diagnose the most common disorders of the central nervous system: neurodegenerative diseases, brain cancer, and neuroAIDS. The properties of these magnetic materials have been described, placing emphasis on iron-based nanoparticles due to their potential as theranostic agents and their biocompatibility. MNPs have been shown to be attractive nanosystems for controlling the spatial and temporal delivery and co-delivery of drugs, thus achieving a multiple therapeutic or synergistic effect at the specific target, improving therapeutic outcome, and preventing drug degradation, among other properties. However, despite significant advances in therapy and diagnosis derived from tailoring MNPs, several obstacles still remain that hinder their universal use and commercialization. These limitations are primarily related to the impenetrable nature of the BBB. Several strategies have been employed to overcome the difficulty of drug penetration imposed by the complex organization and structure of the brain and its physiological barriers. Most share a common approach consisting in the use of multifunctional particles, mainly by combining the magnetic properties of the particles with specific molecules grafted onto their surface. Thus, magnetic hyperthermia, magnetic targeting, and opening the BBB by means of SPIONs have all yielded encouraging results in the treatment of brain cancer. Moreover, treatment of neuroAIDS using multifunctional systems based on a combination of antiretroviral drugs and latency-breaking agents have also yielded interesting outcomes. For example, the effectiveness of drug-loaded Tat-nanoparticles has been confirmed, as has the increase in release efficacy of encapsulated drugs by means of magnetoelectric nanoparticles. In addition, MNPs play an interesting role in fighting against neurodegenerative diseases due to their capacity to decompose the reactive oxygen species that are abundant in this kind of disorder. Meanwhile, in relation to Alzheimer's disease, conveniently decorated MNPs have been used for the specific detection of Aβ fibrils.

The future success of MNPs requires strategies that increase MNP penetration of the BBB in order to improve therapeutic efficacy. Consequently, research should focus on the design of controllable surfaces with the capacity to guide MNPs toward key sites. In conclusion, the exploration of the therapeutic potential of MPNs for molecular targeting to the central nervous system will become a less utopian challenge in the near future.

References

1. Lu AH, Salabas EL, Schuth F (2007) Magnetic nanoparticles: synthesis, protection, functionalization, and application. Angew Chem Int Ed Engl 46(8):1222–1244

2. Jun YW, Seo JW, Cheon J (2008) Nanoscaling laws of magnetic nanoparticles and their applicabilities in biomedical sciences. Acc Chem Res 41(2):179–189

3. Chen G, Roy I, Yang C, Prasad PN (2016) Nanochemistry and nanomedicine for nanoparticle-based diagnostics and therapy. Chem Rev 116:2826–2885

4. Bogart LK, Pourroy G, Murphy CJ, Puntes V, Pellegrino T, Rosenblum D, Peer D, Levy R (2014) Nanoparticles for imaging, sensing, and therapeutic intervention. ACS Nano 8 (4):3107–3122

5. Arruebo M, Fernández-Pacheco R, Ibarra M, Santamaría J (2007) Magnetic nanoparticles for drug delivery. NanoToday 2(3):22–32

6. Hanini A, Lartigue L, Gavard J, Schmitt A, Kacem K, Wilhelm C, Gazeau F, Chau F, Ammar S (2016) Thermosensitivity profile of malignant glioma U87-MG cells and human endothelial cells following γ-Fe2O3 NP internalization and magnetic field application. RSC Adv 6:15415–15425

7. Busquets MA, Espargaro A, Sabate R, Estelrich J (2015) Magnetic nanoparticles cross the blood-brain barrier: when physics rises a challenge. Nano 5:2231–2248

8. Wui W, Wu Z, Yu T, Jiang C, Kim W-S (2015) Recent progress on magnetic iron oxide nanoparticles: synthesis, surface functional strategies and biomedical applications. Sci Technol Adv Mater 16:023501–023543

9. Gupta AK, Gupta M (2005) Synthesis and surface engineering of iron oxide nanoparticles for biomedical applications. Biomaterials 26(18):3995–4021

10. Garcia-Jimeno S, Escribano E, Queralt J, Estelrich J (2012) External magnetic field-induced selective biodistribution of magnetoliposomes in mice. Nanoscale Res Lett 7 (1):452

11. Estelrich J, Sanchez-Martin MJ, Busquets MA (2015) Nanoparticles in magnetic resonance imaging: from simple to dual contrast agents. Int J Nanomedicine 10:1727–1741

12. Sun Z, Worden M, Thliveris JA, Hombach-Klonisch S, Klonisch T, van Lierop J, Hegmann T, Miller DW (2016) Biodistribution of negatively charged iron oxide nanoparticles (IONPs) in mice and enhanced brain delivery using lysophosphatidic acid (LPA). Nanomedicine 12:1775

13. Lajoie JM, Shusta EV (2015) Targeting receptor-mediated transport for delivery of biologics across the blood-brain barrier. Annu Rev Pharmacol Toxicol 55:613–631

14. Singh D, McMillan JM, Kabanov AV, Sokolsky-Papkov M, Gendelman HE (2014) Bench-to-bedside translation of magnetic nanoparticles. Nanomedicine (Lond) 9 (4):501–516

15. Dilnawaz F, Singh A, Mewar S, Sharma U, Jagannathan NR, Sahoo SK (2012) The transport of non-surfactant based paclitaxel loaded magnetic nanoparticles across the blood brain barrier in a rat model. Biomaterials 33 (10):2936–2951

16. De Cuyper M, Joniau M (1988) Magnetoliposomes. Formation and structural characterization. Eur Biophys J 15(5):311–319

17. Sagar V, Pilakka-Kanthikeel S, Pottathil R, Saxena SK, Nair M (2014) Towards nanomedicines for neuroAIDS. Rev Med Virol 24 (2):103–124

18. Kong SD, Lee J, Ramachandran S, Eliceiri BP, Shubayev VI, Lal R, Jin S (2012) Magnetic targeting of nanoparticles across the intact blood-brain barrier. J Control Release 164 (1):49–57

19. Thomsen LB, Thomsen MS, Moos T (2015) Targeted drug delivery to the brain using magnetic nanoparticles. Ther Deliv 6 (10):1145–1155

20. Stepp P, Thomas F, Lockman PR, Chen H, Rosengart AJ (2009) In vivo interactions of magnetic nanoparticles with the blood-brain barrier. J Magn Magn Mater 321:1591–1593

21. Thomsen LB, Linemann T, Pondman KM, Lichota J, Kim KS, Pieters RJ, Visser GM, Moos T (2013) Uptake and transport of

superparamagnetic iron oxide nanoparticles through human brain capillary endothelial cells. ACS Chem Neurosci 4(10):1352–1360

22. Ding H, Sagar V, Agudelo M, Pilakka-Kanthikeel S, Atluri VS, Raymond A, Samikkannu T, Nair MP (2014) Enhanced blood-brain barrier transmigration using a novel transferrin embedded fluorescent magneto-liposome nanoformulation. Nanotechnology 25(5):055101

23. Huang YF, Zhang B, Xie S, Yang B, Xu Q, Tan J (2016) Superparamagnetic iron oxide nanoparticles modified with tween 80 pass through the intact blood-brain barrier in rats under magnetic field. ACS Appl Mater Interfaces 8:11336–11341

24. Jones AR, Shusta EV (2007) Blood-brain barrier transport of therapeutics via receptor-mediation. Pharm Res 24(9):1759–1771

25. Chen R, Romero G, Christiansen MG, Mohr A, Anikeeva P (2015) Wireless magnetothermal deep brain stimulation. Science 347 (6229):1477–1480

26. Chiti F, Dobson CM (2007) Protein misfolding, functional amyloid, and human disease. Annu Rev Biochem 5:333–366

27. Sahni JK, Doggui S, Ali J, Baboota S, Dao L, Ramassamy C (2011) Neurotherapeutic applications of nanoparticles in Alzheimer's disease. J Control Release 152(2):208–231

28. Busquets MA, Sabate R, Estelrich J (2014) Potential applications of magnetic particles to detect and treat Alzheimer's disease. Nanoscale Res Lett 9(1):538

29. Mirsadeghi S, Shanehsazzadeh S, Atyabi F, Dinarvand R (2016) Effect of PEGylated superparamagnetic iron oxide nanoparticles (SPIONs) under magnetic field on amyloid beta fibrillation process. Mater Sci Eng C Mater Biol Appl 59:390–397

30. Brambilla D, Le Droumaguet B, Nicolas J, Hashemi SH, Wu LP, Moghimi SM, Couvreur P, Andrieux K (2011) Nanotechnologies for Alzheimer's disease: diagnosis, therapy, and safety issues. Nanomedicine 7 (5):521–540

31. Skaat H, Corem-Slakmon E, Grinberg I, Last D, Goez D, Mardor Y, Margel S (2013) Antibody-conjugated, dual-modal, near-infrared fluorescent iron oxide nanoparticles for antiamyloidogenic activity and specific detection of amyloid-beta fibrils. Int J Nanomedicine 8:4063–4076

32. Mahmoudi M, Quinlan-Pluck F, Monopoli MP, Sheibani S, Vali H, Dawson KA, Lynch I (2013) Influence of the physiochemical properties of superparamagnetic iron oxide nanoparticles on amyloid beta protein fibrillation in solution. ACS Chem Neurosci 4 (3):475–485

33. Do TD, Ul Amin F, Noh Y, Kim MO, Yoon J (2016) Guidance of magnetic Nanocontainers for treating Alzheimer's disease using an electromagnetic, targeted drug-delivery actuator. J Biomed Nanotechnol 12(3):569–574

34. Phani S, Loike JD, Przedborski S (2012) Neurodegeneration and inflammation in Parkinson's disease. Parkinsonism Relat Disord 18(Suppl 1):S207–S209

35. Edmundson M, Thanh NT, Song B (2013) Nanoparticles based stem cell tracking in regenerative medicine. Theranostics 3 (8):573–582

36. Yang S-Y, Chiu M-J, Lin C-H, Horng H-E, Yang C-C, Chien J-J, Chen H-H, Liu BH (2016) Development of an ultra-high sensitive immunoassay with plasma biomarker for differentiating Parkinson disease dementia from Parkinson disease using antibody functionalized magnetic nanoparticles. J Nanobiotechnol 14(1):41

37. Ramos-Gomez M, Seiz EG, Martinez-Serrano A (2015) Optimization of the magnetic labeling of human neural stem cells and MRI visualization in the hemiparkinsonian rat brain. J Nanobiotechnol 13:20

38. Ramos-Gomez M, Martinez-Serrano A (2016) Tracking of iron-labeled human neural stem cells by magnetic resonance imaging in cell replacement therapy for Parkinson's disease. Neural Regen Res 11(1):49–52

39. Chen Z, Yin JJ, Zhou YT, Zhang Y, Song L, Song M, Hu S, Gu N (2012) Dual enzyme-like activities of iron oxide nanoparticles and their implication for diminishing cytotoxicity. ACS Nano 6(5):4001–4012

40. Zhang Y, Wang Z, Li X, Wang L, Yin M, Wang L, Chen N, Fan C, Song H (2016) Dietary iron oxide nanoparticles delay aging and ameliorate neurodegeneration in Drosophila. Adv Mater 28(7):1387–1393

41. Chen Y, Carter RL, Cho IK, Chan AW (2014) Cell-based therapies for Huntington's disease. Drug Discov Today 19(7):980–984

42. Moraes L, Vasconcelos-dos-Santos A, Santana FC, Godoy MA, Rosado-de-Castro PH, Jasmin A-PRL, Cintra WM, Gasparetto EL, Santiago MF, Mendez-Otero R (2012) Neuroprotective effects and magnetic resonance imaging of mesenchymal stem cells labeled with SPION in a rat model of Huntington's disease. Stem Cell Res 9 (2):143–155

43. Hadjipanayis CG, Van Meir EG (2009) Brain cancer propagating cells: biology, genetics and targeted therapies. Trends Mol Med 15 (11):519–530

44. Wankhede M, Bouras A, Kaluzova M, Hadjipanayis CG (2012) Magnetic nanoparticles: an emerging technology for malignant brain tumor imaging and therapy. Expert Rev Clin Pharmacol 5(2):173–186

45. Heimberger AB, Hlatky R, Suki D, Yang D, Weinberg J, Gilbert M, Sawaya R, Aldape K (2005) Prognostic effect of epidermal growth factor receptor and EGFRvIII in glioblastoma multiforme patients. Clin Cancer Res 11 (4):1462–1466

46. Painbeni T, Venier-Julienne MC, Benoit JP (1998) Internal morphology of poly(D,L-lactide-co-glycolide) BCNU-loaded microspheres. Influence on drug stability. Eur J Pharm Biopharm 45(1):31–39

47. Kim JO, Kabanov AV, Bronich TK (2009) Polymer micelles with cross-linked polyanion core for delivery of a cationic drug doxorubicin. J Control Release 138(3):197–204

48. Moore A, Marecos E, Bogdanov A Jr, Weissleder R (2000) Tumoral distribution of long-circulating dextran-coated iron oxide nanoparticles in a rodent model. Radiology 214 (2):568–574

49. Zimmer C, Weissleder R, Poss K, Bogdanova A, Wright SC Jr, Enochs WS (1995) MR imaging of phagocytosis in experimental gliomas. Radiology 197(2):533–538

50. Mamot C, Drummond DC, Greiser U, Hong K, Kirpotin DB, Marks JD, Park JW (2003) Epidermal growth factor receptor (EGFR)-targeted immunoliposomes mediate specific and efficient drug delivery to EGFR- and EGFRvIII-overexpressing tumor cells. Cancer Res 63(12):3154–3161

51. Kohler N, Sun C, Fichtenholtz A, Gunn J, Fang C, Zhang M (2006) Methotrexate-immobilized poly(ethylene glycol) magnetic nanoparticles for MR imaging and drug delivery. Small 2(6):785–792

52. Chang J, Jallouli Y, Kroubi M, Yuan XB, Feng W, Kang CS, Pu PY, Betbeder D (2009) Characterization of endocytosis of transferrin-coated PLGA nanoparticles by the blood-brain barrier. Int J Pharm 379 (2):285–292

53. Saiyed ZM, Gandhi NH, Nair MP (2010) Magnetic nanoformulation of azidothymidine 5'-triphosphate for targeted delivery across the blood-brain barrier. Int J Nanomedicine 5:157–166

54. Qiao R, Jia Q, Huwel S, Xia R, Liu T, Gao F, Galla HJ, Gao M (2012) Receptor-mediated delivery of magnetic nanoparticles across the blood-brain barrier. ACS Nano 6 (4):3304–3310

55. Kenzaoui BH, Bernasconi CC, Hofmann H, Juillerat-Jeanneret L (2012) Evaluation of uptake and transport of ultrasmall superparamagnetic iron oxide nanoparticles by human brain-derived endothelial cells. Nanomedicine (Lond) 7(1):39–53

56. Philosof-Mazor L, Dakwar GR, Popov M, Kolusheva S, Shames A, Linder C, Greenberg S, Heldman E, Stepensky D, Jelinek R (2013) Bolaamphiphilic vesicles encapsulating iron oxide nanoparticles: new vehicles for magnetically targeted drug delivery. Int J Pharm 450(1–2):241–249

57. Chen G-J, Su Y-Z, Hsu C, Lo Y-L, Huang S-J, Ke J-H, Kuo Y-C, Wang L-F (2014) Angiopep-pluronic F127-conjugated superparamagnetic iron oxide nanoparticles as nanotheranostic agents for BBB targeting. J Mater Chem B 2:5666–5675

58. Cui Y, Xu Q, Chow PK, Wang D, Wang CH (2013) Transferrin-conjugated magnetic silica PLGA nanoparticles loaded with doxorubicin and paclitaxel for brain glioma treatment. Biomaterials 34(33):8511–8520

59. Yan F, Wang Y, He S, Ku S, Gu W, Ye L (2013) Transferrin-conjugated, fluorescein-loaded magnetic nanoparticles for targeted delivery across the blood-brain barrier. J Mater Sci Mater Med 24:2371–2379

60. Jiang W, Xie H, Ghoorah D, Shang Y, Shi H, Liu F, Yang X, Xu H (2012) Conjugation of functionalized SPIONs with transferrin for targeting and imaging brain glial tumors in rat model. PLoS One 7(5):e37376

61. Hu K, Li J, Shen Y, Lu W, Gao X, Zhang Q, Jiang X (2009) Lactoferrin-conjugated PEG--PLA nanoparticles with improved brain delivery: in vitro and in vivo evaluations. J Control Release 134(1):55–61

62. Fang JH, Chiu TL, Huang WC, Lai YH, Hu SH, Chen YY, Chen SY (2016) Dual-targeting lactoferrin-conjugated polymerized magnetic polydiacetylene-assembled nanocarriers with self-responsive fluorescence/magnetic resonance imaging for in vivo brain tumor therapy. Adv Healthc Mater 5 (6):688–695

63. Liu H, Zhang J, Chen X, Du XS, Zhang JL, Liu G, Zhang WG (2016) Application of iron oxide nanoparticles in glioma imaging and therapy: from bench to bedside. Nanoscale 8 (15):7808–7826

64. Remsen LG, McCormick CI, Roman-Goldstein S, Nilaver G, Weissleder R, Bogdanov A, Hellstrom I, Kroll RA, Neuwelt EA (1996) MR of carcinoma-specific monoclonal antibody conjugated to monocrystalline iron oxide nanoparticles: the potential for noninvasive diagnosis. AJNR Am J Neuroradiol 17(3):411–418

65. Ruoslahti E, Bhatia SN, Sailor MJ (2010) Targeting of drugs and nanoparticles to tumors. J Cell Biol 188(6):759–768

66. Hadjipanayis CG, Machaidze R, Kaluzova M, Wang L, Schuette AJ, Chen H, Wu X, Mao H (2010) EGFRvIII antibody-conjugated iron oxide nanoparticles for magnetic resonance imaging-guided convection-enhanced delivery and targeted therapy of glioblastoma. Cancer Res 70(15):6303–6312

67. Allard E, Passirani C, Benoit JP (2009) Convection-enhanced delivery of nanocarriers for the treatment of brain tumors. Biomaterials 30(12):2302–2318

68. Shevtsov MA, Nikolaev BP, Yakovleva LY, Marchenko YY, Dobrodumov AV, Mikhrina AL, Martynova MG, Bystrova OA, Yakovenko IV, Ischenko AM (2014) Superparamagnetic iron oxide nanoparticles conjugated with epidermal growth factor (SPION-EGF) for targeting brain tumors. Int J Nanomedicine 9:273–287

69. Kaluzova M, Bouras A, Machaidze R, Hadjipanayis CG (2015) Targeted therapy of glioblastoma stem-like cells and tumor non-stem cells using cetuximab-conjugated iron-oxide nanoparticles. Oncotarget 6(11):8788–8806

70. Lyons SA, O'Neal J, Sontheimer H (2002) Chlorotoxin, a scorpion-derived peptide, specifically binds to gliomas and tumors of neuroectodermal origin. Glia 39(2):162–173

71. Sun C, Veiseh O, Gunn J, Fang C, Hansen S, Lee D, Sze R, Ellenbogen RG, Olson J, Zhang M (2008) In vivo MRI detection of gliomas by chlorotoxin-conjugated superparamagnetic nanoprobes. Small 4:372–379

72. Kievit FM, Veiseh O, Fang C, Bhattarai N, Lee D, Ellenbogen RG, Zhang M (2010) Chlorotoxin labeled magnetic nanovectors for targeted gene delivery to glioma. ACS Nano 4(8):4587–4594

73. Christian S, Pilch J, Akerman ME, Porkka K, Laakkonen P, Ruoslahti E (2003) Nucleolin expressed at the cell surface is a marker of endothelial cells in angiogenic blood vessels. J Cell Biol 163(4):871–878

74. Reddy GR, Bhojani MS, McConville P, Moody J, Moffat BA, Hall DE, Kim G, Koo YE, Woolliscroft MJ, Sugai JV, Johnson TD,

Philbert MA, Kopelman R, Rehemtulla A, Ross BD (2006) Vascular targeted nanoparticles for imaging and treatment of brain tumors. Clin Cancer Res 12(22):6677–6686

75. Perlstein B, Finniss SA, Miller C, Okhrimenko H, Kazimirsky G, Cazacu S, Lee HK, Lemke N, Brodie S, Umansky F, Rempel SA, Rosenblum M, Mikkelsen T, Margel S, Brodie C (2013) TRAIL conjugated to nanoparticles exhibits increased anti-tumor activities in glioma cells and glioma stem cells in vitro and in vivo. Neuro-Oncology 15(1):29–40

76. Agemy L, Friedmann-Morvinski D, Kotamraju VR, Roth L, Sugahara KN, Girard OM, Mattrey RF, Verma IM, Ruoslahti E (2011) Targeted nanoparticle enhanced proapoptotic peptide as potential therapy for glioblastoma. Proc Natl Acad Sci U S A 108 (42):17450–17455

77. Shapiro B, Kulkarni S, Nacev A, Muro S, Stepanov PY, Weinberg IN (2014) Open challenges in magnetic drug targeting. WIREs Nanomed Nanobiotechnol 7(3):446–457

78. Cherry EM, Maxim PG, Eaton JK (2010) Particle size, magnetic field, and blood velocity effects on particle retention in magnetic drug targeting. Med Phys 37(1):175–182

79. Chertok B, David AE, Huang Y, Yang VC (2007) Glioma selectivity of magnetically targeted nanoparticles: a role of abnormal tumor hydrodynamics. J Control Release 122 (3):315–323

80. Al-Jamal KT, Bai J, Wang JT, Protti A, Southern P, Bogart L, Heidari H, Li X, Cakebread A, Asker D, Al-Jamal WT, Shah A, Bals S, Sosabowski J, Pankhurst QA (2016) Magnetic drug targeting: preclinical in vivo studies, mathematical modeling, and extrapolation to humans. Nano Lett 16 (9):5652–5660

81. Chertok B, Moffat BA, David AE, Yu F, Bergemann C, Ross BD, Yang VC (2008) Iron oxide nanoparticles as a drug delivery vehicle for MRI monitored magnetic targeting of brain tumors. Biomaterials 29 (4):487–496

82. Chertok B, David AE, Yang VC (2010) Polyethyleneimine-modified iron oxide nanoparticles for brain tumor drug delivery using magnetic targeting and intracarotid administration. Biomaterials 31:6317–6324

83. Chertok B, David AE, Yang VC (2011) Brain tumor targeting of magnetic nanoparticles for potential drug delivery: effect of administration route and magnetic field topography. J Control Release 155(3):393–399

84. Driscoll CF, Morris RM, Senyei AE, Widder KJ, Heller GS (1984) Magnetic targeting of microspheres in blood flow. Microvasc Res 27 (3):353–369

85. Marie H, Lemaire L, Franconi F, Lajnef S, Frapart I-M, Nicolas V, Frébourg G, Trichet M, Ménager C, Lessieur S (2015) Superparamagnetic liposomes for MRI monitoring and external magnetic field-induced selective targeting of malignant brain tumors. Adv Funct Mater 25:1258–1269

86. Zhao M, Liang C, Li A, Chang J, Wang H, Yan R, Zhang J, Tai J (2010) Magnetic paclitaxel nanoparticles inhibit glioma growth and improve the survival of rats bearing glioma xenografts. Anticancer Res 30(6):2217–2223

87. Chae GS, Lee JS, Kim SH, Seo KS, Kim MS, Lee HB, Khang G (2005) Enhancement of the stability of BCNU using self-emulsifying drug delivery systems (SEDDS) and in vitro antitumor activity of self-emulsified BCNU-loaded PLGA wafer. Int J Pharm 301 (1–2):6–14

88. Hua MY, Liu HL, Yang HW, Chen PY, Tsai RY, Huang CY, Tseng IC, Lyu LA, Ma CC, Tang HJ, Yen TC, Wei KC (2011) The effectiveness of a magnetic nanoparticle-based delivery system for BCNU in the treatment of gliomas. Biomaterials 32(2):516–527

89. Yang HW, Hua MY, Liu HL, Huang CY, Tsai RY, Lu YJ, Chen JY, Tang HJ, Hsien HY, Chang YS, Yen TC, Chen PY, Wei KC (2011) Self-protecting core-shell magnetic nanoparticles for targeted, traceable, long half-life delivery of BCNU to gliomas. Biomaterials 32(27):6523–6532

90. Akilo OD, Choonara YE, Strydom AM, du Toit LC, Kumar P, Modi G, Pillay V (2016) AN in vitro evaluation of a carmustine-loaded Nano-co-Plex for potential magnetic-targeted intranasal delivery to the brain. Int J Pharm 500(1–2):196–209

91. Pardridge WM (2002) Drug and gene delivery to the brain: the vascular route. Neuron 36(4):555–558

92. Hynynen K, McDannold N, Vykhodtseva N, Raymond S, Weissleder R, Jolesz FA, Sheikov N (2006) Focal disruption of the blood-brain barrier due to 260-kHz ultrasound bursts: a method for molecular imaging and targeted drug delivery. J Neurosurg 105(3):445–454

93. Muldoon LL, Soussain C, Jahnke K, Johanson C, Siegal T, Smith QR, Hall WA, Hynynen K, Senter PD, Peereboom DM, Neuwelt EA (2007) Chemotherapy delivery issues in central nervous system malignancy: a reality check. J Clin Oncol 25 (16):2295–2305

94. Liu HL, Hua MY, Yang HW, Huang CY, Chu PC, Wu JS, Tseng IC, Wang JJ, Yen TC, Chen PY, Wei KC (2010) Magnetic resonance monitoring of focused ultrasound/magnetic nanoparticle targeting delivery of therapeutic agents to the brain. Proc Natl Acad Sci U S A 107(34):15205–15210

95. Fan Z, Chen D, Deng CX (2013) Improving ultrasound gene transfection efficiency by controlling ultrasound excitation of microbubbles. J Control Release 170(3):401–413

96. Chen PY, Liu HL, Hua MY, Yang HW, Huang CY, Chu PC, Lyu LA, Tseng IC, Feng LY, Tsai HC, Chen SM, Lu YJ, Wang JJ, Yen TC, Ma YH, Wu T, Chen JP, Chuang JI, Shin JW, Hsueh C, Wei KC (2010) Novel magnetic/ultrasound focusing system enhances nanoparticle drug delivery for glioma treatment. Neuro-Oncology 12 (10):1050–1060

97. Ranney DF, Huffaker HH (1987) Magnetic microspheres for the targeted controlled release of drugs and diagnostic agents. Ann N Y Acad Sci 507:104–119

98. Jain S, Mishra V, Singh P, Dubey PK, Saraf DK, Vyas SP (2003) RGD-anchored magnetic liposomes for monocytes/neutrophils-mediated brain targeting. Int J Pharm 261 (1–2):43–55

99. Bellizzi G, Bucci OM, Chirico G (2016) Numerical assessment of a criterion for the optimal choice of the operative conditions in magnetic nanoparticle hyperthermia on a realistic model of the human head. Int J Hyperth 32(6):688–703

100. Silva AC, Oliveira TR, Mamani JB, Malheiros SM, Malavolta L, Pavon LF, Sibov TT, Amaro E Jr, Tannus A, Vidoto EL, Martins MJ, Santos RS, Gamarra LF (2011) Application of hyperthermia induced by superparamagnetic iron oxide nanoparticles in glioma treatment. Int J Nanomedicine 6:591–603

101. Bellizzi G, Bucci OM (2010) On the optimal choice of the exposure conditions and the nanoparticle features in magnetic nanoparticle hyperthermia. Int J Hyperth 26 (4):389–403

102. Jordan A, Scholz R, Maier-Hauff K, van Landeghem FK, Waldoefner N, Teichgraeber U, Pinkernelle J, Bruhn H, Neumann F, Thiesen B, von Deimling A, Felix R (2006) The effect of thermotherapy using magnetic nanoparticles on rat malignant glioma. J Neuro-Oncol 78(1):7–14

103. van Landeghem FK, Maier-Hauff K, Jordan A, Hoffmann KT, Gneveckow U, Scholz R, Thiesen B, Bruck W, von Deimling A (2009) Post-mortem studies in

glioblastoma patients treated with thermo-therapy using magnetic nanoparticles. Biomaterials 30(1):52–57

104. Maier-Hauff K, Rothe R, Scholz R, Gneveckow U, Wust P, Thiesen B, Feussner A, von Deimling A, Waldoefner N, Felix R, Jordan A (2007) Intracranial thermo-therapy using magnetic nanoparticles combined with external beam radiotherapy: results of a feasibility study on patients with glioblastoma multiforme. J Neuro-Oncol 81 (1):53–60

105. Maier-Hauff K, Ulrich F, Nestler D, Niehoff H, Wust P, Thiesen B, Orawa H, Budach V, Jordan A (2011) Efficacy and safety of intratumoral thermotherapy using magnetic iron-oxide nanoparticles combined with external beam radiotherapy on patients with recurrent glioblastoma multiforme. J Neuro-Oncol 103(2):317–324

106. Muller S (2009) Magnetic fluid hyperthermia therapy for malignant brain tumors—an ethical discussion. Nanomedicine 5(4):387–393

107. Juratli TA, Schackert G, Krex D (2013) Current status of local therapy in malignant gliomas—a clinical review of three selected approaches. Pharmacol Ther 139 (3):341–358

108. Tabatabaei SN, Duchemin S, Girouard H, Martel S (2012) Towards MR-navigable nanorobotic carriers for drug delivery into the brain. IEEE Int Conf Robot Autom 2012:727–732

109. Tabatabaei SN, Girouard H, Carret AS, Martel S (2015) Remote control of the permeability of the blood-brain barrier by magnetic heating of nanoparticles: a proof of concept for brain drug delivery. J Control Release 206:49–57

110. Dan M, Bae Y, Pittman TA, Yokel RA (2015) Alternating magnetic field-induced hyperthermia increases iron oxide nanoparticle cell association/uptake and flux in blood-brain barrier models. Pharm Res 32(5):1615–1625

111. Pang S, Koyanagi Y, Miles S, Wiley C, Vinters HV, Chen IS (1990) High levels of unintegrated HIV-1 DNA in brain tissue of AIDS dementia patients. Nature 343(6253):85–89

112. Spencer DC, Price RW (1992) Human immunodeficiency virus and the central nervous system. Annu Rev Microbiol 46:655–693

113. Nair M, Jayant RD, Kaushik A, Sagar V (2016) Getting into the brain: potential of nanotechnology in the management of neuroAIDS. Adv Drug Deliv Rev 103:202–217

114. Dou H, Destache CJ, Morehead JR, Mosley RL, Boska MD, Kingsley J, Gorantla S, Poluektova L, Nelson JA, Chaubal M, Werling J, Kipp J, Rabinow BE, Gendelman HE (2006) Development of a macrophage-based nanoparticle platform for antiretroviral drug delivery. Blood 108(8):2827–2835

115. Dou H, Grotepas CB, McMillan JM, Destache CJ, Chaubal M, Werling J, Kipp J, Rabinow B, Gendelman HE (2009) Macrophage delivery of nanoformulated antiretroviral drug to the brain in a murine model of neuroAIDS. J Immunol 183(1):661–669

116. Saiyed ZM, Gandhi NH, Nair MPN (2009) AZT 5-triphosphate nanoformulation suppresses HIV-1 replication in peripheral blood mononuclear cells. J Neurovirol 15:343–347

117. Nair M, Guduru R, Liang P, Hong J, Sagar V, Khizroev S (2013) Externally controlled on-demand release of anti-HIV drug using magneto-electric nanoparticles as carriers. Nat Commun 4:1707

118. Raymond AD, Diaz P, Chevelon S, Agudelo M, Yndart-Arias A, Ding H, Kaushik A, Jayant RD, Nikkhah-Moshaie R, Roy U, Pilakka-Kanthikeel S, Nair MP (2015) Microglia-derived HIV Nef+ exosome impairment of the blood-brain barrier is treatable by nanomedicine-based delivery of Nef peptides. J Neurovirol 22(2):129–139

119. Wen X, Wang K, Zhao Z, Zhang Y, Sun T, Zhang F, Wu J, Fu Y, Du Y, Zhang L, Sun Y, Liu Y, Ma K, Liu H, Song Y (2014) Brain-targeted delivery of trans-activating transcriptor-conjugated magnetic PLGA/lipid nanoparticles. PLoS One 9(9):e106652

120. Decher G (1997) Fuzzy nanoassemblies: toward layered polymeric multicomposites. Science 277:1232–1237

121. Jayant RD, Atluri VS, Agudelo M, Sagar V, Kaushik A, Nair M (2015) Sustained-release nanoART formulation for the treatment of neuroAIDS. Int J Nanomedicine 10:1077–1093

122. Fiandra L, Colombo M, Mazzucchelli S, Truffi M, Santini B, Allevi R, Nebuloni M, Capetti A, Rizzardini G, Prosperi D, Corsi F (2015) Nanoformulation of antiretroviral drugs enhances their penetration across the blood brain barrier in mice. Nanomedicine 11:1387–1397

123. Guo CJ, Li Y, Tian S, Wang X, Douglas SD, Ho WZ (2002) Morphine enhances HIV infection of human blood mononuclear phagocytes through modulation of beta-chemokines and CCR5 receptor. J Investig Med 50(6):435–442

124. Itoh K, Mehraein P, Weis S (2000) Neuronal damage of the substantia nigra in HIV-1 infected brains. Acta Neuropathol 99:376–384

125. Wang GJ, Chang L, Volkow ND, Telang F, Logan J, Ernst T, Fowler JS (2004) Decreased brain dopaminergic transporters in HIV-associated dementia patients. Brain 127(Pt 11):2452–2458

126. Nosheny RL, Ahmed F, Yakovlev A, Meyer EM, Ren K, Tessarollo L, Mocchetti I (2007) Brain-derived neurotrophic factor prevents the nigrostriatal degeneration induced by human immunodeficiency virus-1 glycoprotein 120 in vivo. Eur J Neurosci 25 (8):2275–2284

127. Pardridge WM (2007) Blood-brain barrier delivery. Drug Discov Today 12(1–2):54–61

128. Pilakka-Kanthikeel S, Atluri VS, Sagar V, Saxena SK, Nair M (2013) Targeted brain derived neurotropic factors (BDNF) delivery across the blood-brain barrier for neuroprotection using magnetic nano carriers: an in-vitro study. PLoS One 8(4):e62241

Chapter 8

Nose-to-Brain Drug Delivery Enabled by Nanocarriers

Zachary Warnken, Yang Lu, Hugh D. C. Smyth, and Robert O. Williams III

Abstract

There continues to be a growing need for treatments of central nervous system diseases such as neurodegenerative disease. A significant obstacle to the treatment of central nervous system diseases is the blood-brain barrier. Over the last several decades, the intranasal route of delivery has been reported to provide a bypass of the blood-brain barrier, delivering drugs directly from the nose to the brain. In addition to the opportunities this route of delivery offers, nanocarrier delivery systems can further improve the targeting of drugs by this pathway, offering advantages over other formulation strategies such as sustained release and enzymatic protection. In this chapter the current status of nanocarriers for nose-to-brain delivery will be reviewed, including the proposed pathways for nanocarrier and drug transport to the brain, nanocarrier attributes which improve brain targeting, the directions intranasal nanocarrier research has focused in terms of disease treatments, as well as delivery device considerations for translation to human delivery. In addition to this review, essential methods for the development of nanocarrier systems for this delivery modality will be presented and discussed.

Key words Nasal delivery, Brain targeting, Olfactory epithelium, Axonal transport, Neurodegeneration, Epilepsy

1 Introduction

The blood-brain barrier (BBB) is a protective barrier, essential for excluding toxins and infectious agents from entering into the brain from the systemic circulation. While the BBB is vitally important for healthy life, it becomes a significant barrier for the treatment of central nervous system (CNS)-based diseases. In fact, around 98% of small molecule drugs and nearly no large molecules are capable of crossing the BBB [1]. In our current aging population, the need for CNS therapeutics continues to grow. For example, Alzheimer's disease is currently the sixth leading cause of death in the United States, with one out of three senior deaths being a result of Alzheimer's disease or other dementia [2]. The need for new CNS therapeutics is evident, making it one of the top disease areas for drug development; however, as of June 2009, only 8.2% of the drugs developed for CNS diseases have successfully been approved

Javier O. Morales and Pieter J. Gaillard (eds.), *Nanomedicines for Brain Drug Delivery*, Neuromethods, vol. 157,
https://doi.org/10.1007/978-1-0716-0838-8_8, © Springer Science+Business Media, LLC, part of Springer Nature 2021

[3]. A portion of this disconnect between development of drugs for treatment of CNS diseases and their approval for clinical use is due to the narrow requirements and suitability of drugs to reach the brain. In some cases, to achieve therapeutic brain levels of a drug, high systemic concentrations may be required, which may induce adverse drug reactions and toxicities. As many CNS diseases require chronic drug therapy, noninvasive means for delivery of drugs to the brain are required. Over the last several decades, a growing amount of evidence supports intranasal delivery provides pathways for administered drugs to reach the brain without having to cross the BBB, allowing noninvasive and chronic delivery of both small and large molecules for treating CNS diseases.

In order to facilitate enhanced drug delivery to the CNS, nanocarrier drug delivery systems have been designed for both small and large molecules to be administered intranasally. Several of the studies discussed below have shown the advantages of using nanocarrier systems over administration of more conventional suspensions and solutions. The adaptability of the carrier material and surface coatings gives nanocarrier systems unique opportunities for delivering drugs to the brain. Based on literature reports, intranasal drug delivery to bypass the BBB is promising; however, the introduction of new delivery systems to this route of administration brings challenges. In addition to the potential toxicity of the materials used in the nanocarriers, nasal mucosa toxicity and cilia toxicity are also of concern for nasal delivery. Discussed in this book chapter is the ability of nanocarriers to increase brain concentrations and some of the techniques used to characterize their effectiveness. Additionally, formulation changes can play a major role in deposition within the nasal cavity, which will alter the predominant pathways for absorption [4]. Methods for accessing nasal cavity deposition will be discussed in this chapter.

2   Proposed Pathways and Mechanisms of Direct Nose-to-Brain Drug Delivery

The pathways and mechanisms for substances to be transported from the nose into the brain are not yet fully understood. Although several pathways and mechanisms may exist for transport, depending on the placement within the nasal cavity and the properties of the substance, particular pathways will predominate. These mechanisms and pathways have been reviewed in detail elsewhere, and the reader is referred to reference citations [5–7]. This chapter provides an overview of the proposed pathways and a discussion of these pathways in relation specifically to nanocarrier drug transport. One of the reasons for characterizing the possible pathways for a particular nanocarrier system is that the pathways can influence the pharmacokinetic parameters and distribution of the substance in the brain and systemic tissues.

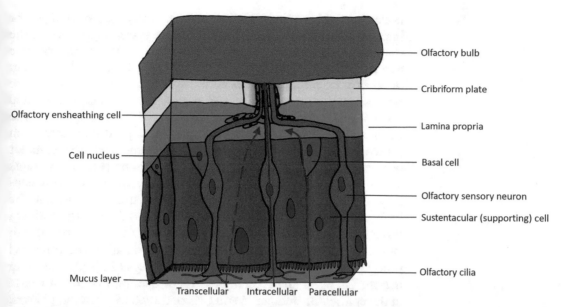

Fig. 1 The olfactory region of the nasal cavity. The olfactory epithelium is depicted including the cell types which play the largest role in direct nose-to-brain drug delivery. The proposed pathways for direct brain transport are illustrated by the red arrows

A majority of the proposed pathways for direct nose-to-brain drug delivery are present in the olfactory membrane due to the anatomical features of the olfactory sensory neurons. The olfactory epithelium is the only location where first-order neurons are in direct contact with the environment (Fig. 1). The olfactory epithelium is a ciliated epithelium; however, unlike in the respiratory region of the nasal cavity, the cilia are nonmotile. In fact, the cilia in the olfactory epithelium are the tips of the dendritic processes of the olfactory sensory neurons, positioned for promoting our sense of smell. The olfactory sensory neurons extend through the olfactory epithelium and converge with the axons of other olfactory neurons to form nerve bundles called fila olfactoria. These bundles are ensheathed by olfactory ensheathing cell within the lamina propria before they travel through the cribriform plate of the skull where they synapse at the olfactory bulb [8, 9]. The architecture of the olfactory epithelium outlines many of the pathways for substances to be directly transported from the nose to the brain.

For drugs to be directly transported to the brain from the nose, the drug must cross the nasal epithelium. This can be accomplished by either intracellular or extracellular mechanisms depending on the properties of the drug or drug carrier system. Intracellular transport can take place at either the neurons (axonal transport) or the supporting cells (transcytosis) in nasal epithelium. Transcytosis and paracellular transport allow access of drugs to the underlying lamina propria, where the drugs may enter the systemic circulation through blood vessels, be transported to the lymph nodes by the lymphatic system, or be directly transported to the brain by olfactory and/or trigeminal nerve systems [5]. One of the proposed pathways for olfactory transport to the brain involves perineural transport of drugs to the olfactory bulb via the bundles of olfactory neurons. Perineural transport has also been shown to be possible in the olfactory and respiratory regions of the nasal cavity by the trigeminal nerve, directing drug transport to the brain stem [10]. Perineural transport has predominately been shown for small molecules and peptides. Some evidence exists for nanocarrier delivery systems to enter the lamina propria by transcytosis across the nasal mucosa, followed by transport extracellularly to the olfactory bulb by the olfactory neuron bundles and the brain stem by the trigeminal nerve. In order to discern the pathway taken for wheat germ agglutinin conjugated PEG-PLA nanoparticles, Liu et al. employed image analysis to visualize the transport process for fluorescently marked nanoparticles [11]. Upon examination, Liu et al. observed high levels of nanoparticle fluorescence in the epithelium and lamina propria of the olfactory region in rats, consistent with transcytosis of the nanoparticles across the olfactory epithelium. Furthermore, fluorescence was seen in olfactory nerve bundles and the associated surrounding connective tissue, with only small amounts colocalized with directing with the nerve bundles, most distributed around them. These results are consistent with transcytosis across the epithelium and extracellular transport around the nerve bundles to the olfactory bulb as the predominating pathway for brain delivery of the nanocarrier.

Direct nose-to-brain transport can be accomplished by intracellular transport through the olfactory neurons to the olfactory bulb. This axonal pathway has been shown transport of proteins, viruses, and even nanoparticles from the nasal epithelium to the olfactory bulb after endocytosis of the substance by the olfactory neurons [12–17]. As the average diameter of olfactory axons is between 100 and 700 nm, it is theoretically possible for any substances smaller than this range to undergo transcellular transport [18, 19]. Axonal transport of nanoparticles has been demonstrated using quantum dots by Hopkins et al. by fluorescence light microscopy and transmission electron microscopy [16]. These microscopy methods allowed Hopkins et al. to visualize the quantum dots within olfactory nerves and sections of the olfactory bulb. In addition to understanding the general pathway for the particles, they were able to discern more about this transport mechanism based on the time to observe brain levels. Axonal transport can be either fast or slow depending on if the components are transported in vesicles or by simple diffusion, respectively. Simple diffusion is slow in axonal transport due to the narrowness of the axons. Hopkins et al. observed quantum dots within the olfactory bulb in less than 3 h, supporting the case that the nanoparticles undergo vesicle-mediated transport within the neurons.

Overall, few studies have explored the specific pathways in which nanocarriers are transported to the brain. Besides Liu et al.'s assessment for wheat germ agglutinin-coated PEG-PLA nanoparticles, there is limited evidence for nanoparticles to travel along the nerve bundles through extracellular mechanisms. Currently, studies observing the pathways of nanoparticle substances are the most translatable to nanocarrier pathways and mechanisms. Several studies have investigated transport of ultrafine particles from the nasal cavity to the brain, such as elemental carbon, manganese oxide, and copper among others [7, 17, 20, 21]. Axonal transportation of nanoparticles is the most commonly reported pathway for solid substances <100 nm. In order to better understand how the physicochemical properties of nanoparticles affect the nose-to-brain transport in mice, Mistry et al. studied chitosan-coated, polysorbate 80-coated, and noncoated polystyrene 100 and 200 nm particles [22]. The uptake of negatively charged polysorbate 80 and uncoated particles into the olfactory epithelium was similar, while the uptake of the positively charged chitosan-coated particles was significantly diminished. Although there were some size differences between the chitosan- and polysorbate 80-coated particles, the largest observed difference was the surface charging. However, no particles were observed in the olfactory bulb after administration with any of the nanoparticle preparations. At the time of the article, 2009, Mistry et al. report only nanoparticles less than 100 nm had shown evidence of axonal transport to the brain and therefore conclude nanoparticles likely need to be less than 100 nm for transport. Upon completion of ex vivo permeation studies, cell uptake of the nanoparticles was observed but could not be detected in the receiver solution [23]. Other nanoparticles tested with in vitro cell monolayers have shown the ability to cross the membrane to the receiver compartment. Gartziandia et al. evaluated the penetration of poly(D,L-lactide-co-glycolide) (PLGA) and nanostructured lipid carrier particles across olfactory cell monolayer and report superior penetration from the lipid carriers. This penetration was further improved with the use of cell-penetrating peptides and chitosan on the surface of the nanocarriers [24]. More recently, nose-to-brain transport has been shown for wheat germ agglutinin PEG-PLA nanoparticles and gelatin nanoparticles with average particle sizes greater than 100 nm [11, 25]. It becomes clear that more research is required to better understand the attributes of nanoparticle systems that allow nose-to-brain transport of the carrier system.

3   Nanocarrier Enhancement of Nose-to-Brain Drug Delivery

Although the nose-to-brain pathways for intranasally administered nanocarrier systems are not fully understood, their ability to increase drug levels in the brain compared to other routes of delivery and formulations has been shown in several examples in the literature. Nanocarriers reported in the literature vary widely in charge, size, and surface characteristics; however, as only a few of these studies report pharmacokinetic information, it is difficult to compare and draw broader conclusions about which characteristics are best for this route of delivery. Several delivery indexes and pharmacokinetic parameters are used to assess the extent and targeting of drug delivery to the brain. The direct targeting efficacy percentage (DTE%; Eq. 1) is a comparative measure between the amount of drug in the brain compared to systemic circulation after nasal delivery compared to systemic delivery. Values above 100% reflect increased targeting to the brain after intranasal administration compared to systemic administration [26]. Another index used for comparing between drug formulations is the direct transport percentage (DTP%; Eq. 2). This quantifies the percentage of drug that accessed the brain utilizing pathways not associated with systemic drug absorption [27]. While DTE% is a standard method of comparing the targeting potential between formulations, it can be greatly influenced by the ability of the systemically administered drug to cross the blood-brain barrier. In addition to these parameters, it is useful to compare the overall extent of drug delivered to the brain by comparing the brain AUC and Cmax values from the formulations. While these values are influenced by the administered dose and properties of the drugs, pharmacologic activity requires sufficiently high concentrations. Limitation of comparing DTE% between formulations and studies includes the inability of the parameter to reflect if the values are meaningful for treatment, and differences in the systemically administered formulation can influence the value. However, as it is a standard parameter used in several nanocarrier studies and demonstrates the ability of the administered formulation to target the brain compared to systemic administration, we will use it below to compare nanocarrier attributes. Of the many nanocarrier-mediated nose-to-brain transport papers in the literature, many report pharmacological evidence of improved delivery by nasal delivery with a nanocarrier, with only a relative few presenting pharmacokinetic evidence which we have used in the following discussion and are presented in Table 1.

$$\text{DTE (\%)} = \frac{[\text{AUC}_{\text{brain}}/\text{AUC}_{\text{blood}}]_{\text{i.n.}}}{[\text{AUC}_{\text{brain}}/\text{AUC}_{\text{blood}}]_{\text{i.v.}}} \times 100\%. \tag{1}$$

$$\text{DTP} = \left(\frac{B_{\text{in}} - B_{\text{x}}}{B_{\text{in}}}\right) \times 100.$$

$$B_{\text{x}} = \frac{B_{\text{iv}}}{P_{\text{iv}}} \times P_{\text{in}}. \tag{2}$$

where B_{in} is the AUC in the brain after intranasal administration, B_{iv} is the AUC in the brain after intravenous administration, and P_{iv} is the plasma AUC after intravenous administration. P_{in} is the plasma

Table 1
Nanocarriers for nose-to-brain delivery presenting pharmacokinetic data

API	Carrier type	Base material	DTE%	DTP (%)	Charge (mv)	Size (nm)	Animal model	Citation
Clonazepam	Polymeric micelle	Polymeric	144	99.3	−25	124	Mouse	[30]
Tarenflurbil	Solid lipid nanoparticle	Lipid	183	45.4	−23	169	Rat	[31]
Diazepam	PLGA nanoparticles	Polymeric	258	61.3	−29	183	Rat	[32]
Tarenflurbil	PLGA nanoparticles	Polymeric	287	65.2	−30	133	Rat	[31]
Rivastigmine	Chitosan nanoparticle	Polymeric	355	71.8	38.4	185	Rat	[33]
Ondansetron HCl	Nanostructured lipid carrier	Lipid	506	97.1	−11.5	92	Rat	[34]
Olanzapine	Polymeric micelle	Polymeric	520	80.8	N/R	59	Rat	[35]
Duloxetine HCl	Nanostructured lipid carrier	Lipid	758	86.8	N/R	130	Rat	[36]
Clonazepam	Organogel	Lipid	758	86.8	N/R	62	Rat	[37]
Rutin	Chitosan nanoparticle	Polymeric	1443	93	31	92	Rat	[38]
Haloperidol	Solid lipid nanoparticle	Lipid	2362	95.8	−20	240	Rat	[39]
Thymoquinone	Chitosan nanoparticle	Polymeric	3318	97.0	30	172	Rat	[40]
Albendazole sulfoxide	Microemulsion	Lipid	10,029	98.2	−10	16	Rat	[41]

DTE% direct targeting efficiency percentage, *DTP* direct transport percentage, *N/R* not reported

AUC after intranasal administration. The defined term B_x is calculated to reflect the amount of drug in the brain that is attributed to systemic absorption after nasal delivery.

3.1 Effect of Nanocarrier Size Brain Targeting

Nanocarrier sizes reported in the literature for nose-to-brain directed formulations range from 15 nm to around 230 nm [28, 29]. Figure 2 depicts the various DTE% values found for nanocarrier formulations of various sizes. A majority of the reported nanocarrier sizes with pharmacokinetic evidence were between 100 and 200 nm. No general trends can be determined between brain targeting and efficiency and size based on the data available currently. The smallest nanocarrier size presented has the highest DTE%; however, it is only a single study making it difficult

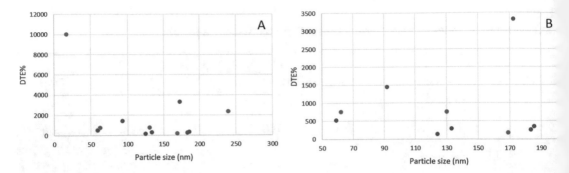

Fig. 2 Effect of nanocarrier particle size on DTE%. Blue circles represent lipid-based nanocarriers, while orange circles depict polymeric-based nanocarriers. A comparison of studies that reported the necessary pharmacokinetic data (**a**). A zoomed-in graph focused on nanocarriers with particle sizes between 50 and 200 nm (**b**)

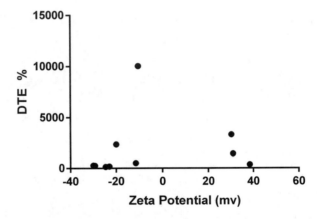

Fig. 3 Comparison of the direct targeting efficiency to the brain from nanocarriers with varying surface charges

to discern if this increased targeting can be attributed to the size of the nanocarrier or some other property of its formulation or administration.

3.2 Effect of Nanocarrier Surface Charge on Brain Targeting

The charge on the nanocarrier can have potential differences in the association with the olfactory epithelium environment as well as transcytosis across the membrane. Figure 3 shows the range of surface charges depicted as zeta potentials reported in the literature for nose-to-brain drug delivery of nanocarriers. As the zeta potential reaches higher magnitude values, either negative or positive, there appears to be some decrease in the targeting efficiency of the nanocarrier systems. The positive zeta potential nanocarrier systems were formulated as chitosan nanoparticles. Unlike the chitosan-coated polystyrene nanoparticles discussed previously, these nanoparticles were formed by similar ionic gelation techniques, which encapsulated the drugs. In several studies, Mistry et al. have shown

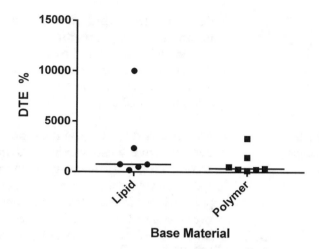

Fig. 4 Effect of nanocarrier base material on the direct targeting efficiency to the brain

the inability of their chitosan-coated nanoparticles to physically cross the nasal epithelium [22, 23]. Whether or not the transport of the drug to the brain from the ionic gelation-produced chitosan nanocarriers arises from intact nanocarrier transport to the brain was not discerned in the studies.

3.3 Effect of Nanocarrier Material Composition on Brain Targeting

While the materials for the nanocarriers range from a wide variety of excipients, the general base materials can be separated into two main categories, lipid and polymer based. With the exception of the high DTE% of 10,000% from a docosahexaenoic acid-containing microemulsion in the lipid-based category, the values obtained were similar for both lipid- and polymeric-based nanocarrier systems (Fig. 4). The choice for nanocarrier system may then rely more heavily on the encapsulation efficiency for the given drug, known toxicity of the nanocarrier excipients, and adaptability of the nanocarrier system.

4 Intranasal Nanocarriers for Diseases of the Central Nervous System

As the previously discussed studies are limited to those which presented pharmacokinetic evidence in order to fairly compare the targeting ability of the nanocarriers in the studies, a current status update of nanocarrier systems separated by disease states is provided to demonstrate the potential for nanocarrier systems for nose-to-brain delivery. Research into the nose-to-brain treatment using nanocarriers has been reported for several diseases, the majority of the studies focusing on neurological disorders and psychiatric diseases. Both categories of diseases are often chronic requiring continued noninvasive therapy. Nanocarriers are versatile drug delivery systems which may enhance drug delivery to the brain after intranasal delivery, providing a method for chronic treatment that can be self-administered by patients . Nanocarriers can be used as a delivery system for both small and large molecules, improving the delivery of peptide- and nucleic acid-based therapeutics to their intended site of action.

4.1 Neuro-degenerative Disorders

Alzheimer's disease is the most prevalent neurodegenerative disorder target in the nanocarrier-mediated nose-to-brain literature and accounts for a vast majority of age-related dementia [42]. Alzheimer's disease is a dementia, which increases in prevalence with older populations. Due to the aging population, the number of patients over the age of 65 with Alzheimer's diseases is projected to increase from 5.2 million Americans to 13.8 million by the year 2050 [2]. The disease is characterized by the accumulation of amyloid β deposits and neurofibrillary tangles in the brain, resulting in cognitive decline. Currently, there are no therapies approved for curing the disease, and current available treatments target symptoms of the disease but are unable to slow the progression of the disease [18, 43].

Drugs that have poor BBB penetration and require chronic dosing for effective treatment are prime candidates for nanocarrier-mediated nose-to-brain drug delivery systems. For example, tarenflurbil is a drug which showed activity in reducing amyloid β in vitro and in vivo and even in a phase II trial [44]. However, upon entering one of the largest Alzheimer's disease clinical trials ever conducted, it failed to slow down the progression of cognitive decline, which was partially attributed to its poor BBB penetration [45]. In order to improve the targeting of tarenflurbil into the brain, Muntimadugu et al. formulated the drug into nanocarrier systems. In order to discern the best nanocarrier system for tarenflurbil for nose-to-brain delivery, both PLGA and solid lipid nanoparticle preparations were formulated and tested in rats. Interestingly, Muntimadugu et al. report higher DTE% from the PLGA nanoparticle compared to the solid lipid nanoparticles, each performing better than a solution formulation. Utilizing nanocarrier systems in combination with the nasal route of delivery may overcome the barriers tarenflurbil faced during its phase III clinical trial and slow the progression of Alzheimer's disease. Another Alzheimer's medication, galantamine hydrobromide is a cholinesterase inhibitor currently approved for treatment. Cholinesterase inhibitors provide symptomatic treatment for Alzheimer's disease but are often accompanied with adverse effects such as nausea and diarrhea due to the increase in acetylcholine in systemic areas. Direct nose-to-brain drug delivery is an attractive option for reducing systemic drug concentrations while improving brain concentrations which can be improved through the use of nanocarriers [46, 47]. Hanafy et al. formulated galantamine hydrobromide encapsulated in chitosan nanoparticles which when labeled with rhodamine and administered intranasally to rats showed distribution within the olfactory bulb, hippocampus, and orbitofrontal and parietal cortices [46]. Furthermore, after delivery to rats, there was significantly lower acetylcholine esterase protein and activity from the intranasal chitosan nanoparticle formulation compared to oral solution and even intranasal solution [48]. Nanocarriers can improve the brain delivery after nasal administration for a range of different molecules with varying properties including peptides. Gao et al. tested the ability of poly (ethylene glycol)-poly (lactic acid) nanoparticles surface decorated with wheat germ agglutinin to deliver vasoactive intestinal peptide to the brain in hopes of improving the neuroprotective effects of the peptide [49]. In order to assess the effectiveness of the nanocarrier system, Gao et al. radiolabeled the peptide to quantify brain distribution in addition to performing a water maze test to assess the neuroprotective potential of the therapy and observing acetylcholinesterase activity after therapy completion. Significantly higher brain levels in the olfactory regions, cerebrum, and cerebellum were reported after dosing with the surface-decorated nanoparticles compared to naked nanoparticles and peptide solution formulations. The increased brain levels translated into a significant pharmacological response as measured by the water maze test. Surface modification of the nanoparticles resulted in increased accumulation at the olfactory epithelium compared to the respiratory epithelium, which assists in greater brain targeting and transport of the modified nanoparticles. Localization of drug to the olfactory epithelium is an important aspect of nose-to-brain drug targeting which is often not addressed in early studies.

Another neurodegenerative disorder that has received the attention of intranasal nanocarrier research is Parkinson's disease. Like Alzheimer's disease, Parkinson's disease is growing in prevalence with the world's aging population [50]. It is caused by the loss of dopaminergic cells within the substantia nigra region of the brain. Treatments for Parkinson's disease are often designed to replace or promote the exogenous dopamine levels in this section of the brain. Substance P is a neuropeptide which has been reported to be regulated via a positive feedback mechanism with dopamine, causing increased dopamine release upon increased substance P levels. Pharmacologic evidence for substance P treatment in Parkinson's disease originated from intracerebroventricular administration studies, which can result in high local concentration of the neuropeptide in the brain. In order to obtain brain levels high enough to show efficacy in a hemiparkinsonian rat model, Lu et al. formulated substance P in phospholipid-based gelatin nanoparticles [51]. The nanocarrier formulation outperformed a substance P solution when both were administered intranasally. Additionally, intranasal administration demonstrated better behavioral improvement compared to intravenous substance P nanocarrier administrations [51, 52]. Research into several nanocarrier types continues for intranasal administration; without standardized methods for assessing the capabilities of these nanocarriers, it may be difficult to select a preferred formulation technique as there are not clear superior nanocarrier systems presented for brain targeting.

4.2 Psychiatric Disorders

Psychiatric disorders such as schizophrenia, bipolar disorder, and depression are chronic central nervous system disorders which require continuous therapy for treatment. These disorders are not immediately life-threatening and as such do not warrant the risks associated with invasive methods of drug delivery [53]. These attributes of psychiatric diseases make them good candidates for targeted brain delivery by the intranasal route. Interestingly, many of the studies for nose-to-brain delivery with nanocarriers for psychiatric disease utilized drugs which are already known to be clinically effective in treating disorders [28, 35, 36, 39, 54]. The versatility of nanocarrier systems can do more than promoting targeting into the brain tissues though and can be used for applications such as sustaining release of the therapeutic agent. For example, Seju et al. produced and assessed the drug delivery of olanzapine formulated in biodegradable nanocarriers using PLGA [54]. Utilizing ex vivo permeation studies, Seju et al. report a similar permeation value and drug release, meaning the main mechanism of drug getting into the brain is likely from drug releasing from the nanocarriers within the nasal cavity and then being transported to the brain. Although brain levels in the first 30 min were higher from the intravenously administered solution, the PLGA nanocarriers resulted in higher levels at 1 h and maintained those levels until the end of the study (3 h). Intranasally applied solution of olanzapine had a similar Tmax compared to the nanocarrier but lower Cmax and was eliminated quickly afterward, which mismatch with the in vitro and ex vivo results. The disconnect between these results may be explained by differences in the clearance of the formulations from the nasal cavities [54]. Mucoadhesive systems with prolonged olanzapine release can improve the drug delivery to the brain when given nasally [55]. Nanocarrier systems can also be used to improve brain delivery for poorly water-soluble drugs. Haloperidol, another antipsychotic medication, is hindered in its bioavailability due to its poor water solubility. Katare et al. explored the ability of polyamidoamine dendrimers to entrap haloperidol, increasing its water solubility and brain targeting. They report comparable behavior responses between haloperidol administered by intraperitoneal injection and haloperidol dendrimers administered intranasally at a dose 6.7 times lower than the former demonstrating the ability of dendrimers to improve the brain delivery of poorly water-soluble drugs when given nasally [28].

4.3 Epilepsy

Epilepsy is a neurological disorder with recurrent episodes associated with electrical abnormalities in the brain, referred to as seizures. Quick resolution of the epileptic episodes is required to prevent permanent damage to the brain caused by the seizure. Intravenous delivery can provide rapid onset of action and subside a seizure; however, this is often difficult to administer during a seizure episode due to convulsions and treatment is often not administered by a trained healthcare professional. The nasal route of delivery can provide rapid onset of action to treat the episode without extensive knowledge or training on how to administer the formulation. In the case of epilepsy, the nanocarrier delivery systems can improve nose-to-brain drug delivery to further minimize the time required to subside a seizure.

Several researchers have evaluated the delivery of benzodiazepines using nanocarriers intranasally [30, 32, 37, 56]. Consistently they have found greater drug levels after intranasal administration of the nanocarrier formulation than with systemic delivery showing the possible application of these delivery systems for epilepsy treatment. In addition to benzodiazepine drug delivery, nanocarriers have allowed improved delivery of neuropeptides for the treatment of seizures. Kubek et al. evaluated the anticonvulsant ability of thyrotropin-releasing hormone when given encapsulated in poly (lactic acid) nanoparticles. A significantly better response was found with administration of the neuropeptide nanoparticles in their animal model than blank nanoparticles. While the neuropeptide alone had an effect on the seizures, Kubek et al. discovered the sustained release properties of the nanocarriers were required for a feasible therapeutic effect [57]. Another example of sustained and increased anticonvulsant activity after intranasal administration of a nanocarrier system is depicted by Sharma et al. [58]. Sharma et al. used a nanolipid carrier delivery system to entrap embelin, a drug reported to have anticonvulsant activity, to provide improved and sustained brain levels after intranasal administration. They report higher brain levels after administration as well as improved attenuation of seizures in animal models after chemical induction. These nanocarrier systems demonstrate the wide variety of platforms which can be incorporated for different drugs, both small and large molecule, to improve the treatments for epilepsy. Utilizing this platform for drug delivery, it is likely that we will continue to see new anticonvulsant agents which when given by another route would not have the same therapeutic benefits as when given nasally.

5 Targeting the Olfactory Region in Humans

A unique aspect of nanocarrier-mediated delivery by the intranasal route is the need for targeting the delivery of the formulation to a specific area of the body. For effective nose-to-brain targeting, the olfactory epithelium is the most important region of the nasal cavity to target [59]. The nasal cavity is designed to disrupt air flows in order to filter out debris and increase the humidity of the air before it reaches the lower airways. This is made possible by the turbinate structures and the nasal valve area within the nasal cavity. These same structures also complicate the delivery pathways for drugs to get into the olfactory region of the nasal cavity.

Many of the current marketed nasal preparations are delivered using meter-dosed pump sprays. These devices allow accurate and reproducible delivery of small volumes (25–200 μL) into the nose in a patient-friendly manner [60]. Although these devices are sufficient for currently marketed products, their deposition patterns are typically isolated to the anterior regions of the nasal cavity [61–65]. In vivo studies with nasal sprays that characterize the amount of spray delivered to the region corresponding with the olfactory epithelium report around 2.5% deposition in this region of the relatively small administered volume [66].

When administered properly, nasal drop delivery systems are capable of covering a larger region of the nasal cavity compared to nasal sprays; however, they are often cleared faster [62]. Charlton et al. compared the distribution as well as the retention times of nasal drops and nasal sprays in human subjects by endoscopic examination [67]. Greater distribution within the olfactory region was observed after administration of nasal drops. The clearance from the olfactory region can be reduced by the use of mucoadhesive agents. Charlton et al. report a residence time of 14 min within the olfactory region after administration of nasal drops with mucoadhesive agents compared to 1.3 min for control solution (no mucoadhesives) [67]. While nasal drops can obtain relatively high distribution at the olfactory epithelium, they are limited by their strong dependence on accurate administration by the patient, requiring complex positioning to obtain proper head positioning [60].

Novel nasal delivery devices have been developed to overcome the limitations found with conventional nasal delivery systems in targeting the olfactory region. Nanocarriers are often formulated as suspended in a liquid medium but may also be lyophilized or dried into powder dosage forms for administrations. The preparation type may have implications on the type of delivery device selected to effectively deliver the product. ViaNase™ is a nasally delivered product used for liquid dosage forms developed using the Kurve Technology®. The device consists of a nebulizer and vortex chamber, causing nebulized drops to leave the device with vortex flow, increasing the deposition area within the nasal cavity. This device has been used in one of the few clinical trials with nose-to-brain drug delivery [68]. One limitation of using nebulized particles with nasal delivery is the need for the patient to control breathing in order to prevent deposition within the lower airways. Optinose®

has developed a device which helps improve deposition to the olfactory region while overcoming this limitation of lung delivery of the formulation. The device comes in two configurations, one for powders and one for liquids. The devices are bi-directional, meaning that the actuation force of the formulation is provided by exhalation of the patient. During exhalation, the soft palate, which connects the nasal cavity with the lower airways closes, making the nasal cavity an enclosed compartment except for the nostrils. Djupesland and Skretting compared the deposition from Optinose's® powder device with a conventional nasal spray in human patients and report 18% of the powder in the upper region of the nasal cavity and only 2.4% of the spray in the same region, showing the improved ability of the Optinose® device to reach the olfactory area [66]. Another device that has been developed for the delivery of powder or liquid dosage form is the POD device by Impel NeuroPharma. This power for drug atomization in this device comes from pressurized instead of the patients' own exhalation force. Hoekman et al. report significantly higher deposition within the upper region of the nasal cavity with the POD device compared to conventional nasal spray, with 45% of the dose being deposited in the upper region with the POD device and only 12% from the nasal spray [69]. To note, the regions defined as the upper region of the nasal cavity were not the same between the two studies, thus making a direct comparison impossible.

6 Approaches for Developing Nanocarriers for Direct Nose-to-Brain Delivery

Provided in this section are what have been reported to be important steps in the development process for nose-to-brain nanocarrier systems (Table 2). The major steps in the preclinical development pathway are firstly characterization of the nanocarrier system often followed by permeation studies, in vivo studies in animal models, and deposition studies in human-based models. The characterization of nanocarrier systems, such as particle size, charge, and drug release, is reported elsewhere in nearly every publication on the development of nanocarrier systems and is not specific for nose-to-brain delivery and therefore will not be discussed in this book chapter.

6.1 In Vitro/Ex Vivo Permeation Studies

Permeation studies of nanocarrier systems allow relatively quick comparison between formulations as well as preliminary toxicity testing to the nasal mucosa. An immediate limitation of permeation studies with regard to nanocarrier systems is the dependence on nose-to-brain transport pathway of the nanocarrier for it to translate to in vivo studies. Permeation studies are useful for formulations that adhere to mucosal surface and release drug, which then permeates across the epithelium for further transport. Additionally,

Table 2
Research approaches needed for the development of novel nanocarrier systems for nose-to-brain drug delivery

Approach	Rationale
Nanocarrier characterization	• Understanding and controlling the attributes of the nanocarrier system is essential to effectively study the system
Permeation studies	• Help elucidate transport pathway • Allow quick comparison between formulations
In vivo studies – Pharmacokinetic studies – Pharmacological studies – Brain transport studies	• Quantitatively assess brain targeting from route and delivery system • Understand timescale for delivery • Effect of drug on treating the disease • Help elucidate transport pathway
In vitro nasal deposition studies	• Confirms the formulation is depositing where it can effectively promote nose-to-brain delivery
Toxicity studies	• Assess irritation to the nasal mucosa • Understand toxicity to nasal cilia • Assess toxicity and accumulation in brain

these studies may translate well for nanocarrier systems that undergo transcytosis across the olfactory epithelium and transport by extracellular methods, discussed previously. The permeation studies will likely not reflect well the formulations that undergo axonal transport. Permeation studies of nanocarriers can be accomplished with either in vitro-cultured olfactory cell monolayers or excised olfactory epithelium from mammals [23, 24, 54, 70]. Gartziandia et al. have established an in vitro cell monolayer model derived from rat olfactory epithelium for the assessment of nanoparticle transport [24]. Briefly, after extraction of olfactory mucosal cells, the cells are isolated from the connective tissue and cultured. For transport studies, cell monolayers are prepared on Transwell® inserts. Gartziandia et al. evaluated the cell number and growth time required to form a monolayer with consistent integrity before performing transport studies. The authors report 500,000 cells per insert showed the highest and most consistent transepithelial electrical resistance (TEER) at 21 days. They concluded 500,000 cells/insert with growth time of 21 days and TEER values above 160 Ω cm^2 is required for the monolayer to be used in transport studies.

The use of olfactory cell monolayers for studying formulations for nose-to-brain transport is a relatively more recent development. Several studies have resorted to ex vivo studies, using excised olfactory epithelium from recently diseased mammals [32, 54, 56,

71, 72]. The translation of studies looking at permeability and transport across the membrane from ex vivo animal epithelium to humans is expected to be very similar, as the olfactory epithelium is structurally similar between all vertebrates with some minor exceptions [73]. In addition to drug permeation across the epithelium, ex vivo studies have been used to measure active transport across the epithelium as well as nanocarrier uptake into cells [23, 74]. Ex vivo studies of nasal diffusion are often completed in Franz diffusion cells. Freshly excised olfactory epithelium is typically immediately placed in an appropriate buffer solution for storage until the permeation study begins. The viability of excised nasal membranes has been established to be at least 4 h from the time of harvesting from the animal, typically the limiting time for measuring the permeation [74]. In order to ensure the integrity of the excised membrane through the permeation experiment, it is important to take TEER measurements before and have the study period, discarding any data that have changes in TEER greater than 20%, as this would show significant changes in permeability due to tissue integrity differences which would likely skew the data. Permeation data can help elucidate the mechanism of drug transport to the brain when compared with drug release data from the nanocarrier. It is often the case that the release of the drug from the nanocarrier reflects the permeation across the olfactory epithelium, resulting in slower diffusion into the acceptor compartment than solution dosage forms and showing the dependence for drug to release from the carrier before transport in the in vitro setting [54, 56, 71]. There is a disagreement between the in vitro and in vivo data for some of these cases which may be explained by either axonal transport in the in vivo situation or the in vitro tests for drug release and permeation are not accurate depictions of the in vivo environment [56]. In these cases, the permeation and drug release are similar; however, the brain levels in vivo are more rapid and to a greater extent than with the comparator solution formulations. Permeation across olfactory epithelium experiments are a quick way to test how changes in formulation, such as surface coatings, can affect the penetration of drug or intact nanocarriers across the membrane for direct brain transport, reducing the number of animals required for in vivo testing.

6.2   In Vivo Brain Accumulation Studies

In order to study the direct targeting capabilities of an intranasally delivered nanocarrier drug delivery system, in vivo studies in animals must be completed. Most often these studies are completed in rat or mice models. Endpoints for measuring brain targeting vary between studies, often either accessing pharmacokinetic differences between formulations and route of administration or measuring the change in a pharmacodynamic response after drug administration. Rats and mice are often the model of choice because the pharmacological tests are already developed and validated in the animals and pharmacokinetic brain levels typically require animals to be sacrificed and their brains removed to quantitate their contents. Although there are advantages to these animal models, they also bring their own limitations. The area of the nasal cavity which comprises olfactory epithelium in mice and rats is around 50%, much greater than the 8–10% reported for human nasal cavities [59]. Because of these anatomical differences, targeting and transport will likely be overestimated from in vivo studies. These studies are still helpful however, proving whether or not the formulation enhances brain targeting when applied to the olfactory epithelium. Ruigrok and de Lange have proposed using pharmacokinetic-pharmacodynamic models to bridge the results from in vivo data to human data which can improve our understanding of brain levels in humans, since they cannot readily be attained in a safe manner [59].

Pharmacokinetic studies in animal models are performed after single-dose administration. Animals are anesthetized and placed in the supine position for nasal administration although some reports explain methods for dosing awake mice [75]. Administration of nanocarriers for animal studies is often performed as a nanosuspension with a micropipette dropwise into each nostril. The volume reported for intranasal administration in mice ranges from 5 to 25 µL [76, 77], while that in rats varies between 10 and 200 µL [39, 54] with a majority of studies administering around 20 µL divided between both nostrils in rats. At each time point, three or more animals are sacrificed and their brains removed and processed for drug quantitation. In order to compare the targeting improvement from the nanocarrier delivery system, it is necessary to test the formulation against a comparator formulation, such as a solution, nasally as well as intravenously. With exception to nanocarriers specifically designed to cross the BBB, they are typically not expected to cross into the brain to an appreciable degree [78–81]. When possible, it is preferred to use solution formulations for systemic delivery as intravenous nanocarrier systems may limit the amount of drug reaching the brain, falsely elevating the perceived targeting after intranasal administration. Once brain and plasma concentration versus time data is obtained from the in vivo studies, Eqs. 1 and 2 can be used to calculate the DTE% and DTP for a more standardized method of comparing between formulations and routes of delivery.

In vivo studies can also be used as a method to assess the toxicity of formulations to the nasal mucosa. Toxicity studies are typically performed over 1–2 weeks. Mice or rats are dosed daily with drug-containing formulation, blank nanocarriers, or control solution for typically at least 1 week. At this point three animals from each group are sacrificed, and nasal mucosa is harvested and fixed and embedded in paraffin. Sections are then cut and stained with hematoxylin/eosin and compared under light microscopy for indicators of irritation such as inflammation, fibrosis, or atypical findings. The rest of the animals from the treatment group are left to recover for 1 week and then analyzed in the same manner to assess the ability of the nasal mucosa to recover from what damage may have occurred due to the formulation.

The pathway or pathways undertaken for the drug to reach the brain after administration in the nanocarrier should be accessed. As mentioned previously few nanocarrier studies report on the mechanism for the drug to reach the brain, whether in intact nanocarriers or not. This can have relevance for several different reasons. If the nanocarrier materials are reaching the brain, then the toxicity and accumulation of the materials in the brain must be studied to understand the risks of the therapy. It is also essential information to guide future scientists in formulating nanocarriers for this route of delivery. Imaging technologies are invaluable for discerning the possible pathways for specific nanocarrier transport. Detailed methods for visualizing the transport of nanocarriers have been reported previously [11, 82]. Detection of nanocarrier transported around or within olfactory neurons can be performed with fluorescence microscopy. After encapsulation of a fluorescent dye, such as 6-coumarin, release testing should be performed in physiologically relevant conditions to ensure that the dye is maintained within the nanocarrier. The formulation is then dosed to animals in the same manner as pharmacokinetic dosing. At the endpoint one side of the nasal cavity with the olfactory bulb is isolated and fixed in 4% paraformaldehyde. Jansson et al. have reported a schedule for cutting the nasal cavity into slices characterizing different regions of the nasal cavity [82]. The regional slices are dehydrated and embedded followed by further slicing into 5 μm slices using a microtome. Further staining can be performed to mark the cells to better visualize the nanocarrier positions to the olfactory neurons. Looking at the slices from the cribriform region, conclusions can be made about the transport mechanism by assessing the amount in the lamina propria, olfactory neurons, and the connective tissues surrounding the neurons.

6.3 Targeting the Olfactory Region

Evidence from in vivo studies supports the ability of the nanocarrier systems to improve drug targeting to the brain after administration to the target region within the nasal cavity. Since rats are often dosed in the supine position under constant anesthesia for in vivo studies, targeting to the olfactory region is controlled by animal positioning. For most therapeutic applications, it is unreasonable to assume a patient can accurately maintain certain head positioning for the entirety of the transport of the drug across the epithelium. The ability for the formulation to reach the olfactory region needs to be assessed before testing in humans. Estimation of formulation deposition within the nasal cavity can be determined using three-dimensional models derived from medical imaging in order to have an accurate representation of the nasal geometry. Particle deposition within the nasal cavity can be determined during inhalation or device actuation computationally or by physical experimentation [64, 65, 83–89]. As 3D printing becomes increasingly available and a rapid and affordable tool for research, the methods in this section will focus on experimentation in determining distribution of a nanocarrier formulation within the nasal cavity. The 3D models derived from medical imaging can be printed and utilized to quantitate the distribution of drug in the region most associated with the olfactory epithelium. A CT scan or MRI file is uploaded to software capable of segmenting the nasal cavity from the remaining image and developing a three-dimensional model by combining the slices, such as Mimics or 3D Slicer. After creation of the 3D nasal cavity model, the file can be either directly printed or modified so it can be disassembled for easy quantitation of the olfactory region compartment. The region of the nasal cavity delineated for assessing relevant nose-to-brain deposition varies between studies making it difficult to compare between devices and studies [66, 69, 89].

7   Conclusion

Nanocarrier drug delivery systems given by the nasal route have shown improved targeting of drug to the brain compared to other formulations and routes of delivery. There is a clear need for therapeutics to treat central nervous system diseases which may be addressed by the improved delivery characteristics of nose-to-brain drug delivery mediated by nanocarriers. While some studies report pharmacokinetic data for drug transport, the evidence supporting many of the studies is pharmacological, making it difficult to compare between studies and drug delivery systems. More research is needed which more completely characterizes the nanocarriers and their delivery, such as the pathways used by the formulation to enhance drug delivery, pharmacokinetics, distribution within the brain tissues, and toxicity of the nanocarriers to determine the preferred nanocarrier system. Often overlooked in nose-to-brain drug delivery is understanding not only the targeting of the formulation to the brain once administered to the olfactory epithelium but also the targeting of the formulation to the olfactory region in humans. In conclusion, nanocarrier drug delivery systems are a promising formulation approach for nose-to-brain delivery; with improved methodology in analyzing drug delivery from these systems, their role in neurotherapeutics will be better established and tested in humans.

References

1. Pardridge WM (2005) The blood-brain barrier: bottleneck in brain drug development. NeuroRx 2(1):3–14

2. Alzheimer's Association (2016) 2016 Alzheimer's disease facts and figures. Alzheimers Dement 12(4):459–509

3. DiMasi JA, Feldman L, Seckler A, Wilson A (2010) Trends in risks associated with new drug development: success rates for investigational drugs. Clin Pharmacol Ther 87 (3):272–277. https://doi.org/10.1038/clpt.2009.295

4. Warnken ZN, Smyth HDC, Watts AB, Weitman S, Kuhn JG, Williams Iii RO (2016) Formulation and device design to increase nose to brain drug delivery. J Drug Deliv Sci Technol 35:213–222. https://doi.org/10.1016/j.jddst.2016.05.003

5. Lockhead J, Thorne RG (2013) Intranasal drug delivery to the brain. In: Hammarlund-Udenaes M, Thorne RG (eds) Drug delivery to the brain: physiological concepts, methodologies and approaches. Springer Science & Business Media, New York, NY

6. Dhuria SV, Hanson LR, Frey WH (2010) Intranasal delivery to the central nervous system: mechanisms and experimental considerations. J Pharm Sci 99:1654–1673. https://doi.org/10.1002/jps.21924

7. Genter MB, Krishan M, Prediger RD (2015) The olfactory system as a route of delivery for agents to the brain and circulation. In: Doty R (ed) Handbook of olfaction and gustation. John Wiley & Sons Inc, Hoboken, NJ, pp 453–484. https://doi.org/10.1002/9781118971758.ch19

8. Dennis JC, Aono S, Vodyanoy VJ, Morrison EE (2015) Development, morphology, and functional anatomy of the olfactory epithelium. In: Doty R (ed) Handbook of olfaction and gustation. John Wiley & Sons Inc, Hoboken, NJ, pp 93–108. https://doi.org/10.1002/9781118971758.ch4

9. Lochhead JJ, Thorne RG (2012) Intranasal delivery of biologics to the central nervous system. Adv Drug Deliv Rev 64:614–628. https://doi.org/10.1016/j.addr.2011.11.002

10. Thorne RG, Pronk GJ, Padmanabhan V, Frey WH II (2004) Delivery of insulin-like growth factor-I to the rat brain and spinal cord along olfactory and trigeminal pathways following intranasal administration. Neuroscience 127:481–496. https://doi.org/10.1016/j.neuroscience.2004.05.029

11. Liu Q, Shen Y, Chen J, Gao X, Feng C, Wang L, Zhang Q, Jiang X (2012) Nose-to-brain transport pathways of wheat germ agglutinin conjugated PEG-PLA nanoparticles. Pharm Res 29(2):546–558. https://doi.org/10.1007/s11095-011-0641-0

12. Kristensson K, Olsson Y (1971) Uptake of exogenous proteins in mouse olfactory cells. Acta Neuropathol 19(2):145–154

13. Buchner K, Seitz-Tutter D, Schönitzer K, Weiss DG (1987) A quantitative study of anterograde and retrograde axonal transport of exogenous proteins in olfactory nerve C-fibers. Neuroscience 22(2):697–707. https://doi.org/10.1016/0306-4522(87)90366-6

14. Balin BJ, Broadwell RD, Salcman M, El-Kalliny M (1986) Avenues for entry of peripherally administered protein to the central nervous system in mouse, rat, and squirrel monkey. J Comp Neurol 251:260–280. https://doi.org/10.1002/cne.902510209

15. Majde JA, Bohnet SG, Ellis GA, Churchill L, Leyva-Grado V, Wu M, Szentirmai E, Rehman A, Krueger JM (2007) Detection of mouse-adapted human influenza virus in the olfactory bulbs of mice within hours after intranasal infection. J Neurovirol 13(5):399–409. https://doi.org/10.1080/13550280701427069

16. Hopkins LE, Patchin ES, Chiu P-L, Brandenberger C, Smiley-Jewell S, Pinkerton KE (2014) Nose-to-brain transport of aerosolised quantum dots following acute exposure. Nanotoxicology 8(8):885–893. https://doi.org/10.3109/17435390.2013.842267

17. Oberdörster G, Sharp Z, Atudorei V, Elder A, Gelein R, Kreyling W, Cox C (2004) Translocation of inhaled ultrafine particles to the brain. Inhal Toxicol 16(6–7):437–445

18. Md S, Mustafa G, Baboota S, Ali J (2015) Nanoneurotherapeutics approach intended for direct nose to brain delivery. Drug Dev Ind Pharm 41(12):1922–1934. https://doi.org/10.3109/03639045.2015.1052081

19. Mistry A, Stolnik S, Illum L (2009) Nanoparticles for direct nose-to-brain delivery of drugs. Int J Pharm 379:146–157. https://doi.org/10.1016/j.ijpharm.2009.06.019

20. Elder A, Gelein R, Silva V, Feikert T, Opanashuk L, Carter J, Potter R, Maynard A, Ito Y, Finkelstein J (2006) Translocation of inhaled ultrafine manganese oxide particles to the central nervous system. Environ Health Perspect 114:1172–1178

21. Liu Y, Gao Y, Zhang L, Wang T, Wang J, Jiao F, Li W, Liu Y, Li Y, Li B (2009) Potential health impact on mice after nasal instillation of nano-sized copper particles and their translocation in mice. J Nanosci Nanotechnol 9 (11):6335–6343

22. Mistry A, Glud SZ, Kjems J, Randel J, Howard KA, Stolnik S, Illum L (2009) Effect of physicochemical properties on intranasal nanoparticle transit into murine olfactory epithelium. J Drug Target 17(7):543–552

23. Mistry A, Stolnik S, Illum L (2015) Nose-to-brain delivery: investigation of the transport of nanoparticles with different surface characteristics and sizes in excised porcine olfactory epithelium. Mol Pharm 12(8):2755–2766. https://doi.org/10.1021/acs. molpharmaceut.5b00088

24. Gartziandia O, Egusquiaguirre SP, Bianco J, Pedraz JL, Igartua M, Hernandez RM, Préat V, Beloqui A (2016) Nanoparticle transport across in vitro olfactory cell monolayers. Int J Pharm 499(1–2):81–89. https://doi. org/10.1016/j.ijpharm.2015.12.046

25. Kim I-D, Sawicki E, Lee H-K, Lee E-H, Park HJ, Han P-L, Kim K, Choi H, Lee J-K (2016) Robust neuroprotective effects of intranasally delivered iNOS siRNA encapsulated in gelatin nanoparticles in the postischemic brain. Nanomedicine 12(5):1219–1229. https://doi.org/ 10.1016/j.nano.2016.01.002

26. Sherry Chow H-H, Chen Z, Matsuura GT (1999) Direct transport of cocaine from the nasal cavity to the brain following intranasal cocaine administration in rats. J Pharm Sci 88:754–758. https://doi.org/10.1021/ js9900295

27. Zhang Q, Jiang X, Jiang W, Lu W, Su L, Shi Z (2004) Preparation of nimodipine-loaded microemulsion for intranasal delivery and evaluation on the targeting efficiency to the brain. Int J Pharm 275:85–96. https://doi.org/10. 1016/j.ijpharm.2004.01.039

28. Katare YK, Daya RP, Sookram Gray C, Luckham RE, Bhandari J, Chauhan AS, Mishra RK (2015) Brain targeting of a water insoluble antipsychotic drug haloperidol via the intranasal route using PAMAM Dendrimer. Mol Pharm 12:3380–3388. https://doi.org/10. 1021/acs.molpharmaceut.5b00402

29. Elnaggar YSR, Etman SM, Abdelmonsif DA, Abdallah OY (2015) Intranasal piperine-loaded chitosan nanoparticles as brain-targeted therapy in Alzheimer's disease: optimization, biological efficacy, and potential toxicity. J Pharm Sci 104(10):3544–3556. https://doi. org/10.1002/jps.24557

30. Nour SA, Abdelmalak NS, Naguib MJ, Rashed HM, Ibrahim AB (2016) Intranasal brain-targeted clonazepam polymeric micelles for immediate control of status epilepticus: in vitro optimization, ex vivo determination of cytotoxicity, in vivo biodistribution and pharmacodynamics studies. Drug Delivery 23 (9):3681–3695. https://doi.org/10.1080/ 10717544.2016.1223216

31. Muntimadugu E, Dhommati R, Jain A, Challa VGS, Shaheen M, Khan W (2016) Intranasal delivery of nanoparticle encapsulated tarenflurbil: a potential brain targeting strategy for Alzheimer's disease. Eur J Pharm Sci 92:224–234. https://doi.org/10.1016/j.ejps.2016.05.012

32. Sharma D, Sharma RK, Sharma N, Gabrani R, Sharma SK, Ali J, Dang S (2015) Nose-to-brain delivery of PLGA-diazepam nanoparticles. AAPS PharmSciTech 16(5):1108–1121. https://doi.org/10.1208/s12249-015-0294-0

33. Fazil M, Md S, Haque S, Kumar M, Baboota S, Jk S, Ali J (2012) Development and evaluation of rivastigmine loaded chitosan nanoparticles for brain targeting. Eur J Pharm Sci 47:6–15. https://doi.org/10.1016/j.ejps.2012.04.013

34. Devkar TB, Tekade AR, Khandelwal KR (2014) Surface engineered nanostructured lipid carriers for efficient nose to brain delivery of ondansetron HCl using *Delonix regia* gum as a natural mucoadhesive polymer. Colloids Surf B Biointerfaces 122:143–150. https:// doi.org/10.1016/j.colsurfb.2014.06.037

35. Abdelbary GA, Tadros MI (2013) Brain targeting of olanzapine via intranasal delivery of core–shell difunctional block copolymer mixed nanomicellar carriers: in vitro characterization, ex vivo estimation of nasal toxicity and in vivo biodistribution studies. Int J Pharm 452:300–310. https://doi.org/10.1016/j. ijpharm.2013.04.084

36. Alam MI, Baboota S, Ahuja A, Ali M, Ali J, Sahni JK, Bhatnagar A (2014) Pharmacoscintigraphic evaluation of potential of lipid nano-carriers for nose-to-brain delivery of antidepressant drug. Int J Pharm 470:99–106. https://doi.org/10.1016/j. ijpharm.2014.05.004

37. Abdel-Bar HM, Abdel-Reheem AY, Awad GA, Mortada ND (2013) Evaluation of brain targeting and mucosal integrity of nasally administrated nanostructured carriers of a CNS active drug, clonazepam. J Pharm Pharm Sci 16 (3):456–469

38. Ahmad N, Ahmad R, Naqvi AA, Alam MA, Ashafaq M, Samim M, Iqbal Z, Ahmad FJ (2016) Rutin-encapsulated chitosan

nanoparticles targeted to the brain in the treatment of cerebral ischemia. Int J Biol Macromol 91:640–655. https://doi.org/10.1016/j.ijbiomac.2016.06.001

39. Yasir M, Sara U, Som I (2016) Haloperidol loaded solid lipid nanoparticles for nose to brain delivery: stability and in vivo studies. J Nanomed Nanotechnol S7:1–9

40. Alam S, Khan ZI, Mustafa G, Kumar M, Islam F, Bhatnagar A, Ahmad FJ (2012) Development and evaluation of thymoquinone-encapsulated chitosan nanoparticles for nose-to-brain targeting: a pharmacoscintigraphic study. Int J Nanomedicine 7:5705–5718. https://doi.org/10.2147/IJN.S35329

41. Shinde RL, Bharkad GP, Devarajan PV (2015) Intranasal microemulsion for targeted nose to brain delivery in neurocysticercosis: role of docosahexaenoic acid. Eur J Pharm Biopharm 96:363–379. https://doi.org/10.1016/j.ejpb.2015.08.008

42. Bertram L, Tanzi RE (2005) The genetic epidemiology of neurodegenerative disease. J Clin Investig 115(6):1449–1457. https://doi.org/10.1172/JCI24761

43. Di Stefano A, Iannitelli A, Laserra S, Sozio P (2011) Drug delivery strategies for Alzheimer's disease treatment. Expert Opin Drug Deliv 8(5):581–603

44. Wilcock GK, Black SE, Hendrix SB, Zavitz KH, Swabb EA, Laughlin MA, Investigators TPIS (2008) Efficacy and safety of tarenflurbil in mild to moderate Alzheimer's disease: a randomised phase II trial. Lancet Neurol 7 (6):483–493

45. Green RC, Schneider LS, Amato DA, Beelen AP, Wilcock G, Swabb EA, Zavitz KH, Group TPS (2009) Effect of tarenflurbil on cognitive decline and activities of daily living in patients with mild Alzheimer disease: a randomized controlled trial. JAMA 302(23):2557–2564

46. Hanafy AS, Farid RM, ElGamal SS (2015) Complexation as an approach to entrap cationic drugs into cationic nanoparticles administered intranasally for Alzheimer's disease management: preparation and detection in rat brain. Drug Dev Ind Pharm 41 (12):2055–2068. https://doi.org/10.3109/03639045.2015.1062897

47. Bhavna MS, Ali M, Ali R, Bhatnagar A, Baboota S, Ali J (2014) Donepezil nanosuspension intended for nose to brain targeting: in vitro and in vivo safety evaluation. Int J Biol Macromol 67:418–425. https://doi.org/10.1016/j.ijbiomac.2014.03.022

48. Hanafy AS, Farid RM, Helmy MW, ElGamal SS (2016) Pharmacological, toxicological and

neuronal localization assessment of galantamine/chitosan complex nanoparticles in rats: future potential contribution in Alzheimer's disease management. Drug Deliv 23:1–12. https://doi.org/10.3109/10717544.2016.1153748

49. Gao X, Wu B, Zhang Q, Chen J, Zhu J, Zhang W, Rong Z, Chen H, Jiang X (2007) Brain delivery of vasoactive intestinal peptide enhanced with the nanoparticles conjugated with wheat germ agglutinin following intranasal administration. J Control Release 121:156–167. https://doi.org/10.1016/j.jconrel.2007.05.026

50. Dorsey ER, Constantinescu R, Thompson JP, Biglan KM, Holloway RG, Kieburtz K, Marshall FJ, Ravina BM, Schifitto G, Siderowf A, Tanner CM (2007) Projected number of people with Parkinson disease in the most populous nations, 2005 through 2030. Neurology 68(5):384–386. https://doi.org/10.1212/01.wnl.0000247740.47667.03

51. Lu C-T, Jin R-R, Jiang Y-N, Lin Q, Yu W-Z, Mao K-L, Tian F-R, Zhao Y-P, Zhao Y-Z (2015) Gelatin nanoparticle-mediated intranasal delivery of substance P protects against 6-hydroxydopamine-induced apoptosis: an in vitro and in vivo study. Drug Des Devel Ther 9:1955–1962. https://doi.org/10.2147/DDDT.S77237

52. Zhao Y-Z, Jin R-R, Yang W, Xiang Q, Yu W-Z, Lin Q, Tian F-R, Mao K-L, Lv C-Z, Wáng Y-XJ LC-T (2016) Using gelatin nanoparticle mediated intranasal delivery of neuropeptide substance P to enhance neuro-recovery in hemiparkinsonian rats. PLoS One 11(2): e0148848. https://doi.org/10.1371/journal.pone.0148848

53. Singh D, Rashid M, Hallan SS, Mehra NK, Prakash A, Mishra N (2016) Pharmacological evaluation of nasal delivery of selegiline hydrochloride-loaded thiolated chitosan nanoparticles for the treatment of depression. Artif Cells Nanomed Biotechnol 44(3):865–877. https://doi.org/10.3109/21691401.2014.998824

54. Seju U, Kumar A, Sawant KK (2011) Development and evaluation of olanzapine-loaded PLGA nanoparticles for nose-to-brain delivery: in vitro and in vivo studies. Acta Biomater 7:4169–4176. https://doi.org/10.1016/j.actbio.2011.07.025

55. Fonseca FN, Betti AH, Carvalho FC, Gremiao MPD, Dimer FA, Guterres SS, Tebaldi ML, Rates SMK, Pohlmann AR (2015) Mucoadhesive amphiphilic methacrylic copolymer-functionalized poly(epsilon-caprolactone) nanocapsules for nose-to-brain delivery of

olanzapine. J Biomed Nanotechnol 11:1472–1481. https://doi.org/10.1166/jbn.2015.2078

56. Sharma D, Maheshwari D, Philip G, Rana R, Bhatia S, Singh M, Gabrani R, Sharma SK, Ali J, Sharma RK, Dang S (2014) Formulation and optimization of polymeric nanoparticles for intranasal delivery of lorazepam using Box-Behnken design: *in vitro* and *in vivo* evaluation. Biomed Res Int 1:e156010. https://doi.org/10.1155/2014/156010

57. Kubek MJ, Domb AJ, Veronesi MC (2009) Attenuation of kindled seizures by intranasal delivery of neuropeptide-loaded nanoparticles. Neurotherapeutics 6(2):359–371. https://doi.org/10.1016/j.nurt.2009.02.001

58. Sharma N, Bhandari S, Deshmukh R, Yadav AK, Mishra N (2016) Development and characterization of embelin-loaded nanolipid carriers for brain targeting. Artif Cells Nanomed Biotechnol 45(3):409–413. https://doi.org/10.3109/21691401.2016.1160407

59. Ruigrok MJR, de Lange ECM (2015) Emerging insights for translational pharmacokinetic and pharmacokinetic-pharmacodynamic studies: towards prediction of nose-to-brain transport in humans. AAPS J 17:493. https://doi.org/10.1208/s12248-015-9724-x

60. Kublik H, Vidgren MT (1998) Nasal delivery systems and their effect on deposition and absorption. Adv Drug Deliv Rev 29:157–177. https://doi.org/10.1016/S0169-409X(97)00067-7

61. Pennington AK, Ratcliffe JH, Wilson CG, Hardy JG (1988) The influence of solution viscosity on nasal spray deposition and clearance. Int J Pharm 43:221–224. https://doi.org/10.1016/0378-5173(88)90277-3

62. Hardy JG, Lee SW, Wilson CG (1985) Intranasal drug delivery by spray and drops. J Pharm Pharmacol 37(5):294–297

63. Shah SA, Dickens CJ, Ward DJ, Banaszek AA, George C, Horodnik W (2014) Design of experiments to optimize an in vitro cast to predict human nasal drug deposition. J Aerosol Med Pulm Drug Deliv 27(1):21–29. https://doi.org/10.1089/jamp.2012.1011

64. Foo MY, Cheng YS, Su WC, Donovan MD (2007) The influence of spray properties on intranasal deposition. J Aerosol Med 20(4):495–508. https://doi.org/10.1089/jam.2007.0638

65. Cheng YS et al (2001) Characterization of nasal spray pumps and deposition pattern in a replica of the huan nasal airway. J Aerosol Med 14(2):267–280

66. Djupesland PG, Skretting A (2012) Nasal deposition and clearance in man: comparison of a bidirectional powder device and a traditional liquid spray pump. J Aerosol Med Pulm Drug Deliv 25(5):280–289. https://doi.org/10.1089/jamp.2011.0924

67. Charlton S, Jones NS, Davis SS, Illum L (2007) Distribution and clearance of bioadhesive formulations from the olfactory region in man: effect of polymer type and nasal delivery device. Eur J Pharm Sci 30:295–302. https://doi.org/10.1016/j.ejps.2006.11.018

68. Craft S et al (2012) Intranasal insulin therapy for alzheimer disease and amnestic mild cognitive impairment: a pilot clinical trial. Arch Neurol 69:29–38. https://doi.org/10.1001/archneurol.2011.233

69. AAPS (2013) SPECT imaging of direct nose-to-brain transfer of MAG-3 in man (trans: Hoekman J, Brunelle A, Hite M, Kim P, Fuller C). AAPS, Arlington, VA

70. Schmidt MC, Peter H, Lang SR, Ditzinger G, Merkle HP (1998) In vitro cell models to study nasal mucosal permeability and metabolism. Adv Drug Deliv Rev 29(1–2):51–79. https://doi.org/10.1016/S0169-409X(97)00061-6

71. Pardeshi CV, Belgamwar VS (2016) Ropinirole-dextran sulfate nanoplex for nasal administration against Parkinson's disease: in silico molecular modeling and in vitro–ex vivo evaluation. Artif Cells Nanomed Biotechnol 45:1–14. https://doi.org/10.3109/21691401.2016.1167703

72. Madane RG, Mahajan HS (2016) Curcumin-loaded nanostructured lipid carriers (NLCs) for nasal administration: design, characterization, and in vivo study. Drug Deliv 23(4):1326–1334. https://doi.org/10.3109/10717544.2014.975382

73. Menco B (2003) Morphology of the mammalian olfactory epithelium: form, fine structure, function, and pathology. In: Doty RL (ed) Handbook of olfaction and gustation, vol 2nd ed. CRC Press, New York, NY

74. Chemuturi NV, Donovan MD (2007) Role of organic cation transporters in dopamine uptake across olfactory and nasal respiratory tissues. Mol Pharm 4(6):936–942. https://doi.org/10.1021/mp070032u

75. Hanson LR, Fine JM, Svitak AL, Faltesek KA (2013) Intranasal administration of CNS therapeutics to awake mice. J Vis Exp 74:4440. https://doi.org/10.3791/4440

76. Patel S, Chavhan S, Soni H, Ak B, Mathur R, Ak M, Sawant K (2011) Brain targeting of risperidone-loaded solid lipid nanoparticles by intranasal route. J Drug Target 19:468–474.

https://doi.org/10.3109/1061186X.2010.
523787

77. Abd-Elal RMA, Shamma RN, Rashed HM, Bendas ER (2016) Trans-nasal zolmitriptan novasomes: in-vitro preparation, optimization and in-vivo evaluation of brain targeting efficiency. Drug Deliv 23:1–13. https://doi.org/10.1080/10717544.2016.1183721

78. Alyautdin RN, Petrov VE, Langer K, Berthold A, Kharkevich DA, Kreuter J (1997) Delivery of loperamide across the blood-brain barrier with polysorbate 80-coated polybutyl-cyanoacrylate nanoparticles. Pharm Res 14 (3):325–328

79. Alyautdin R, Tezikov E, Ramge P, Kharkevich D, Begley D, Kreuter J (1998) Significant entry of tubocurarine into the brain of rats by adsorption to polysorbate 80-coated polybutylcyanoacrylate nanoparticles: an in situ brain perfusion study. J Microencapsul 15 (1):67–74

80. Lockman PR, Oyewumi MO, Koziara JM, Roder KE, Mumper RJ, Allen DD (2003) Brain uptake of thiamine-coated nanoparticles. J Control Release 93(3):271–282. https://doi.org/10.1016/j.jconrel.2003.08.006

81. Lu W, Zhang Y, Tan Y-Z, Hu K-L, Jiang X-G, Fu S-K (2005) Cationic albumin-conjugated pegylated nanoparticles as novel drug carrier for brain delivery. J Control Release 107 (3):428–448.   https://doi.org/10.1016/j.jconrel.2005.03.027

82. Jansson B, Björk E (2002) Visualization of in vivo olfactory uptake and transfer using fluorescein dextran. J Drug Target 10(5):379–386

83. Si XA, Xi J (2016) Modeling and simulations of olfactory drug delivery with passive and active controls of nasally inhaled pharmaceutical aerosols. J Vis Exp 111:e53902. https://doi.org/10.3791/53902

84. Garcia GJM, Schroeter JD, Kimbell JS (2015) Olfactory deposition of inhaled nanoparticles in humans. Inhal Toxicol 27(8):394–403. https://doi.org/10.3109/08958378.2015.1066904

85. Inthavong K, Tian ZF, Tu JY, Yang W, Xue C (2008) Optimising nasal spray parameters for efficient drug delivery using computational fluid dynamics. Comput Biol Med 38 (6):713–726.   https://doi.org/10.1016/j.compbiomed.2008.03.008

86. Inthavong K, Ge Q, Se CMK, Yang W, Tu JY (2011) Simulation of sprayed particle deposition in a human nasal cavity including a nasal spray device. J Aerosol Sci 42(2):100–113. https://doi.org/10.1016/j.jaerosci.2010.11.008

87. Kimbell JS, Segal RA, Asgharian B, Wong BA, Schroeter JD, Southall JP, Dickens CJ, Brace G, Miller FJ (2007) Characterization of deposition from nasal spray devices using a computational fluid dynamics model of the human nasal passages. J Aerosol Med 20 (1):59–74.   https://doi.org/10.1089/jam.2006.0531

88. Kelly JT, Asgharian B, Kimbell JS, Wong BA (2004) Particle deposition in human nasal airway replicas manufactured by different methods. Part I: inertial regime particles. Aerosol Sci Technol 38:1063–1071. https://doi.org/10.1080/027868290883360

89. Xi J, Yuan JE, Zhang Y, Nevorski D, Wang Z, Zhou Y (2016) Visualization and quantification of nasal and olfactory deposition in a sectional adult nasal airway cast. Pharm Res 33 (6):1527–1541.   https://doi.org/10.1007/s11095-016-1896-2

Chapter 9

In Vitro Models of Central Nervous System Barriers for Blood-Brain Barrier Permeation Studies

Sounak Bagchi, Behnaz Lahooti, Tanya Chhibber, Sree-pooja Varahachalam, Rahul Mittal, Abhijeet Joshi, and Rahul Dev Jayant

Abstract

One of the biggest challenging diseases are the neurodegenerative diseases which are not easy to target due to the presence of a complex semipermeable, dynamic, and adaptable barrier between the central nervous system (CNS) and the systemic circulation termed as the blood-brain barrier (BBB), which controls the exchange of molecules. Its semipermeable nature restricts the movement of bigger molecules, like drugs, across it and leads to minimal bioavailability of drugs in the CNS. This poses the biggest shortcoming in the development of therapeutics for CNS disorders. Although the complexity of the BBB muddles the drug delivery approaches into the CNS and can promote disease progression, understanding the composition and functions of BBB provides a platform for unraveling the way toward drug development. The BBB is comprised of brain microvascular endothelial CNS cells which communicate with other CNS cells (astrocytes, pericytes) and behave according to the state of the CNS, by retorting against pathological environments and modulating disease progression. This chapter discusses the fundamentals of BBB, permeation mechanisms, an overview of different in vitro BBB models with their advantages and disadvantages, and rationale of selecting penetration prediction methods toward the important role in the development of CNS therapeutics.

Key words Blood-brain barrier (BBB), Brain microvascular endothelial cells (BMECs), Astrocytes, Pericytes, Tight junctions, Central nervous system, In vitro BBB models

1 Introduction

The first evidence for existence of a barrier between the (central nervous system) CNS and the systemic circulation was shown in the studies done by Paul Ehrlich in 1885 and Edwin Goldmann in 1913, and the term "blood-brain barrier" (BBB) was coined by Stern and Gaultier in 1922 [1, 2]. The nature of the BBB is semipermeable as it restricts incoming of the detrimental molecules and cells from the blood side and allows uptake of selective nutrients and hormones only. The major cells comprising the BBB are

Javier O. Morales and Pieter J. Gaillard (eds.), *Nanomedicines for Brain Drug Delivery*, Neuromethods, vol. 157,
https://doi.org/10.1007/978-1-0716-0838-8_9, © Springer Science+Business Media, LLC, part of Springer Nature 2021

brain microvascular endothelial cells (BMECs), which are supported by astrocytes and pericytes [3]. To discuss in vitro BBB models, we briefly introduce the biological properties and functions of individual BBB components [1, 4, 5].

The assurance of the access of nutrients to the brain is due to the big surface area and the diminutive diffusion distance from the BBB capillaries to the neurons in the CNS, and the chemicals/molecules penetrate the BBB by utilizing intra- and intercellular routes. The intracellular route is regulated by tight junctions (TJs) to facilitate the passage of molecules based on their lipophilicity, ionization, polarity, and other physicochemical properties, whereas the intercellular route is controlled by passive diffusion, endocytosis, and the ratio of influx and efflux transporters [6]. The inefficacy of drugs to reach across the BBB is mainly due to poor pharmacokinetic properties, i.e., inefficient ADME (absorption, distribution, metabolism, and excretion). In addition to ADME, toxicity of the CNS drugs is also one among the major shortcomings [7]. Drug exposure is controlled by plasma pharmacokinetic properties of the drug which are different from the brain pharmacokinetics of the drug. Studying drug pharmacokinetics for CNS specifically involves understanding the correlation of physicochemical properties of drug compound and physiologic function of the BBB [1, 6, 7]. Therefore, in this chapter, we will discuss the fundamentals of BBB focusing on the permeation mechanisms, penetration measurement, prediction methods, and disease patterns which have been changing in recent times so that better understanding of BBB can be obtained.

In case of CNS infectious diseases or aging disorders, the biggest factor is the incapability of the BBB to maintain its integrity and open temporarily allowing the access of the drugs into the CNS. Unable to maintain the brain homeostasis and allowing minimal bioavailability of the drug into the CNS, the BBB directly contributes toward the progression of the pathological conditions which makes BBB as a potent therapeutic target for designing the drugs which can cross the BBB and help restore the stability of BBB, in CNS disorders. Therefore, in order to study the drug transmigration across the BBB, simplified in vitro BBB models have been developed, including the monolayer models, co-culture models, dynamic models, and microfluidic BBB models, to understand the dynamics and role of the BBB. The in vitro BBB models come with the shortcoming of not being able to be replicated in in vivo scenarios; therefore understanding the limitations of the in vitro BBB models would be critical for experimental design and data interpretation [1].

2   Fundamentals and Composition of BBB

The BBB consists of a monolayer of brain microvascular endothelial cells (BMVEC) joined together by much tighter junctions than peripheral vessels and form a cellular membrane which is known as the physical backbone of the BBB [7, 8]. The key characteristics of the BBB are its uniform thickness, absence of fenestrae, least pinocytotic activity, and negative surface charge.

In the BBB composition, BMVECs are supported by the capillary basement membrane, pericytes, astrocytes, and microglial cells forming the neurovascular unit. The basement membrane is made of collagen and elastin structural proteins, fibronectin and laminin which are specialized proteins, and finally proteoglycans, which give the structural specificity and membrane stability. Pericytes are the cellular constituents of microvessels, including capillaries and post-capillary venules, which cover 22–32% of the capillaries and share the same basement membrane, helping in various structural and nonstructural tasks of the BBB. They synthesize structural and signaling proteins and contribute to the BMVECs proliferation, migration, and differentiation processes [8].

The next important cells are the astrocytes, whose end feet from the lamellae which are in close contact with the outer surface of capillary endothelium and basement membrane, contributing as a part of neurovascular unit. Additionally, the presence of immuno-competent brain microglial cells is crucial, as they examine the local microenvironment with their motile extensions and are capable of changing their phenotype according to the homeostatic distur-bance in the CNS [9]. The interactions of the BMVECs with the basement membrane, glial cells (microglia and astrocytes), neurons, and perivascular pericytes lead to specific brain microvascular biol-ogy. The presence of matrix adhesion receptors and signaling pro-teins forms an extensive and complex matrix which is essential for maintenance of the BBB [8, 10] (Fig. 1).

2.1   Molecular Properties of BBB

The BMVECs assembly is regulated by the molecular constituents of TJs, adherence junctions, and signaling pathways. TJs are highly dynamic structures which are responsible for the selective perme-ability property of the BBB as the apical region of the endothelial cells is sealed together by TJs allowing limited paracellular perme-ability. Structurally, TJs are formed by the interaction of integral transmembrane proteins with the neighboring plasma membrane [11]. Among these proteins, junction adhesion molecules (JAM), claudins, and occludins (intermembrane) bind to cytoplasmic pro-teins (e.g., zonula occludens, cinguline) and are well studied for their role in the TJs constituting the BBB (Fig. 2) [12, 13]. In addition to contributing in the physical restriction property of the BBB, other functions such as control of gene expression, cell

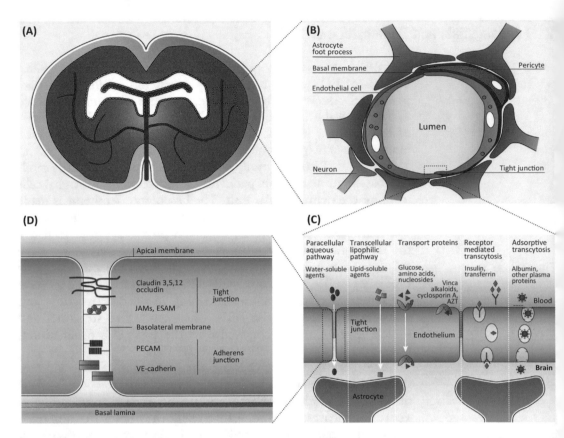

Fig. 1 Structure and functionality of the blood-brain barrier (BBB). (**a**) *Brain barriers*, the brain has several barriers, including (1) the BBB, (2) the outer blood-cerebrospinal fluid (CSF)-brain barrier, and (3) the blood-CSF barrier. (**b**) *BBB structure*, the BBB is formed by endothelial cells (ECs) that are in close association with astrocyte end feet and pericytes, forming a physical barrier. (**c**) *BBB transport*, routes for molecular traffic across the BBB are shown. Some transporters are energy-dependent (e.g., P-glycoprotein, P-gp) and act as efflux transporters. (**d**) *Tight junctions*, tight junctions are typically located on the apical region of ECs. The tight junctions form complex networks that result in multiple barriers that restrict the penetration of polar drugs or other biomolecules into the brain (Reproduced from Aday et al., 2016 [10], with permission)

proliferation, and differentiation are also the properties of the TJs. Below the TJs, actin filaments (including cadherins and catenins) link together and form a belt of adherence junctions [6]. These adherence junctions contribute to the barrier property along with additional roles such as promoting BMVECs adhesions, contact inhibition throughout vascular growth, cell polarity, and controlling paracellular permeability regulations. Dynamic interactions between TJs and adherence junctions, through signaling pathways, regulate the BBB permeability. These signaling pathways include mitogen-activated protein kinases, endothelial nitric oxide synthases, and G-proteins, and interaction between these pathways regulates the paracellular route. The signal transduction includes signals from the interior of the cells to the TJs (facilitating the assembly and regulating the permeability) and signals from the

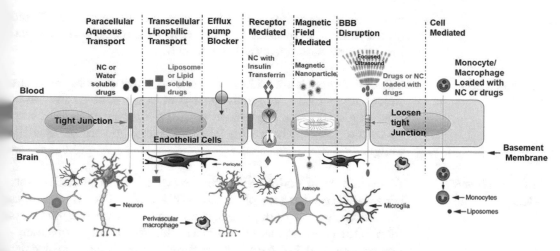

Fig. 2 Schematic representation of mechanisms available for drugs/biomolecules or nanoparticle-based formulation to cross the BBB. The above schematic shows several ways in which transport across the BBB works. For the nanoparticle delivery across the BBB, the most common mechanisms are receptor-mediated transcytosis and adsorptive transcytosis (passive transport). Also, diffusion and active transport (magnetic field or ultrasound) are the other main types of transport mechanisms (Reproduced from Nair et al., 2016 [13], with permission)

TJs to the interior of the cell for modulating gene expression, proliferation, and differentiation [7, 11]. In addition to the proteins with enzymatic activities, there are other specific proteins (drug efflux transporters, multidrug resistance proteins, organic anion transporting polypeptides) which work as the BBB transporters, responsible for the rapid efflux of xenobiotics from the CNS [14] and for the delivery of the essential nutrients and transmitters into the brain, resulting in the specific barrier functions of the BBB, important for protecting CNS against harmful xenobiotics. BBB is complex, and transmigration of essential drugs is a big challenge, even though limited BBB permeation is attributed to passive diffusion, active transport, and endocytosis [15] (Fig. 2).

3   In Vitro BBB Permeation Measurement Methods

To expedite the brain research and accelerate the R&D of novel drugs for numerous neurological diseases, different types of in vitro BBB models have been established. However, as none of these in vitro models entirely reproduce the in vivo conditions, thus, there is no perfect in vitro BBB model. Therefore, it is utmost important to carefully choose the in vitro BBB model according to the requirement of the study and to interpret the data efficiently [16–18]. Here, we are summarizing the most widely used in vitro BBB models, including the recently developed microfluidic BBB models, and analyzing their advantages and disadvantages. Based on the simulation of shear stress, in vitro BBB models are classified into the static and dynamic BBB models.

3.1 Static BBB Models

Static BBB models are commonly used, but they do not imitate the sheer stress which is generally generated in vivo due to the blood flow. Static BBB models are divided into further monolayer and co-culture models, based on types of cells involved in the BBB design.

3.1.1 Monolayer BBB Models

A monolayer of endothelial cells grown in the Transwell insert (Fig. 3a-i) is used as a simple in vitro BBB model. The insert mimics the blood (luminal) side, whereas the well in which the insert sits mimics the parenchymal (abluminal) side. The microporous membrane support (0.2–0.4 μm) allows the exchange of small molecules and cell-secreted growth factors but prevents the migration of cells between the two compartments. To mimic the unique properties of BMECs, primary or low passage number cells are used for the BBB model preparation. This process is challenging, because isolation and culturing the primary BMECs are critical and have a great risk of getting contaminated by the mural cells and secondly low yield of primary BMECs after isolation is also a shortcoming as the brain vasculature accounts only for 0.1% (v/v) of the brain; therefore a large number of rodents are needed to isolate enough primary BMECs to continue with the cell culture studies [1, 19]. To overcome this limitation, larger species and nonhuman primates are utilized for the experiment for isolating larger amounts of BMECs, for further experiments [20]. Human cells are used when the studies focus on human-specific transporters/receptors or immunological aspects but are not usually available due to ethical issues. To evade this issue, many immortalized human cell lines, such as human cerebral microvascular endothelial cell line (hCMEC/D3) and immortalized human cerebral endothelial cells, have been generated, which are useful, but have lower expression of some of the BBB-specific transporters and enzymes, leading to decreased generation of a tight monolayer and thus having inadequate barrier function [21]. This inadequacy is counteracted by addition of BBB-modulating compounds like cAMP and glucocorticoids, which increase the endothelial monolayer tightness and stability [22, 23]. Monolayer models are employed in studying signaling pathways, transporter kinetics, binding affinity, and high-throughput screenings. However, the monolayer model is not ideal for BBB integrity studies, as it has only one type of cells (BMECs) and lacks to imitate the brain microenvironment in which cell-to-cell communication is essential between different cell CNS cell types [24, 25]. Therefore, for the study of BBB integrity, a more vivid and complex BBB model is required, such as co-culture and dynamic models.

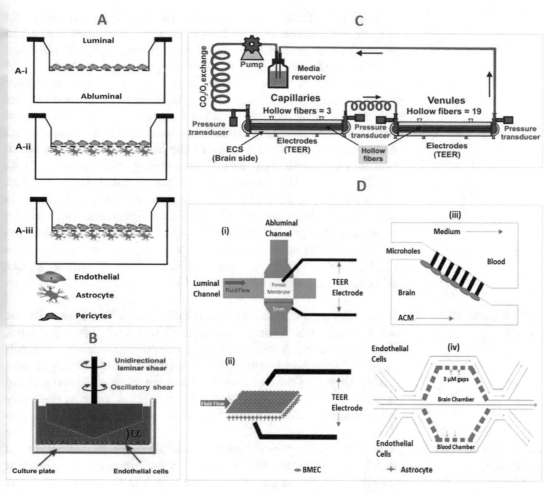

Fig. 3 Schematic representation of different in vitro BBB models. (**a**) (i) Monolayer models constructed using brain capillary endothelial cells (BCECs) on the upper side of the microporous semipermeable membrane (Transwell). (ii) 2D co-culture contact models, endothelial cells are grown on porous cell culture inserts and co-cultured with primary astrocytes. (iii) 3D co-culture models, triple cultures of endothelial cells with astrocytes and pericytes; (**b**) cone-plate apparatus; (**c**) schematic representation of the dynamic in vitro blood-brain barrier (DIV-BBB) model setup; (**d**) microfluidic-based in vitro blood-brain barrier (BBB) models. (i) Microfluidic BBB (μBBB) model. (ii) 3D view of the porous membrane at the intersection of the flow channels in the μBBB model. (iii) Diagram of the microfluidic device containing microholes. (iv) Structure of the synthetic microvasculature model of the BBB model. *ACM* astrocyte-conditioned medium, *BMEC* brain microvascular endothelial cell, *TEER* transendothelial electric resistance (Reproduce from He et al. 2014 [1], with permission)

3.1.2 Co-culture BBB Models

In order to mimic the anatomic structure of the in vivo BBB, BMECs are co-cultured with other CNS cells that directly contribute to the barrier properties of BBB. Interaction between BMECs and other brain cells increases the expression of transporters; TJs in BMECs induce the cell polarity in BMECs promoting a phenotype more closely mimicking the in vivo BBB. In this co-culture model (BMECs with astrocytes or pericytes) [7, 26–37], BMECs are seeded in the Transwell insert, and astrocytes are grown either on the underside of the Transwell insert or at the bottom of the well in which the insert sits (Fig. 3a-ii). Since pericytes also have a key role in BBB regulation, a BMEC-astrocyte-pericyte co-culture model has also been developed, which is termed as a triple co-culture system (BMEC-pericyte-astrocyte). Addition of pericytes enhances the quality of the co-culture model compared to the monolayer model. In this model, BMECs are plated in the Transwell insert with pericytes on the underside of the insert and astrocytes at the bottom of the well (Fig. 3a-iii). This arrangement although lacks the direct cell-to-cell communication between BMECs, astrocytes, and pericytes, it utilizes indirect cell-to-cell communication via secreted soluble factors, which promotes BBB regulation. BMEC-pericyte-astrocyte triple co-culture model is a more reliable in vitro BBB model due to the higher transendothelial electric resistance (TEER) value and lower permeability, which generates tighter BBB, ideal for permeability studies [19, 38].

3.2 Dynamic BBB Models

In physiological conditions, the regular blood flow generates the shear stress, which regulates transporters and TJs expression, donating toward efficacious barrier function. Shear stress developed by blood flow increases ZO-1 expression and reduces permeability; therefore the dynamic BBB models, with shear stress, have been developed, which are of three major types, namely, the cone-plate, dynamic, and microfluidic-based models [39, 40].

3.2.1 Cone-Plate BBB Apparatus

The cone-plate apparatus was used initially to construct shear force, in which a rotating cone produces shear force and the angular velocity and the angle of the cone regulate the produced shear stress. The sheer stress then reaches the endothelial monolayer via the medium (Fig. 3b), but it is not evenly distributed along the radius of the plates, and therefore the endothelial monolayer receives varying shear stress depending upon the location. This model does not include astrocytes and pericytes; therefore, it has a limited application, low reliability, and less significant to be used in the BBB studies [41, 42].

3.2.2 Dynamic In Vitro BBB Model

To incorporate both the components, i.e., shear stress and various cell types, microporous hollow fibers are used (Fig. 3c). In this model, BMECs and astrocytes are plated in the inner (luminal) and outer (abluminal) sides of the porous hollow fibers, respectively [1, 42]. The culture medium is then pumped into the system via a variable-speed pump to generate shear stress comparable with that seen in physiological conditions in vivo (5–23 dynes/cm^2) [43]. To maintain the stable microenvironment, a gas-permeable tubing system is used for the exchange of O_2 and CO_2. This dynamic in vitro BBB model has been used to study the pathophysiology of various CNS diseases, including ischemia-reperfusion-induced injury and epilepsy [44, 45]. Recently, a revised model with hollow fibers with transmural microholes of 2–4 µm has been developed to facilitate transmigration/trafficking studies. However, the dynamic BBB model has many disadvantages like (1) it does not allow direct visualization of the endothelial morphology in the luminal side; (2) the cell numbers ($>1 \times 10^6$) required to build this model are relatively high; and (3) the time required to reach steady-state TEER is longer (9–12 days) compared to co-culture models (3–4 days). These shortcomings prevent the use of dynamic in vitro BBB model in large-scale screens. This model, however, is useful in lead compound validation/optimization in new drug R&D [45].

3.2.3 Microfluidic-Based BBB Models

Microfluidic device-based in vitro BBB models have been developed to overcome the shortcomings found in dynamic BBB models [1, 46, 47]. The microfluidic BBB (µ-BBB) comprises two perpendicularly crossing channels, allowing the dynamic flow to generate shear stress; a polycarbonate porous membrane is placed over the intersection of the perpendicular channels, enabling the co-culture of BMECs and astrocytes (on the luminal and abluminal sides, respectively). The channels are 200 µm tall, 2 mm (luminal), and 5 mm (abluminal) wide. It also contains multiple built-in Ag/AgCl electrodes for facilitating the TEER measurement (Fig. 3d-i, ii) [48]. Pumps and a gas-permeable tubing system are used to generate shear stress and allow O_2-CO_2 exchange, respectively. This µ-BBB model is further improved by replacing the oxidation-sensitive AgCl electrodes with inert platinum ones and decreasing the cross-sectional area. These modifications allow accurate measurement of TEER and reduce the amount of cells needed. Additionally, the microfluidic-based BBB model with the microhole has been designed to study the BBB permeability of drugs, which is composed of two horizontally aligned chambers connected by a microhole structure (Fig. 3d-iii). However, the shortcomings of the model are that (1) it lacks cell-cell contact and (2) it does not replicate the dimensions of microvasculature in vivo. Therefore, currently the development of a new version of microfluidic device, i.e., synthetic microvasculature model of the BBB (SyM-BBB), is progressing [47]. The SyM-BBB model contains two microchannels separated by microfabricated pillars with 3 µm gaps (Fig. 3d-iv), which mimic the porous membrane of the µBBB model. Endothelial cells get infused to the blood chamber via ports 1 and 2, whereas astrocyte-conditioned medium or astrocytes are infused to the brain chamber from port 3. The flow speed of medium in these chambers determines the shear stress. This design better mimics the in vivo microcirculatory system by including the diverging and converging bifurcations.

In comparison with the dynamic BBB model, the microfluidic models are closer replicates of the in vivo BBB structure, as they have more membrane thickness (<10 µm) which allows efficacious transmigration studies conditions. With the microfluidic model, nondestructive microscopy is possible because of the transparency of the materials; it takes less time (3–4 days) to reach steady-state TEER and requires only a small amount of cells and is less demanding in terms of technical skills [46].

Nonetheless, the microfluidic models have limitations too, i.e.: (1) TEER value is not high enough (250–300 Ω cm^2), and (2) they only incorporate two cell types, given that the membrane and pillars (hollow fibers in the dynamic in vitro BBB model) have only two sides. With the growth of research data, these microfluidic models may be used to aid neurovascular research and new drug R&D in the future because of its small size, short time to reach steady-state TEER, low-to-moderate technical skill requirement, and low cost. The advantages and disadvantages of these in vitro BBB models are summarized in Table 1, and comparison between each among different in vitro BBB models is summarized in Table 2.

Table 1
Advantages and disadvantages of different in vitro BBB models

Model type	Advantages	Disadvantages
Epithelial cells overexpressing Transporters model	• Inexpensive • Easy to standardize	• Differences between epithelial and endothelial cells • Non-physiologically high levels of transporter
Transwell monoculture model – Cerebral endothelial cells on microporous membranes	• Uses brain endothelial cells • Inexpensive	• Effect of other cellular components of the neurovascular unit (NVU-astrocytes, pericytes) is neglected • No shear stress
Co-cultures models – Co- culture of cerebral microvascular endothelial cells with astrocytes – Co-culture models using pericytes – Triple cell co-culture models (astrocytes, endothelial, and pericytes) – Co-culture of brain endothelial cells with neuronal precursors	• Takes into account the influence of other elements of the neurovascular unit (NVU)	• Relatively expensive and time-consuming • No shear stress
Dynamic in vitro (DIV) model	• Mimics in vivo situation possibility of co-culture.	• Expensive • No possibility to optically monitor the cells • Special skills required to culture cells in these conditions
Microfluidic model	• Mimics in vivo situation possibility of co-culture	• Not well-established models presently expensive

Table 2
Comparison among different in vitro BBB models for biomolecules/drugs/nanoformulations transport

Model type	Other brain cell required	Sheer stress produced	Time to stable TEER (days)	Appropriate for migration assay	Cost	Technical requisite
Monolayer	No	No	3–4	Yes	Low	Low
Co-culture	Yes	No	3–4	Yes	Low to moderate	Moderate
Cone-plate apparatus	No	Yes	3–4d	No	Low	Low to moderate
Dynamic in vitro BBB	Yes	Yes	9–12	No	High	High
Microfluidic based model	Yes	Yes	3–4	Yes	High	Moderate

4 BBB Permeation Prediction Methods (In Silico Methods)

The widely used in silico prediction of the BBB permeability is an inexpensive, less time-consuming, and high-throughput filtering method for novel compounds in the drug discovery process. Although this method is based on several molecular descriptors or physicochemical characteristics of the molecule, it has its own strengths and weaknesses [49–52]. These computational models are typically based on the previous in vivo and in vitro experimental data. Therefore, for the predictive power of estimations, selection of datasets is a critical component. The assumption of passive diffusion of a compound as a major route of transport through the BBB is the base for the in silico predictions which does not consider various BBB transport pathways, e.g., nanoparticle-based transport/carrier-mediated, receptor-mediated, and active efflux or influx transport methods [53]. Recently, cerebrospinal fluid (CSF) penetration is also considered in in silico model while analyzing the brain penetration of the molecule [54].

To increase the predictive values of these computational models, new approaches have become more sophisticated. Table 3 shows different computational models to predict newly designed or synthesized compound's BBB penetration. In general in brain penetration studies, brain-to-plasma ratios will be measured, and consequently most of the in silico predictions are based on the available logBB data (Table 3), which represents the most readily available experimental data [52, 55, 56]. In the training set, several molecular descriptors of the compounds are calculated with already known log BB values which are experimentally determined. To derive the equation to give the relationship between log BB and

Table 3

In silico models and their parameters for estimating drug penetrability of the BBB

Model	Description	Parameters involved
Brain penetrability parameters		
log BB	Brain-to-plasma ratio (log C_{brain}/log C_{blood})	Correlation with quantitative structure-activity relationship data
log PS	BBB permeability surface area product	Correlation with quantitative structure-activity relationship data
log CSF	Cerebrospinal fluid to plasma ratio ((log C_{CSF}/log C_{blood})	Correlation with quantitative structure-activity relationship data
Molecular descriptors		
log P_{oct}	Octanol/water partition coefficient	Hydrophobicity, H-bond donor potential
Δ log P	The difference in octanol/water and cyclohexane/water partition coefficients (log P_{oct} – log P_{cyc})	Low overall H-bonding ability
log D	Log distribution coefficient	Lipophilicity ($0 <$ log $D < 3$)
Classical descriptors	Physicochemical parameters	Polar surface area, molecular weight, molecular size, shape, and flexibility charge
P-glycoprotein substrate	High-affinity P-glycoprotein substrate probability	Efflux transport through the BBB
Rule-based models		
Hansch's rule of 2	Prediction based on octanol/water partition coefficient	Compounds having log $P_{oct} \approx 2.0$ have optimal brain penetration
Modified Lipinski's rules for CNS penetration	Prediction based on selected molecular descriptors	H-bond donors ≤ 3; H-bond acceptors ≤ 7; molecular weight ≤ 400 Da; log $P_{oct} \leq 5.0$; $7.5 <$ p$K_a < 10.5$
CNS-active drugs	Prediction based on selected molecular descriptors	Polar surface area < 90 Å^2; H-bond donors <3; 2.0 log $P_{oct} < 5.0$; molecular weight < 450 Da
Quantitative structure-activity relationship (Qsar)		
Linear QSAR	Prediction based on selected molecular descriptors	Multiple linear regression (MLR); partial least squares (PLS) methods; variable selection and modelling method based on the prediction (VSMP); linear discriminant analysis (LDA); comprehensive descriptors for structural and statistical analysis (CODESSA)
Nonlinear QSAR	Prediction based on selected molecular descriptors	Neural networks (NN); Bayesian modelling; support vector machine (SVM); Gaussian processes; k nearest neighbor method; recursive partitioning; substructure analysis

H-bond hydrogen bond, *log BB* brain-to-plasma ratio, *log CSF* cerebrospinal fluid to plasma ratio, *log D* log distribution coefficient, *log P* log octanol/water partition coefficient, *log PS* blood-brain-barrier permeability surface area product, *pKa* log of acidic dissociation constant the compound's computed descriptors, typically, regression methods are used. Because of its physiological relevance, the permeability surface area product (PS value/log PS) would be an impressive method of determining the BBB permeability for a specific molecule both in vivo and in vitro, compared with the currently more popular log BB [57, 58]. Unfortunately, the limited availability of log PS is due to the complicated measurement of log PS than that of log BB.

Based on Lipinski's rule, molecules which have not more than 5 H-bond donors and not more than ten H-bond acceptors with a MW of <500 Da and an octanol/water partition coefficient log P under 5 can be the drug candidates [59]. Molecules with these physicochemical characteristics have good aqueous solubility and intestinal permeability. Approximately, 90% of the orally active drug molecules which are under Phase II clinical trials have these characteristics [60]. Specifically, guidelines for the properties of new molecules that can be of potential CNS-active drug have been proposed [60–63]. The relationship between the experimental data computationally available parameters of a new compound and its blood-brain barrier penetration has been studied for long. Among different datasets, the octanol/water partitioning coefficient (log P_{oct}) is one of the earliest predictive factors available for BBB permeability. For compounds with MW <400 Da, it is possible to predict the relationship of the capillary permeability coefficient (log PC) to the log P_{oct} [64]. In 1988, when a linear correlation of antihistamines and Δ log P between the brain-to-blood ratios was established, the computational prediction of BBB penetration for these compounds began [65]. The observation of an inverse relation between the hydrogen bonding activity of the compound and BBB permeability provided a theoretical concept for designing BBB-permeable drugs.

To obtain the descriptors for the general linear free energy response (LFER) equation, calculations of log P_{oct}, -cyclohexane, and -dichloromethane systems were used [66]. To estimate the blood-to-brain distribution ratio, calculation of compound's physicochemical and biochemical properties is useful [66]. For the first time, universal quantitative scales of solute's hydrogen-bond acidity and basicity have been standardized, and along with other descriptors, these descriptors have been used in equations to calculate, predict, analyze, and correlate various solute properties. At the same time, these equations can be used for the analysis of various compound's physicochemical LFERs and biological properties, such as quantitative structure-activity relationships (QSARs) set up for blood-brain distribution.

Although these rule-based models can be used for the qualitative BBB permeability estimation using various physicochemical descriptors, these will not be useful to predict the active BBB transport through efflux pumps, carriers, and receptor-mediated transmigration. Overall, these in silico quantitative models as classification tools have more than 70% accuracy in predicting log BB [49, 52]. For the analysis of various molecular descriptors listed in Table 3, both linear and nonlinear statistical methods can be used.

These in silico models for predicting log BB and log PS have become increasingly reliable and popular than in vitro and in vivo BBB models in the drug discovery process as these methods are economic and faster. However, size and quality of the training set play a major role in accuracy of predicting the passive permeability. Similar to the recently developed model for P-glycoprotein substrate properties, if new models for the active transport mediated by carriers, receptors, and efflux pumps are developed, the predictive power of the in silico models will be tremendously increased [51, 52, 67].

5   Rationale for BBB Model Selection

In vitro BBB models are extensively used for the initial stages of novel drug development, which includes lead identification, hit identification, and target identification/optimization as shown in Fig. 4 [4]. After a target (enzyme, receptors, etc.) is identified, high-throughput screening (HTS) is employed to identify probable drug contenders. At this phase, a large number of compounds need to be screened and thus require easy and fast in vitro BBB screening model. Selecting suitable in vitro models not only enables accurate interpretation of the data but also saves time and money.

The key criteria in model selection is the purpose of the study, in the case of monolayer or co-culture models, which generally take 3–4 days to reach steady-state TEER value and are reasonably easy to construct, can be used. Previous studies have shown that the use of different immortalized cell lines provides the best correlation between in vitro and in vivo data for permeability assays. The co-culture models and dynamic in vitro BBB models are best models to study the drug permeability [1, 17]. Many multi-culture models (2D or 3D) are commercially available now, which considerably reduce the efforts and time but increase the cost. For trafficking/migration studies, microfluidic BBB model is the right option due to the incorporation of the shear stress component and that it mimics real in vivo conditions [1].

To study signaling pathways/transporter kinetics or quantifying binding affinities, the monolayer model is the best option due to intrinsic simplicity. For the lead identification/optimization phase studies, validation, structure-activity relationships (SAR), and toxicological profile, more sensitive in vitro models that replicate most of the in vivo conditions are required, i.e., static co-culture and dynamic models or newly developed microfluidic-based BBB models can serve as an alternative. Primary human-

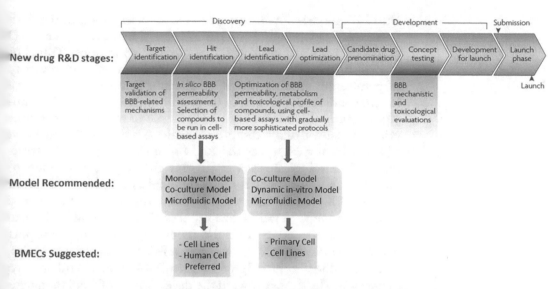

Fig. 4 Suggestions on model selection at early stages of new drug research and development (R&D). During hit identification stage, simple models, including the monolayer and co-culture models, are recommended. Microfluidic-based dynamic models may also be used given their incorporation of shear stress and low cost. Because of the large-scale nature of this stage, immortalized endothelial cell lines, especially human cells, should be used. When it comes to the lead identification and optimization stage, the number of compounds is dramatically reduced. Thus, sophisticated sensitive models that better replicate the in vivo blood-brain barrier (BBB) conditions are strongly recommended, such as the co-culture models and the dynamic in vitro BBB model. Microfluidic-based dynamic models may also be used at this stage. It is advised to use primary cells at this stage, although cell lines may also be used. *BMECs* brain microvascular endothelial cells (Adapted and reproduced from He et al., 2014 [1], with permission)

derived cells are a better option than immortalized cell lines due to closeness to biological properties of the BMECs in vivo. Authentication using human-derived primary cells is highly recommended for the generation of in vitro BBB model to evade species-based differences that may lead to the failure of the product in the later R&D stages. Based on the above problems and suggestions, Fig. 4 summarizes the selection of in vitro BBB models at different R&D stages for the therapeutic development [20, 49, 55, 68, 69].

6 Conclusion and Future Aspects

In vitro models of the BBB have been proven to be exceptionally valued for investigations of endothelial cell properties and mechanistic studies of drug transport via brain endothelial cells. The early effort of the pioneers Joo, Bowmann, Borchardt, Audus, Dehouck, and Cecchelli has been followed up by a large community of investigators and has resulted in a range of in vitro model configurations. In general, the cell culture models of the brain endothelium are believed to reflect the properties of the healthy BBB. In vitro BBB models are important to our understanding of the BBB functions in physiological and pathological situations and the R&D of novel drugs for different neurological conditions. Different in vitro BBB models have been established and used for a variety of permeation studies, yet no single model can imitate the real in vivo conditions. Further, knowing the advantages and disadvantages of each of these models and rationale of selecting the appropriate model will allow the precise understanding of the data and will significantly empower the R&D of novel drugs for the treatment of neurological diseases. In this chapter, we attempted to provide the overview of regularly used and newly advanced in vitro BBB models, equated their strengths and weaknesses, and made an attempt to rationalize the model selection. Model selection parameters are critical for predicting the drug transport because the disease in question may affect the barrier properties. A combinatorial approach of in vitro BBB models and in vivo methods will be the key toward the development of CNS therapeutics with improved pharmacokinetic properties and better BBB penetrability.

Acknowledgments

Rahul Dev Jayant would like to acknowledge the financial support from School of Pharmacy, Texas Tech University Health Sciences Center (TTUHSC) start-up funds, and The Campbell Foundation (Florida). Abhijeet Joshi acknowledges the INSPIRE Fellowship provided by Department of Science and Technology, Government of India.

References

1. He Y et al (2014) Cell-culture models of the blood–brain barrier. Stroke 45(8):2514–2526

2. Stern L, Gautier R (1922) II.–Les Rapports Entre Le Liquide Céphalo-Rachidien Et Les éléments Nerveux De L'axe Cerebrospinal. Arch Int Physiol 17(4):391–448

3. Guillemin GJ, Brew BJ (2004) Microglia, macrophages, perivascular macrophages, and pericytes: a review of function and identification. J Leukoc Biol 75(3):388–397

4. Cecchelli R et al (2007) Modelling of the blood–brain barrier in drug discovery and development. Nat Rev Drug Discov 6(8):650–661

5. Engelhardt B, Ransohoff RM (2012) Capture, crawl, cross: the T cell code to breach the blood–brain barriers. Trends Immunol 33(12):579–589

6. Jouyban A, Soltani S (2012) Blood brain barrier permeation. In: Acree W (ed) Toxicity and drug testing. InTechOpen, London

7. Abbott NJ (2005) Physiology of the blood–brain barrier and its consequences for drug transport to the brain. In: International congress series. Elsevier, Amsterdam

8. Cardoso FL, Brites D, Brito MA (2010) Looking at the blood–brain barrier: molecular anatomy and possible investigation approaches. Brain Res Rev 64(2):328–363

9. Prinz M, Mildner A (2011) Microglia in the CNS: immigrants from another world. Glia 59(2):177–187

10. Aday S et al (2016) Stem cell-based human blood–brain barrier models for drug discovery and delivery. Trends Biotechnol 34(5):382–393

11. Ballabh P, Braun A, Nedergaard M (2004) The blood–brain barrier: an overview: structure,

regulation, and clinical implications. Neurobiol Dis 16(1):1–13

12. Helms HC et al (2016) In vitro models of the blood–brain barrier: an overview of commonly used brain endothelial cell culture models and guidelines for their use. J Cereb Blood Flow Metab 36(5):862–890

13. Nair M et al (2016) Getting into the brain: potential of nanotechnology in the management of NeuroAIDS. Adv Drug Deliv Rev 103:202–217

14. Löscher W, Potschka H (2005) Role of drug efflux transporters in the brain for drug disposition and treatment of brain diseases. Prog Neurobiol 76(1):22–76

15. Alavijeh MS et al (2005) Drug metabolism and pharmacokinetics, the blood-brain barrier, and central nervous system drug discovery. NeuroRx 2(4):554–571

16. Czupalla CJ, Liebner S, Devraj K (2014) In vitro models of the blood–brain barrier. Methods Mol Biol 1135:415–437

17. Garberg P et al (2005) In vitro models for the blood–brain barrier. Toxicol In Vitro 19 (3):299–334

18. Wilhelm I, Fazakas C, Krizbai IA (2011) In vitro models of the blood-brain barrier. Acta Neurobiol Exp (Wars) 71(1):113–128

19. Nakagawa S et al (2009) A new blood–brain barrier model using primary rat brain endothelial cells, pericytes and astrocytes. Neurochem Int 54(3):253–263

20. Lippmann ES et al (2013) Modeling the blood–brain barrier using stem cell sources. Fluids Barriers CNS 10(1):2

21. Daniels BP et al (2013) Immortalized human cerebral microvascular endothelial cells maintain the properties of primary cells in an in vitro model of immune migration across the blood brain barrier. J Neurosci Methods 212 (1):173–179

22. Franke H, Galla H-J, Beuckmann CT (1999) An improved low-permeability in vitro-model of the blood–brain barrier: transport studies on retinoids, sucrose, haloperidol, caffeine and mannitol. Brain Res 818(1):65–71

23. Hurst R, Fritz I (1996) Properties of an immortalised vascular endothelial/glioma cell co-culture model of the blood-brain barrier. J Cell Physiol 167(1):81–88

24. Hori S et al (2004) A pericyte-derived angiopoietin-1 multimeric complex induces occludin gene expression in brain capillary endothelial cells through Tie-2 activation in vitro. J Neurochem 89(2):503–513

25. Toimela T et al (2004) Development of an in vitro blood–brain barrier model—cytotoxicity of mercury and aluminum. Toxicol Appl Pharmacol 195(1):73–82

26. Armulik A et al (2010) Pericytes regulate the blood-brain barrier. Nature 468 (7323):557–561

27. Daneman R et al (2010) Pericytes are required for blood-brain barrier integrity during embryogenesis. Nature 468(7323):562–566

28. Atluri VSR et al (2016) Development of TIMP1 magnetic nanoformulation for regulation of synaptic plasticity in HIV-1 infection. Int J Nanomedicine 11:4287

29. Jayant R (2014) Layer-by-layer (LbL) assembly of anti-HIV drug for sustained release to brain using magnetic nanoparticle. J Neuroimmun Pharmacol 9(1):25–25

30. Jayant R, Nair M (2016) Nanotechnology for the Treatment of NeuroAIDS. J Nanomed Res 3(1):00047

31. Jayant R, Nair M (2016) Role of biosensing technology for neuroAIDS management. J Biosensors Bioelectron 7(1):pii: e141

32. Jayant RD, Madhavan N (2016) Materials and methods for sustained release of active compounds. US Patent App. 15/082611

33. Kaushik A, Jayant RD, Nair M (2016) Advancements in nano-enabled therapeutics for neuroHIv management. Int J Nanomedicine 11:4317

34. Tomitaka A et al (2017) Development of magneto-plasmonic nanoparticles for multimodal image-guided therapy to the brain. Nanoscale 9(2):764–773

35. Nair M et al (2013) Externally controlled on-demand release of anti-HIV drug using magneto-electric nanoparticles as carriers. Nat Commun 4:1707

36. Pilakka-Kanthikeel S et al (2013) Targeted brain derived neurotropic factors (BDNF) delivery across the blood-brain barrier for neuro-protection using magnetic nano carriers: an in-vitro study. PLoS One 8(4):e62241

37. Ding H et al (2014) Enhanced blood–brain barrier transmigration using a novel transferrin embedded fluorescent magneto-liposome nanoformulation. Nanotechnology 25 (5):055101

38. Nakagawa S et al (2007) Pericytes from brain microvessels strengthen the barrier integrity in primary cultures of rat brain endothelial cells. Cell Mol Neurobiol 27(6):687–694

39. Siddharthan V et al (2007) Human astrocytes/astrocyte-conditioned medium and shear stress enhance the barrier properties of human brain microvascular endothelial cells. Brain Res 1147:39–50

40. Tarbell JM (2010) Shear stress and the endothelial transport barrier. Cardiovasc Res 87 (2):320–330

41. Bussolari SR, Dewey CF Jr, Gimbrone MA Jr (1982) Apparatus for subjecting living cells to fluid shear stress. Rev Sci Instrum 53 (12):1851–1854

42. Naik P, Cucullo L (2012) In vitro blood–brain barrier models: current and perspective technologies. J Pharm Sci 101(4):1337–1354

43. Koutsiaris AG et al (2007) Volume flow and wall shear stress quantification in the human conjunctival capillaries and post-capillary venules in vivo. Biorheology 44(5-6):375–386

44. Cucullo L et al (2007) Development of a humanized in vitro blood–brain barrier model to screen for brain penetration of antiepileptic drugs. Epilepsia 48(3):505–516

45. Cucullo L et al (2011) A dynamic in vitro BBB model for the study of immune cell trafficking into the central nervous system. J Cereb Blood Flow Metab 31(2):767–777

46. Booth R, Kim H (2012) Characterization of a microfluidic in vitro model of the blood-brain barrier (μBBB). Lab Chip 12(10):1784–1792

47. Prabhakarpandian B et al (2013) SyM-BBB: a microfluidic blood brain barrier model. Lab Chip 13(6):1093–1101

48. Booth R, Kim H (2011) A multi-layered microfluidic device for in vitro bloodbrain barrier permeability studies. In: International conference on miniaturized systems for chemistry and life sciences

49. Vastag M, Keseru GM (2009) Current in vitro and in silico models of blood-brain barrier penetration: a practical view. Curr Opin Drug Discov Devel 12(1):115–124

50. Abbott NJ (2004) Prediction of blood–brain barrier permeation in drug discovery from in vivo, in vitro and in silico models. Drug Discov Today Technol 1(4):407–416

51. Goodwin JT, Clark DE (2005) In silico predictions of blood-brain barrier penetration: considerations to "keep in mind". J Pharmacol Exp Ther 315(2):477–483

52. Mensch J et al (2009) In vivo, in vitro and in silico methods for small molecule transfer across the BBB. J Pharm Sci 98 (12):4429–4468

53. Deli MA (2011) Drug transport and the blood-brain barrier. In: Tihanyi K, Vastag M (eds) Solubility, delivery, and ADME problems of drugs and drug-candidates. Bentham Science Publication Ltd, Washington, DC, pp 144–165

54. Bendels S et al (2008) In silico prediction of brain and CSF permeation of small molecules using PLS regression models. Eur J Med Chem 43(8):1581–1592

55. Garg P, Verma J, Roy N (2008) In silico modeling for blood–brain barrier permeability predictions. In: Drug absorption studies. Springer, New York, NY, pp 510–556

56. Konovalov DA et al (2007) Benchmarking of QSAR models for blood-brain barrier permeation. J Chem Inf Model 47(4):1648–1656

57. Liu X et al (2004) Development of a computational approach to predict blood-brain barrier permeability. Drug Metab Dispos 32 (1):132–139

58. Abraham MH (2004) The factors that influence permeation across the blood–brain barrier. Eur J Med Chem 39(3):235–240

59. Lipinski CA et al (1997) Experimental and computational approaches to estimate solubility and permeability in drug discovery and development settings. Adv Drug Deliv Rev 23 (1-3):3–25

60. Lipinski CA (2004) Lead-and drug-like compounds: the rule-of-five revolution. Drug Discov Today Technol 1(4):337–341

61. Glave W, Hansch C (1972) Relationship between lipophilic character and anesthetic activity. J Pharm Sci 61(4):589–591

62. Pajouhesh H, Lenz GR (2005) Medicinal chemical properties of successful central nervous system drugs. NeuroRx 2(4):541–553

63. Hitchcock SA (2008) Blood–brain barrier permeability considerations for CNS-targeted compound library design. Curr Opin Chem Biol 12(3):318–323

64. Levin VA et al (1984) Relationship of octanol/water partition coefficient and molecular weight to cellular permeability and partitioning in S49 lymphoma cells. Pharm Res 1 (6):259–266

65. Young RC et al (1988) Development of a new physicochemical model for brain penetration and its application to the design of centrally acting H2 receptor histamine antagonists. J Med Chem 31(3):656–671

66. Abraham MH, Takács-Novák K, Mitchell RC (1997) On the partition of ampholytes: application to blood–brain distribution. J Pharm Sci 86(3):310–315

67. Liu X, Chen C, Smith BJ (2008) Progress in brain penetration evaluation in drug discovery and development. J Pharmacol Exp Ther 325 (2):349–356

68. Sakolish CM et al (2016) Modeling barrier tissues in vitro: methods, achievements, and challenges. EBioMedicine 5:30–39

69. Veszelka S, Kittel Á, Deli MA (2011) Tools of modelling blood–brain barrier penetrability. In: Tihanyi K, Vastag M (eds) Solubility, delivery, and ADME problems of drugs and drug-candidates. Bentham Science Publication Ltd, Washington, DC, pp 166–188

Chapter 10

Safety and Nanotoxicity Aspects of Nanomedicines for Brain-Targeted Drug Delivery

Johanna Catalan-Figueroa and Javier O. Morales

Abstract

Nanotechnology for brain drug delivery comprises the promise for future possibilities of successful treatment in several central nervous system pathologies currently deficient of curative treatments, such as neurodegenerative disorders and malignant glioblastoma. Nevertheless, the neurotoxic effects exerted by several types of nanomaterials are the same as those involved in the pathology of neurodegeneration; thus it is important to have a deep knowledge of these mechanisms, so that a proper approach can be taken into consideration. On the other hand, cancer cells usually respond differently to normal cells, being this characteristic a potential advantage for brain cancer therapy. In this chapter we analyze the mechanisms behind neurotoxic effects, from a multidisciplinary perspective, aiming to highlight the disadvantages of nanomaterials for the development of brain-targeted nanocarriers.

Key words Neurotoxicity, Neurodegeneration, Nanoparticle toxicity, Brain, Autophagy, Oxidative stress

1 Introduction

Drugs aiming for brain diseases' treatment must overcome the physiological barrier imposed by the blood-brain barrier (BBB), which separates the central nervous system (CNS) from systemic circulation [1]. The main function of the BBB is to protect the brain against: i) the effects of peripheral immune system cells and derivatives; ii) endogenous metabolites present in the bloodstream; and iii) toxins and potentially harmful xenobiotics [2]. Because of BBB, nowadays there is a dearth of effective treatments for brain-associated disorders, being most of them just symptomatic [1]. Lately, many efforts from the pharmaceutical industries and academic research have focused on the potential that nanotechnology may offer [3]. Nevertheless, to attain clinical applications, nanomedicine for brain-targeted drug delivery still requires a deeper knowledge and understanding of the potential neurotoxic effects that nanoparticles may exert over brain tissue. Health risks

Javier O. Morales and Pieter J. Gaillard (eds.), *Nanomedicines for Brain Drug Delivery*, Neuromethods, vol. 157, https://doi.org/10.1007/978-1-0716-0838-8_10, © Springer Science+Business Media, LLC, part of Springer Nature 2021 are determined as much as the intrinsic effect of the specific xeno-biotic (in this case, nanomaterials), plus the burden and lasting of the exposition; thus biocompatibility, biodegradability, and nano-materials' clearance capacity of the organism are the main charac-teristics to consider [4]. Nanomaterials' physicochemical parameters such as shape, size, zeta potential, and surface hydro-phobicity are determinants for their toxicity, being the main critical parameters size and zeta potential [5]. Size is a key factor that determines the extent of the total surface of the different nanoma-terial systems' (nanosystems) mass that may interact with biological structures (providing larger areas for nanosystems to react), as well as one of the parameters that define whether the organisms' immune response is triggered or not [6, 7]. On the other hand, zeta potential is involved in nanosystem stability (i.e., dispersion or agglomeration), and the interaction with charged biological mole-cules (e.g., proteins, nucleic acids, etc.) and with cellular organelles (e.g., lysosomes).

The necessity to assess and predict nanosystems' neurotoxic effects has led to a new nanotechnology discipline, the nanoneur-otoxicology. However, in general, the broad variety of nanomater-ials available for the development of engineered nanosystems makes it difficult to standardize nanotoxicological effects; and, in particu-lar, neurotoxicity is also hard to generalize, because of the different kind of pathologies that affect the CNS. Regardless, oxidative stress, inflammation, and proteolytic pathways' impairment are the main mechanisms that have been described as triggers for the toxicity exerted by nanoparticles [8]. Thus, these effects must be carefully assessed, since they may alter brain homeostasis, as well as increase deleterious processes involved in some CNS pathologies, such as neurodegenerative disorders; in contrast, they may provide advantages in the case of brain cancer treatment.

Human brain is a very complex organ, which has a morpho-functional organization and extensive metabolic needs (~20% of the total body mass energy demand) [9]. Given brain morphofunction-ality, the different neuroanatomic regions display diverse levels of physical and/or chemical susceptibility; in addition, high metabolic rates in the brain make it more vulnerable than other organs to oxidative stress and inflammation. The latter implies the need for intracellular structures to have a constant quality control and a proper turnover, processes that are mediated by mechanisms such as autophagy. Moreover, it is important to consider that neuronal regenerative capacity is very limited, whereas brain restoration by new neurons is impossible in most of the CNS neuroanatomic areas [10].

The knowledge about engineered nanosystems' toxic effects over the brain is still not as abundant as the data available for nanotoxicity on other tissues. Nevertheless, nanoscale systems' toxic effects have been studied for a long time, since the recognition of health problems induced by environmental pollution (i.e., nano-particulate matter) [11].

In this chapter we will discuss first general nanotoxicity mechanisms to provide wide-ranging overview that may facilitate a personal examination for the reader. Finally, we will present the neurotoxic effects for representative nanomaterials, as well as their potential biomedical applications, aiming for a proper and balanced analysis.

2   General Nanotoxicity Mechanisms

In general, the effects of free radicals on biomolecules, such as lipids, proteins, and nucleic acids, are well-known, being the brain particularly vulnerable to the effects of oxidative stress. This is mainly because of its high energy demand; thus, a high number of healthy mitochondria are necessary for brain functions to remain intact. In mitochondria, the electron cascade provides the demanded energy, but it also suffers a small physiological leak of electrons, which are precursors for the physiological amounts of the reactive oxygen species (ROS) [12]. Because of this constant ROS exposure, old or distressed mitochondria require a continuous turnover, process known as mitophagy (a specific type of autophagy) [13]. When there is an overproduction of oxygen free radicals, they trigger oxidative stress and the failure of cell redox homeostasis maintenance [14]. In turn, this may lead to cell dysfunction by oxidative stress induction of lipoperoxidation, proteotoxicity, and genotoxicity, as well as potential modifications on gene expression, and immune system inflammation response modulation [15].

Oxidative stress is one of the most characterized nanotoxicity mechanisms [16] and, as nanosystems' total surface area increases, also its capacity to react with cell membrane biomolecules increases [17]. Nanomaterials' ROS production can be direct (because of physicochemical properties) or indirect (due to the organism's reaction to a foreign body). In addition, even apparently unharmful substances can exert a toxic response when cells and tissues are exposed to their nanometric scale variants [17]. Excessive ROS production can induce DNA fragmentation, oxidative stress, micronucleus formation, or homolog chromatid exchange; therefore, it can turn into a carcinogenic cascade [18]. In addition, transition metals, such as iron, manganese, gold, and copper [19], have been extensively studied for their participation in the pathogenesis of neurodegenerative disorders, principally since they are well-known triggers for oxidative toxicity, which is self-perpetuated by the Haber-Weiss/Fenton reaction (Fig. 1) [20]. Oxidative stress is also related to the inflammation response, by the activation of

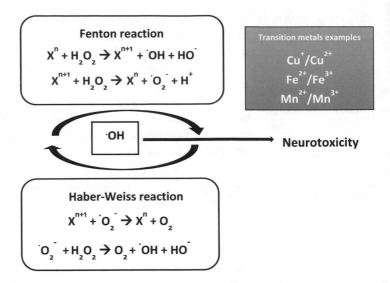

Fig. 1 Transition metals induction of Fenton/Haber-Weiss reaction (X^n/X^{n+1}). Highly reactive hydroxyl radical ($\cdot$OH) induces severe oxidative damage in susceptible tissues, as brain under neurodegenerative processes. Hydrogen peroxide, even though not highly reactive, has a high capacity of diffusion across cellular membranes, thus promoting oxidative damage propagation and autoperpetuation

immune cells such as neutrophils, which in turn can cause oxidative DNA damage in rat lung epithelial cells, thus potentially inducing carcinogenesis [21, 22]. In addition, macrophages are also known to significantly influence nanosystems-induced inflammatory response, since they play an important role in the uptake and clearance of inhaled nanoparticles (NPs) [23]. Furthermore, it has been shown that NPs' protein corona can activate macrophages through their interaction with surface receptors, inducing the secretion of pro-inflammatory mediators [24]. Also, it has been observed that endotoxin-free single-walled carbon nanotubes (SWCNTs) and graphene oxide nanosystems (GO) can be uptaken by primary human macrophages, but only SWCNTs were shown to exert chemokine production, which was independent from their cellular internalization, mainly regulated by the transcription factor NF-κβ [25]. Moreover, further analysis showed an important role of Toll-like receptor 2 (TLR2)/NF-κβ signaling pathway in the induction of chemokine production by macrophages, evidencing a direct activation of TLRs by SWCNTs protein corona free [25]. The TLRs sense pathogen-associated molecular patterns and induce the secretion of inflammatory cytokines, while NF-κβ positively modulates the immune system and inflammation, as well as apoptosis [26, 27]. In addition, it has been shown that NF-κβ stimulates the production of the beta-amyloid (Aβ) by upregulating the expression of beta-secretase 1 [28]. Interestingly, oxidative stress in neurons triggers the activation of transcriptional factors HIF1α, p53, and NF-κβ, as well as the induction of the reticulum stress response and, finally, the stimulation of macroautophagy through the activation of AMPK or ATG4 (an autophagy-related protein) [29]. Macroautophagy (from here on out called as autophagy) is a proteolytic process that uses lysosomes for the degradation of damaged cellular components, as well as for providing energy substrates in case of nutrient starvation [30], thus preserving the balance between synthesis and degradation. Autophagy flux processes can be summarized as follows [31]: (1) initiation, characterized by the phagophore formation, where the autophagy-related proteins (ATG) are recruited; (2) phagophore elongation and cargo sequestration during the autophagosome formation; (3) autophagosome maturation by its fusion with lysosome, turning into the structure known as autolysosomes; and (4) autolysosome content degradation. Once autophagy is activated, autophagosomes' formation is mediated by the conjugation of the soluble form of the microtubules-associated protein, MAP1LC3 (LC3), to phosphatidylethanolamine (PE), which is incorporated into the membranes of the autophagosomes in its non-soluble form (LC3-II), favoring autophagosomes' elongation and the p62-mediated cargo sequestration for their ulterior degradation by the lysosomes [32, 33]; these are the main autophagy markers most usually analyzed in the literature. Low levels of basal autophagy are required for the turnover of old or damaged cellular components (quality control autophagy), whereas ATP depletion, oxidative stress, and organelle dysfunction may elicit a stress-mediated autophagic response [34]. Interestingly, gold nanoparticles have been shown to colocalize with polyubiquitin protein aggregates [35, 36], suggesting that cells may target them for autophagy degradation, like invading microorganisms [37].

Furthermore, nanomaterials have been observed inside autophagosomes in several cell types, as well as they have been associated with lysosomal dysfunction [8]. In this context, it is important to highlight the fact that an augmentation on autophagosome markers does not necessarily implicate an enhancement of autophagy, but it may be reflecting an impairment of the autophagic flux, which might occur at different stages of autophagy. In different cellular models, diverse nanomaterials have been related to either a disruption in autophagic flux or autophagy induction [8, 38–41]. Regarding this, it is important to have taken into consideration that an augmentation of LC3-II or autophagic vacuoles does not necessarily implicate autophagy induction (further analysis can be found elsewhere) [4]. In addition, it has been shown that iron oxide NPs and gelatin NPs can form adducts with cytoskeleton proteins, potentially disrupting vesicle trafficking, thus impeding the maturation of autophagosomes [42, 43]. As it was aforementioned, autophagosomes' maturation involves their fusion with lysosomes;

thus, lysosomal nanomaterials-induced dysfunction can influence the disruption of autophagic processes. Lysosomal dysfunction can result from the sequestration into lysosomal compartment of non-biodegradable nanomaterials, such as fullerene-derived NPs and poly(amidoamine) (PAMAM) dendrimers [8, 44, 45]. Interestingly, lysosomal dysfunction has been related to the pathogenesis of sphingolipidoses and mucopolysacharidoses, which are lysosomal storage disorders that induce degenerative diseases in nervous and musculoskeletal systems [46–48]. On the other hand, nanosystems with positive zeta potential have been related to lysosomal membrane permeabilization (LMP), probably due to the "proton sponge effect." This could be the result of cationic NPs sequestration of protons in unsaturated amine groups, resulting in an over-activation of the H^+-ATPase (proton pump), which in turn may lead to an osmotic swelling and lysosome rupture. It is widely accepted in the literature that, in general, positively charged NPs are related to this hypothesis, which implies a LMP and a potential lysosomal rupture [49]. Interestingly, it has been reported that the inflammatory response induced by different nanomaterials (titanium dioxide fibers, carbon nanotubes, and amino-functionalized polystyrene NPs) [50–53] might be related to nanomaterials-TLR interaction, leading to an increase in the release of IL-1β, linked to caspase 1 and NLRP3 inflammasome activation [52, 53]. The inflammasome is a protein complex that acts as an activation platform for caspase-1 activation, characteristic for the innate immune response [54]. This response can be triggered by: (a) variations in the concentration of ions and/or intracellular-extracellular ATP; (b) the phagolysosome (phagocytic vesicle fusion with lysosome) destabilization; and (c) redox reaction-mediated mechanisms [54]. Interestingly, it has also been documented to be induced by several types of nanomaterials [55]. On the other hand, inflammasome signaling inhibition has been shown to be autophagy-dependent, probably by avoiding inflammasome ubiquitination and/or by the clearance of pro-IL-1β molecules and free radicals [57]. Interestingly, it has been observed that nanomaterials may exert inflammatory response, in the absence of cytotoxicity [58], probably due to the autophagy-mediated regulator effects.

3　Nanoneurotoxicity

3.1　Inorganic Nanoparticles

A wide variety of inorganic NPs have high potential for their use in biological applications as fluorescent labeling, biosensors, imaging contrast, tissue engineering, diagnostic, and treatment [59–62]. There is growing evidence in the literature supporting that small-sized nanoparticles (~5 nm) can cross the BBB, independently from their route of access into the organism. Moreover, it has also been demonstrated that these NPs can persist up to 4 months in brain tissue, among others [63, 64]. In particular, metal oxide NPs may exert oxidative stress as a common feature, mainly by the Haber-Weiss/Fenton reaction cycle [65] (Fig. 1).

3.1.1 Iron Oxide Nanoparticles

Iron oxide nanoparticles (IONPs) have great potential for nanotechnology-based diagnosis and therapy, also known as theranostic [66]. The IONPs have shown promising results in a pilot clinical assay of patients with recurrent glioblastoma, an aggressive and untreatable type of brain cancer [67, 68]. Nevertheless, in the brains of female rats exposed to acute oral doses of IONPs, it has been observed inhibition of Na^+-K^+, Mg^{2+}, and Ca^{2+}-ATPases that may have deleterious effects over membrane potential and conductivity [69]. There is register in the literature that IONPs can accumulate in brain tissue, by olfactory bulb translocation or by crossing the BBB from systemic circulation to the central nervous system (CNS), forming deposits in the striatum and hippocampus [70]. Iron accumulation may induce oxidative stress and neuroinflammation, since the brain is rich in polyunsaturated lipids and has a high iron content [71]. In addition, iron is known to be one of the triggers for Alzheimer's disease (AD) and Parkinson's disease (PD), both neurodegenerative disorders [72]. The hippocampus, in charge of memory consolidation, is the brain subregion that is mainly affected in AD, while PD is characterized for the impairment of the function of the striatum, a cerebral nucleus responsible for movement control and balance. Memory consolidation is a process that depends on glutamatergic neurotransmission (which when excessive may lead to calcium excitotoxicity), whereas dopamine (prone to oxidation in its free cytoplasm form) regulates striatum neurons triggering [19]. In the hippocampus and striatum of rats treated intraperitoneally for 28 days with different concentrations/day of IONPs (1, low dose (LD), 20.3 mg/kg/; 2, moderate dose (MD), 40.6 mg/kg; and 3, high dose (HD), 81.3 mg/kg), it was shown an increase in lipoperoxidation at day 7. Also, a significant enhancement of superoxide dismutase (SOD) activity was detected in MD and HD treated animals at day 14, which was reduced by day 28 [71]. In addition, in rats exposed to intranasal (i.n.) administration of IONPs, oxidative stress was detected in striatum, but not in the hippocampus, as an increase in hydrogen peroxide and a decrease in glutathione (GSH) levels [73]. On the other hand, i.n. administration to mice was shown to induce microglial proliferation and activation in the striatum, hippocampus, and olfactory bulb [74]. Microglial recruitment in the injured tissue is part of the host defense response and the CNS repair; however, when microglia are overactivated, it may cause an excess in the production of free radicals, impairing regional neural cells and self-perpetuating neuronal death [75].

3.1.2 Gold Nanoparticles   There is scarce bibliography regarding nanoneurotoxicity of gold nanoparticles (AuNPs); this is a concerning fact, since there are several studies showing their potential for therapy and diagnoses [59]. For example, it has been shown the potential advantages of gold-based nanosystems for the enhancement of Aβ clearance [76, 77] as well as for brain cancer treatment [78, 79], since they bioaccumulate in the brain tissue, reaching through either the olfactory nerve or by blood circulation [80, 81]. Nevertheless, there is controversial evidence regarding nanoneurotoxicity, with some studies showing no signs of toxicity upon acute administration to rats and mice of AuNPs [81, 82] or slight cytotoxicity mediated by 3 nm AuNPs, but no the larger ones (5, 7, 10, 30, and 60 nm) [83]; however, others have shown alterations in neural cells' homeostasis and viability as detailed in the following paragraphs. It is important to mention that it has been pinpointed that size, surface modifications, and geometry strongly affect gold-based nanosystems' toxicity [84, 85]. For instance, gold nanospheres (AuNS) have been shown to decrease to a greater extent mitochondrial function than gold nanorods (AuNR) and gold urchins (AuU) [84]. The three kinds of nanosystem were shown to exert a transient activation of microglia after their nasal administration to mice, along with the activation of the TLR2, related to Aβ microglial uptake [86]. In addition, AuNS coated with cetyltrimethylammonium bromide (CTAB, a cationic surfactant) were shown to exert more cytotoxicity toward microglia, regarding those coated with poly-ethylene-glycol (PEG, a nonionic surfactant) [84]. This might be because of (1) CTAB potential toxicity (data controversial in the literature) [87, 88]; (2) an increased cellular uptake, inducing more bioaccumulation; and/or (3) the positive charge of the AuNS. Nevertheless, AuNR showed slight cytotoxic effects [84]; therefore, it also might be the result of shape- and surface-related additive effects. Other kinds of AuNPs alterations over brain homeostasis have been observed, such as an augmentation of neuronal excitability in mice hippocampal brain slices [89]; these authors did not analyze the mechanisms underlying this effect, but it can be hypothesized that this may be due to AuNPs effects over ionic channels, since it has been previously observed that AuNPs are able to clog potassium and nicotinic acetylcholine receptor [90, 91].

On the other hand, it has been shown that AuNPs can induce astrogliosis (an over-increase in the number of reactive astrocytes) [92]. Astrogliosis is a part of a global response to CNS insult, which may potentially lead to a scar formation and permanent tissue rearrangement [93]. Male albino rats that orally received (by gastric tube) 400 µg/kg/day of AuNPs, during 8 weeks, exhibited an increase in reactive astrocytes (augmentation in the number of and length of foot process and the cell body diameter), which was more marked in the hippocampus than in the cerebral cortex.

In addition, they observed an augmentation in the number of caspase-3 (an apoptosis mediator) positive choroidal epithelium cells [92]. In general, in rat brain it has been observed that AuNPs generate lipoperoxidation, decrease the levels of glutathione peroxidase (GPx), as well as considerably increase the amounts of 8-hydroxi-deoxy-guanosine (8-OHdG) [94], a widespread biomarker for carcinogenesis [95]. In addition, caspase-3 and heat shock protein 70 (HsP70) levels were found augmented as well. The HsP70 has shown to exert several protective effects over CNS, such as against oxidative damage and Aβ aggregation [96, 97]. It also has been shown that AuNPs accumulation into rat hippocampus exerts cognitive impairment, on a size-dependent fashion [98].

3.1.3 Silver Nanoparticles

Silver NPs (AgNPs) have been shown to exert sensitizer-like effects in vitro and in vivo, over rat malignant glioma cells [99, 100]. For instance, in vivo studies have strongly suggested that AgNPs treatment is synergic to radiotherapy, since rats that received 10 Gy of radiation post 10–20 µg of AgNPs administration exhibited an augmentation on their life spans of approximately a 500%, while ~40% of rats showed no apparent disease after 200 days posttreatment. Moreover, an enhancement in the antimitotic and proapoptotic effects was observed when rats received AgNPs treatment after radiotherapy [101]. Thus, AgNPs have high potential as coadjutant treatment for the treatment of malignant glioma. Nevertheless, in neural primary cultures, it has been observed that AgNPs exert toxic effects, observing a diminution in cells' viability [102, 103]. Also, it has been observed that AgNPs can access to mice brain and impair the BBB permeability, altering normal cerebral functions [104], by either translocating through the olfactory nerve or by transcytosis across the cerebral endothelial cells [105]. In human neuroblastoma SH-SY5Y cells, AgNPs have shown to have the capacity of interacting with mitochondrial membrane and producing free radicals. Then, these free radicals can interact with the nitric oxide (NO) synthesized in response to the oxidative stress, leading later to the formation of highly toxic peroxynitrite species (RNS), with a potent capacity for lipoperoxidation, DNA oxidation, and proteotoxic stress. This finally can produce mitochondrial membrane permeabilization (MMP) and reticulum stress [106], both processes deeply involved in neurodegeneration [15, 107]. In addition, it has been observed in rats' primary mixed neural cell cultures (neurons and astrocytes) that the first response upon exposure to AgNPs was an increase in calcium signal (even before ROS generation) which may affect signal transduction and synaptic transmission. Interestingly, in this study, they found that astrocytes were more prone to the damage induced by AgNPs than neurons [102]. In addition, Xu et al. observed that in primary mixed cultures from rat cortex exposed to AgNPs, these exerted impairments in cell integrity and cell growth, as well as neurites overlap, at early stages and in full-grown cultures [103]. There were also observed vacuoles in the cytoplasm of cultures at early stages of neural outgrowth, when exposed to 1 μg/mL of AgNPs, whereas their structure was severely compromise at concentrations of 10 μg/mL, observing alterations in the morphological integrity as well as a great extent of neurites' degeneration. In addition, in mature cultures, neuronal branches' overlap was decreased (1 μg/mL), while at 10 μg/mL, cluster aggregation of cell bodies was observed along with neurite disruption and damage of glial cell layers [103]. All these effects appeared to be the consequence of the observed concentration-dependent reduction of neuron and glial cells positively labeled for F-actin and β-tubulin, which are cytoskeleton proteins essential for the correct intracellular trafficking and cellular structure, as well as other functions [108]. Interestingly, it has been observed that AgNPs may alter autophagic flux [109], which can be potentially related to their effects over cytoskeleton proteins, as well as for their effects over lysosome pH and membrane potential [110]. In addition, Huang et al. observed in mice cell lines that AgNPs could induce the secretion of IL-1 while penetrating the cellular membrane in cell lines of astrocytes (ALT), microglia (BV-2), and neuroblastoma (N2a) [105]. Interestingly, they also observed the induction on gene expression of the proteins CXCL13, MARCO, and glutathione synthetase (GSS). The MARCO protein is a scavenger receptor involved in the uptake of Aβ by microglia, while CXCL13 protein is a chemokine that controls monocytes' migration and adhesion; both proteins are strongly related to AD [111, 112]. On the other hand, the expression of the amyloid precursor protein (APP) was increased, while the levels of the LDL receptor (LDL-R) and the neutral endopeptidase (NEP) were decreased. The LDL-R and NEP are involved in the uptake and degradation of Aβ [113, 114]; thus, the exposure to AgNPs may induce an accumulation of Aβ, which is involved in the pathogenesis of AD [115].

In the case of in vivo research, it has been observed the induction of toxic effects in the brains of mice exposed to AgNPs, such as augmentation of ROS levels [95]. In this study, it was shown that brain's glutathione peroxidase (GPx) activity was decreased, while it was enhanced in kidneys and spleen. On the other hand, glutathione transferase (GST) activity was increased in brain tissue and diminished in kidneys and spleen, suggesting that the response varies according to the different analyzed tissues. Changes were more prominent when administrating 2 μM of AgNPs than with 1 μM [95]. It is noteworthy the fact that AgNPs have a high affinity toward thiol groups, which are of great importance in protein assembly and GSH function [116]. Thus, AgNPs may exert deleterious effects over cellular redox status and protein folding, which are among the main mechanisms of neurodegeneration [4]. Moreover, AgNPs are known for their ability for DNA binding, which, even if it may be advantageous in the case of cancer treatment, also can contribute to genotoxicity, by forming adducts of 8-hydroxi-2-deoxyguanosine (8-OHdG) [95], indicating DNA oxidative damage.

3.1.4 Zinc Oxide Nanoparticles

Zinc oxide NPs (ZnO-NPs) have gather plenty attention during the last years, mainly because of their antibacterial and anticancer properties [117, 118]. In addition, it has been recently suggested the possibility to use ZnO nanocluster for the diagnoses of AD at early stages [119].

However, there has been described for this kind of nanoparticles reproductive and developmental toxicity; for example, it has been shown that the ZnO-NPs oral administration of 400–500 mg/kg/day in pregnant rats exerted embryotoxic effects as well as in rats' offspring [120]. Xiaoli et al. [121] observed that when pregnant rats were exposed during 18 days to ZnO-NPs at 500 mg/kg/day, their 2-day-old offspring exhibited significant zinc brain accumulation. Their analysis of antioxidant status showed augmentation in the concentration of ROS and malondial-dialdehyde (MDA), as well as a decrease in the activity of SOD and GPx and the presence of 8-OHdG-positive neurons in the brain of pups exposed to ZnO-NPs. The augmentation of ROS correlates with the increase of MDA (a marker for lipid peroxidation) and of 8-OhdG (indicating nucleic acids' oxidative damage). The enzymes that were shown to be affected play an important role in cells' antioxidant capacity; SOD catalyzes the conversion of superoxide radical to hydrogen peroxide (H_2O_2), while GSH-Px transforms H_2O_2 to H_2O (see Eq. 1). In addition to the oxidative damage of nucleic acids observed, the fetal exposure to ZnO-NPs has shown to alter the expression of genes related to cellular redox homeostasis, such as the upregulation at postnatal day 21 of Gsst1 (subunit of GSH-transferase) and Alox12b (lipid oxidation) [121].

$$M^{(n+1)+}\text{-SOD} + O_2^- \rightarrow M^{n+}\text{-SOD} + O_2.$$

$$M^{n+}\text{-SOD} + O_2^- + 2H^+ \rightarrow M^{(n+1)+}\text{-SOD} + H_2O_2. \qquad (1)$$

$$2GSH + H_2O_2 \rightarrow GS\text{-}SG + 2H_2O.$$

where $M = Cu$ ($n = 1$); Mn ($n = 2$); Fe ($n = 2$); and Ni ($n = 2$).

Cell exposure to ZnO-NP has shown to induce alteration of mitochondrial and cell membranes, resulting in an augmentation of ROS production, followed by the destruction of mitochondria and cell death. Toxicity exerted by ZnO-NPs is mainly determined by their dissolution and zinc homeostatic interference [122]. Zinc can be found as a structural component of proteins and, in its free form, as a neurotransmitter involved in the consolidation of long-term memory, being released along with glutamate [123]. It has been shown that excess of zinc in the synaptic cleft induces the upregulation of the calcium AMPA/kainate channels, favoring the zinc access into neurons, thus increasing its free form in the cytoplasm (where in physiological conditions is kept inside vesicles). This cytoplasmic zinc augmentation has shown to inhibit redox homeostasis enzymes and mitochondrial respiration, inducing an energetic depletion and ROS production [124].

Moreover, ZnO-NPs were observed to accumulate in the neural synapse of the pups from the pregnant rats exposed to ZnO-NPs, meaning that they could be transported across the blood-placental barrier, reaching to the fetus' brain [121]. This could lead to further brain damage, affecting brain function. This is supported by the finding of autophagosomes in the cytoplasm of the offspring neurons, which are structures that have been extensively related to neurodegenerative diseases, as a defense mechanism against oxidative stress and inflammation.

Nevertheless, it seems that ZnO-NPs effects are dose dependent, since biphasic responses (hormesis) [125] have been described in the brain tissue of *O. nicotilus* and *T. zilli*; at low levels (500 µg/L), ZnO-NPs were shown to produce a significant antioxidant response, reflected in the decrease of MDA, the augmentation of GSH levels, and the activity of antioxidant enzymes (catalase (CAT); SOD; glutathione reductase (GR); GST; and GPx), contrary to what was observed at concentrations of 2000 µg/L [127].

3.1.5 Titanium Dioxide Nanoparticles

It has been observed that titanium dioxide nanoparticles (TiO_2-NP) can be used as photosensitizers since, after their photocatalytic activation, TiO_2-NP have been observed to inhibit the in vitro growth of malignant glioma cells [128, 129], as well as prolonging the survival of glioma mice treated with TiO_2-NP photodynamic therapy [129] and inhibiting metastasis [130]. Nevertheless, it has been shown that after the nasal exposure to TiO_2-NPs, they accumulate in the hippocampus [131], which is involved in the pathogenesis of AD and Lewy body dementia [132]. It has also been observed, in rats' primary hippocampal neuronal culture and in mice, that TiO_2-NPs in a concentration-dependent manner may induce augmentation of ROS and intracellular calcium levels, as well as the loss of MMP, the downregulation of Bcl-2 (a protein that inhibits apoptosis and autophagy, by sequestering Beclin-1), and the upregulation of caspase-3 and Bax levels, both proapoptotic proteins [133–135]. Interestingly, Hong et al. observed in primary hippocampal cultures that TiO_2-NPs significantly increase glutamate levels while decreasing glutamine, glutamine synthetase, and ATP levels [136], thus altering excitatory synapsis cycle. These alterations have been related to several CNS illnesses, such as AD and encephalopathy [137–139], and are potentially the cause of the observed neurites' development suppression and the synaptic plasticity detriment induced by the TiO_2-NPs [136]. In fact, in utero exposure to TiO_2-NPs was shown to exert alterations in the redox homeostasis, inducing lipoperoxidation and nucleic acids' oxidative damage in the brain of rats' offspring. Moreover, the rats exposed to TiO_2-NPs in their fetal stage, during their adulthood exhibited depressive-like behavior and the impairment of learning and memory consolidation [140–142]. On the other hand, it has also been observed that TiO_2-NPs exposure induces a decrease in the expression of phosphorylated CREB in mice hippocampus, as well as for their target genes, which are involved in the development of brain synapsis and memory consolidation [143, 144]. It is noteworthy that CREB phosphorylation has shown to be dependent on the NMDA glutamate receptor [145, 146], a calcium ion channel deeply involved in neuroplasticity processes related to learning and memory, whose function has shown to be altered in chronic stress, pain, and neurodegenerative disorders [147–150]. Finally, TiO_2-NPs also have shown to impair the Wnt canonical and non-canonical signaling, involved in expression of microtubule cytoskeleton in the increase of dendritic arborization [151]; however, the effects of however, the effects of however, the effects of TiO_2-NPs over Wnt signaling still need further analysis and evaluations, since the results obtained by Hong et al. are controversial in light of previous literature on the effects of the Wnt pathway over neurodevelopment and neurodegeneration [152–155].

3.2 Carbon-Based Nanosystems

Carbon atoms' unique ability to establish covalent bonds with each other in diverse hybridization states or with other non-metallic elements allows them to form a wide variety of structures; thus, they can be found in different arrangements in natural state such as amorphous carbon, diamond, and graphite. Although these materials are all only carbon-based, their properties are totally different, due to the way that carbon atoms are distributed: diamonds are the hardest material known and work as an electric isolator, while graphite is soft and has a high electroconductivity [156]. In the age of engineered nanomaterials, new carbon-based systems have emerged, such as fullerenes (being the C60 structure the most studied), carbon nanotubes (CNTs), and graphene [156].

Data in the literature suggest that C60 derivates may have neuroprotective effects in the case of AD and prion diseases, by reducing the microglial-mediated inflammatory response [157] and attenuating the effects of Aβ [158, 159]. In addition, C60 derivates also have shown to have anti-HIV effects, by inhibiting the viral reverse-transcriptase and protease [160, 161]. Nevertheless, rats exposed to C60 (size ≤460 nm) were shown to have spatial memory and learning impairments, potentially by a reduction in the brain-derived neurotrophic factor (BDNF) levels [162]. The BDNF induces two survival pathways: one mediated by the class I phosphatidylinositol triphosphate kinase (PI3K-I) and the other by the mitogen-activated protein kinase (MAPK), Raf [163]. The first one is associated with the mTORC1 complex, which is mainly in charge of protein synthesis stimulation, inhibiting autophagy [164, 165]. On the other hand, Raf induces the survival signaling pathway of ERK/CREB/Bcl-2, previously mentioned (Subheading 3.1.4). It is noteworthy that both pathways are essential for synaptic plasticity, since dendritic spines requires protein synthesis for their proper maturation and synapsis increase and consolidation [166].

In turn, CNTs are considered one of the most promising materials for cancer multimodal therapy, mainly because of their physicochemical, electromagnetic, and optical characteristics. They can be applied as (1) highly sensitive labeling agents in imaging; (2) drug delivery systems; (3) and tissue engineering; and (4) nanosurgical agents by thermal ablation [167, 168]; hence, they are potential and strong candidates for multimodal anticancer agents, aiming to an early diagnosis and a broad-spectrum treatment. Moreover, it has been suggested that pegylated CNTs can enhance neurites' outgrowth and may reduce astrocytes' glial scar [169, 170]. However, they have asbestos similar properties (such as extremely high length/diameter ratios and low solubility), while CNTs' in vitro toxicity has shown to be greater than asbestos' [171, 172]; thus a cancerogenic CNTs' effect could be associated. Nevertheless, CNTs and asbestos differ in the mechanisms whereby they enter into mesothelial cells, and no reports in the literature have reported CNTs as carcinogenic [173, 174]. On the other hand, single-walled CNTs (SWCNTs) have shown to inhibit the proliferation of PC12 cells (rat pheochromocytoma), as well as induce apoptosis at concentration between 100 and 400 μg/mL and the MMP reduction, whereby ROS production and lipoperoxidation were promoted along with a decrease in the activities of SOD, GPx, and CAT enzymes and the content GSH, in a dose-dependent manner [175]. Moreover, there are studies that have shown multiwalled CNTs (MWCNTs) capacity to induce malignant glioma cells (RG2) apoptosis and to improve the survival of rats with intracerebral glioma, by the induction of mitochondrial dysfunction [176]. Interestingly, Romano-Feinholz et al. [176] also determined the effects of MWCNTs in healthy rats after their intracerebral injection, observing neither inflammation nor diffusion of MWCNTs into healthy tissue at the site of injection, contrasting with MWCNTs intratumoral injection. This was probably by either tumor cells' increased capacity for endocytosis or because of the tumor infiltration by immune cells [176]. Interestingly, the aforementioned cytotoxic effects mostly appear as advantages for brain cancer treatment. It is noteworthy to mention that CNTs' impurities, such as iron (used as catalyzer during CNTs' synthesis), are able to be solubilized in biological systems, inducing toxic effects, which are reduced when metallic impurities are properly removed [167, 177].

Finally, graphene is the last carbon-based nanomaterial to appear on the road, derived from graphite exfoliation [178]; hence, biomedical applications are not as abundant since graphene research is still at early stages. Interestingly, in the recent years, graphene has been tested for drug and gene delivery, as well as for photothermal and photodynamic ablation and for the improvement of biosensing and bioimaging [179]. Nevertheless, the research about their toxic effects is still controversial, while long-term toxicity is yet to be known, and some research have shown that it may induce PD-like signs and symptoms in zebrafish larvae, such as the formation of α-synuclein/ubiquitin aggregates (Lewy bodies), massive dopaminergic neuron loss, mitochondrial damage, and alterations of motor performance [180]. In addition, it has been shown in PC12 cells graphene induction of apoptosis and cell cycle arrestment, associated with ERK signaling pathway imbalance [181]. Therefore, although graphene is a promising nanomaterial, further studies are still required.

4 Conclusion and Future Perspectives

An increasing number of nanosystems have appear in the recent years, providing promising drug delivery systems, as well as novel probing diagnosis strategies. In this context, nanomedicine may have a significant impact in the treatment of brain illnesses, due to the ability of engineered nanosystems to cross the BBB. However, the toxicity/therapeutic equilibrium is a critical concern to be explored, since nanotoxic effects are mainly determined by the dose applied and the physiological/pathological state of the organism analyzed; therefore, whether oxidative stress increase and triggering and blocking autophagy are desirable effects or not will depend on the specific pathology. For instance, in cancer treatment, the aim is to be able to target cytotoxicity specifically toward cancer cells, without affecting healthy ones. On the other hand, regarding neurodegeneration, autophagy induction is suggested to be one of the first lines of defense, exhibiting in patients the extensive presence of autophagosomes and autolysosomes accumulation [182]. Therefore, in the case of other CNS diseases, such as depressive states, autism, and infections, it must be carefully assessed their individual pathological state for a proper analysis of the potential toxic effects of the nanomaterial of interest.

Acknowledgments

The authors acknowledge the support for this work of ANID/ FONDECYT Regular/1181689, ANID/PIA/ACT192144, and ANID/FONDAP/15130011.

References

1. Gabathuler R (2010) Approaches to transport therapeutic drugs across the blood–brain barrier to treat brain diseases. Neurobiol Dis 37 (1):48–57

2. Dyrna F, Hanske S, Krueger M, Bechmann I (2013) The blood-brain barrier. J Neuroimmune Pharmacol 8(4):763–773

3. Comoglu T, Arisoy S, Akkus ZB (2017) Nanocarriers for effective brain drug delivery. Curr Top Med Chem 17(13):1490–1506

4. Catalan-Figueroa J, Palma-Florez S, Alvarez G, Fritz HF, Jara MO, Morales JO (2016) Nanomedicine and nanotoxicology: the pros and cons for neurodegeneration and brain cancer. Nanomedicine 11(2):171–187

5. Jeevanandam J, Barhoum A, Chan YS, Dufresne A, Danquah MK (2018) Review on nanoparticles and nanostructured materials: history, sources, toxicity and regulations. Beilstein J Nanotechnol 9:1050–1074

6. Bantz C, Koshkina O, Lang T et al (2014) The surface properties of nanoparticles determine the agglomeration state and the size of the particles under physiological conditions. Beilstein J Nanotechnol 5:1774–1786

7. Drasler B, Sayre P, Steinhäuser KG, Petri-Fink A, Rothen-Rutishauser B (2017) In vitro approaches to assess the hazard of nanomaterials. NanoImpact 8:99–116

8. Stern ST, Adiseshaiah PP, Crist RM (2012) Autophagy and lysosomal dysfunction as emerging mechanisms of nanomaterial toxicity. Part Fibre Toxicol 9:20

9. Clarke DD, Sokoloff L (1999) Regulation of cerebral metabolic rate. http://www.ncbi.nlm.nih.gov/books/NBK28194/

10. Barker RA, Götz M, Parmar M (2018) New approaches for brain repair-from rescue to reprogramming. Nature 557(7705):329–334

11. Lovisolo D, Dionisi M, Ruffinatti FA, Distasi C (2018) Nanoparticles and potential neurotoxicity: focus on molecular mechanisms. AIMS Mol Sci 5(1):1–13

12. Rossignol DA, Frye RE (2014) Evidence linking oxidative stress, mitochondrial dysfunction, and inflammation in the brain of individuals with autism. Front Physiol 5:150

13. Imai Y, Lu B (2011) Mitochondrial dynamics and mitophagy in Parkinson's disease: disordered cellular power plant becomes a big deal in a major movement disorder. Curr Opin Neurobiol 21(6):935–941

14. Kiffin R, Bandyopadhyay U, Cuervo AM (2006) Oxidative stress and autophagy. Antioxid Redox Signal 8(1–2):152–162

15. Bhat AH, Dar KB, Anees S et al (2015) Oxidative stress, mitochondrial dysfunction and neurodegenerative diseases; a mechanistic insight. Biomed Pharmacother 74:101–110

16. Marano F, Hussain S, Rodrigues-Lima F, Baeza-Squiban A, Boland S (2011) Nanoparticles: molecular targets and cell signalling. Arch Toxicol 85(7):733–741

17. Oberdörster G, Maynard A, Donaldson K et al (2005) Principles for characterizing the potential human health effects from exposure to nanomaterials: elements of a screening strategy. Part Fibre Toxicol 2:8

18. Pietroiusti A, Stockmann-Juvala H, Lucaroni F, Savolainen K (2018) Nanomaterial exposure, toxicity, and impact on human health. Wiley Interdiscip Rev Nanomed Nanobiotechnol. https://doi.org/10.1002/wnan.1513

19. Paris I, Segura-Aguilar J (2011) The role of metal ions in dopaminergic neuron degeneration in Parkinsonism and Parkinson's disease. Chem. Mon 142(4):365–374

20. Kanti Das T, Wati MR, Fatima-Shad K (2014) Oxidative stress gated by Fenton and Haber Weiss reactions and its association with Alzheimer's disease. Arch Neurosci 2(3):1–8. http://www.archneurosci.com/?page=article&article_id=20078

21. Knaapen AM, Schins RPF, Borm PJA, van Schooten FJ (2005) Nitrite enhances neutrophil-induced DNA strand breakage in pulmonary epithelial cells by inhibition of myeloperoxidase. Carcinogenesis 26 (9):1642–1648

22. Knaapen AM, Güngör N, Schins RPF, Borm PJA, Van Schooten FJ (2006) Neutrophils and respiratory tract DNA damage and mutagenesis: a review. Mutagenesis 21(4):225–236

23. Donaldson K, Schinwald A, Murphy F et al (2013) The biologically effective dose in inhalation nanotoxicology. Acc Chem Res 46(3):723–732

24. Deng ZJ, Liang M, Monteiro M, Toth I, Minchin RF (2011) Nanoparticle-induced unfolding of fibrinogen promotes mac-1 receptor activation and inflammation. Nat Nanotechnol 6(1):39–44

25. Mukherjee SP, Bondarenko O, Kohonen P et al (2018) Macrophage sensing of single-walled carbon nanotubes via Toll-like receptors. Sci Rep 8:1115. https://www.ncbi.nlm.nih.gov/pmc/articles/PMC5773626/

26. Li G, Xia Z, Liu Y et al (2018) SIRT1 inhibits rheumatoid arthritis fibroblast-like synoviocyte aggressiveness and inflammatory response via suppressing NF-κB pathway. Biosci Rep 38(3):pii:BSR20180541

27. Ran J, Ma C, Xu K et al (2018) Schisandrin B ameliorated chondrocytes inflammation and osteoarthritis via suppression of NF-κB and MAPK signal pathways. Drug Des Devel Ther 12:1195–1204

28. Yun J, Yeo IJ, Hwang CJ et al (2018) Estrogen deficiency exacerbates Aβ-induced memory impairment through enhancement of neuroinflammation, amylodogenesis and NF-κB activation in ovariectomized mice. Brain Behav Immun 73:282–293

29. He C, Klionsky DJ (2009) Regulation mechanisms and signaling pathways of autophagy. Annu Rev Genet 43:67–93

30. Mizushima N (2007) Autophagy: process and function. Genes Dev 21(22):2861–2873

31. Nikoletopoulou V, Papandreou ME, Tavernarakis N (2015) Autophagy in the physiology and pathology of the central nervous system. Cell Death Differ 22(3):398–407

32. Abounit K, Scarabelli TM, McCauley RB (2012) Autophagy in mammalian cells. World J Biol Chem 3(1):1–6

33. McEwan DG, Dikic I (2011) The three musketeers of autophagy: phosphorylation, ubiquitylation and acetylation. Trends Cell Biol 21(4):195–201

34. Murrow L, Debnath J (2013) Autophagy as a stress-response and quality-control mechanism: implications for cell injury and human disease. Annu Rev Pathol 8:105–137

35. Brancolini G, Kokh DB, Calzolai L, Wade RC, Corni S (2012) Docking of ubiquitin to gold nanoparticles. ACS Nano 6(11):9863–9878

36. Calzolai L, Franchini F, Gilliland D, Rossi F (2010) Protein--nanoparticle interaction: identification of the ubiquitin—gold nanoparticle interaction site. Nano Lett 10(8):3101–3105

37. Zheng YT, Shahnazari S, Brech A, Lamark T, Johansen T, Brumell JH (2009) The adaptor protein p62/SQSTM1 targets invading bacteria to the autophagy pathway. J Immunol 183(9):5909–5916

38. Chen Y, Yang L, Feng C, Wen L-P (2005) Nano neodymium oxide induces massive vacuolization and autophagic cell death in non-small cell lung cancer NCI-H460 cells. Biochem Biophys Res Commun 337(1):52–60

39. Harhaji L, Isakovic A, Raicevic N et al (2007) Multiple mechanisms underlying the anticancer action of nanocrystalline fullerene. Eur J Pharmacol 568(1–3):89–98

40. Seleverstov O, Zabirnyk O, Zscharnack M et al (2006) Quantum dots for human mesenchymal stem cells labeling. A size-dependent autophagy activation. Nano Lett 6(12):2826–2832

41. Stern ST, Zolnik BS, McLeland CB, Clogston J, Zheng J, McNeil SE (2008) Induction of autophagy in porcine kidney cells by quantum dots: a common cellular response to nanomaterials? Toxicol Sci 106(1):140–152

42. Berry CC, Wells S, Charles S, Curtis ASG (2003) Dextran and albumin derivatised iron oxide nanoparticles: influence on fibroblasts in vitro. Biomaterials 24(25):4551–4557

43. Gupta AK, Gupta M, Yarwood SJ, Curtis ASG (2004) Effect of cellular uptake of gelatin nanoparticles on adhesion, morphology and cytoskeleton organisation of human fibroblasts. J Control Release 95(2):197–207

44. Shcharbin D, Jokiel M, Klajnert B, Bryszewska M (2006) Effect of dendrimers on pure acetylcholinesterase activity and structure. Bioelectrochemistry 68(1):56–59

45. Ueng TH, Kang JJ, Wang HW, Cheng YW, Chiang LY (1997) Suppression of microsomal cytochrome P450-dependent monooxygenases and mitochondrial oxidative phosphorylation by fullerenol, a polyhydroxylated fullerene C60. Toxicol Lett 93(1):29–37

46. Aldenhoven M, Sakkers RJB, Boelens J, de Koning TJ, Wulffraat NM (2009) Musculoskeletal manifestations of lysosomal storage disorders. Ann Rheum Dis 68(11):1659–1665

47. Bellettato CM, Scarpa M (2010) Pathophysiology of neuropathic lysosomal storage disorders. J Inherit Metab Dis 33(4):347–362

48. Ravikumar B, Sarkar S, Davies JE et al (2010) Regulation of mammalian autophagy in physiology and pathophysiology. Physiol Rev 90 (4):1383–1435

49. Nel AE, Mädler L, Velegol D et al (2009) Understanding biophysicochemical interactions at the nano-bio interface. Nat Mater 8 (7):543

50. Franchi L, Eigenbrod T, Muñoz-Planillo R, Nuñez G (2009) The inflammasome: a caspase-1-activation platform that regulates immune responses and disease pathogenesis. Nat Immunol 10(3):241–247

51. Hamilton RF, Wu N, Porter D, Buford M, Wolfarth M, Holian A (2009) Particle length-dependent titanium dioxide nanomaterials toxicity and bioactivity. Part Fibre Toxicol 6:35

52. Lunov O, Syrovets T, Loos C et al (2011) Amino-functionalized polystyrene nanoparticles activate the NLRP3 inflammasome in human macrophages. ACS Nano 5 (12):9648–9657

53. Meunier E, Coste A, Olagnier D et al (2012) Double-walled carbon nanotubes trigger IL-1β release in human monocytes through Nlrp3 inflammasome activation. Nanomedicine 8(6):987–995

54. Suárez R, Buelvas N (2015) Inflammasome: activation mechanisms. Invest Clin 56 (1):74–99

55. Farrera C, Fadeel B (2015) It takes two to tango: understanding the interactions between engineered nanomaterials and the immune system. Eur J Pharm Biopharm 95:3–12

56. Alcocer-Gómez E, Casas-Barquero N, Williams MR et al (2017) Antidepressants induce autophagy dependent-NLRP3-inflammasome inhibition in major depressive disorder. Pharmacol Res 121:114–121

57. Latz E, Xiao TS, Stutz A (2013) Activation and regulation of the inflammasomes. Nat Rev Immunol 13(6):397–411. https://www.ncbi.nlm.nih.gov/pmc/articles/PMC3807999/

58. Bhattacharya K, Kiliç G, Costa PM, Fadeel B (2017) Cytotoxicity screening and cytokine profiling of nineteen nanomaterials enables hazard ranking and grouping based on inflammogenic potential. Nanotoxicology 11 (6):809–826

59. Cabuzu D, Cirja A, Puiu R, Grumezescu AM (2015) Biomedical applications of gold nanoparticles. Curr Top Med Chem 15 (16):1605–1613

60. Derfus AM, Chan WCW, Bhatia SN (2004) Probing the cytotoxicity of semiconductor quantum dots. Nano Lett 4(1):11–18

61. Hope MD, Hope TA, Zhu C et al (2015) Vascular imaging with ferumoxytol as a contrast agent. Am J Roentgenol 205(3):W366–W373

62. Kim J-H, Kim D-K, Lee OJ et al (2016) Osteoinductive silk fibroin/titanium dioxide/hydroxyapatite hybrid scaffold for bone tissue engineering. Int J Biol Macromol 82:160–167

63. Disdier C, Devoy J, Cosnefroy A et al (2015) Tissue biodistribution of intravenously administrated titanium dioxide nanoparticles revealed blood-brain barrier clearance and brain inflammation in rat. Part Fibre Toxicol 12:27

64. Lee JH, Kim YS, Song KS et al (2013) Biopersistence of silver nanoparticles in tissues from Sprague–Dawley rats. Part Fibre Toxicol 10(1):36

65. Liochev SI, Fridovich I (2002) The Haber-Weiss cycle – 70 years later: an alternative view. Redox Rep Commun Free Radic Res 7 (1):55–57. author reply 59–60

66. Xie J, Lee S, Chen X (2010) Nanoparticle-based theranostic agents. Adv Drug Deliv Rev 62(11):1064–1079

67. Maier-Hauff K, Rothe R, Scholz R et al (2007) Intracranial thermotherapy using magnetic nanoparticles combined with external beam radiotherapy: results of a feasibility study on patients with glioblastoma multiforme. J Neuro-Oncol 81(1):53–60

68. Maier-Hauff K, Ulrich F, Nestler D et al (2011) Efficacy and safety of intratumoral thermotherapy using magnetic iron-oxide nanoparticles combined with external beam radiotherapy on patients with recurrent glioblastoma multiforme. J Neuro-Oncol 103 (2):317–324

69. Kumari M, Rajak S, Singh SP et al (2013) Biochemical alterations induced by acute oral doses of iron oxide nanoparticles in Wistar rats. Drug Chem Toxicol 36(3):296–305

70. Wang B, Feng WY, Wang M et al (2007) Transport of intranasally instilled fine Fe2O3 particles into the brain: micro-distribution, chemical states, and histopathological observation. Biol Trace Elem Res 118(3):233–243

71. Vyas K, Bhatt D, Soni I, John PJ (2018) Iron oxide nanoparticle (IONP): chemical synthesis and neurotoxic studies in Wistar rat. Int J Zool Appl Biosci 3(1):24–33

72. Myhre O, Utkilen H, Duale N, Brunborg G, Hofer T (2013) Metal dyshomeostasis and inflammation in Alzheimer's and Parkinson's diseases: possible impact of environmental exposures. Oxid Med Cell Longev 2013:726954. https://www.ncbi.nlm.nih.gov/pmc/articles/PMC3654362/

73. Wu J, Ding T, Sun J (2013) Neurotoxic potential of iron oxide nanoparticles in the rat brain striatum and hippocampus. Neurotoxicology 34:243–253

74. Wang Y, Wang B, Zhu M-T et al (2011) Microglial activation, recruitment and phagocytosis as linked phenomena in ferric oxide nanoparticle exposure. Toxicol Lett 205 (1):26–37

75. Block ML, Zecca L, Hong J-S (2007) Microglia-mediated neurotoxicity: uncovering the molecular mechanisms. Nat Rev Neurosci 8(1):57–69

76. Liao Y-H, Chang Y-J, Yoshiike Y, Chang Y-C, Chen Y-R (2012) Negatively charged gold nanoparticles inhibit Alzheimer's amyloid-β fibrillization, induce fibril dissociation, and mitigate neurotoxicity. Small 8 (23):3631–3639

77. Morales-Zavala F, Arriagada H, Hassan N et al (2017) Peptide multifunctionalized gold nanorods decrease toxicity of β-amyloid peptide in a *Caenorhabditis elegans* model of Alzheimer's disease. Nanomedicine 13 (7):2341–2350

78. Cheng Y, Dai Q, Morshed RA et al (2014) Blood-brain barrier permeable gold nanoparticles: an efficient delivery platform for enhanced malignant glioma therapy and imaging. Small 10(24):5137–5150

79. Hainfeld JF, Smilowitz HM, O'Connor MJ, Dilmanian FA, Slatkin DN (2013) Gold nanoparticle imaging and radiotherapy of brain tumors in mice. Nanomedicine 8 (10):1601–1609

80. Balasubramanian SK, Poh K-W, Ong C-N, Kreyling WG, Ong W-Y, Yu LE (2013) The effect of primary particle size on biodistribution of inhaled gold nano-agglomerates. Biomaterials 34(22):5439–5452

81. Lasagna-Reeves C, Gonzalez-Romero D, Barria MA et al (2010) Bioaccumulation and toxicity of gold nanoparticles after repeated administration in mice. Biochem Biophys Res Commun 393(4):649–655

82. Guerrero S, Araya E, Fiedler JL et al (2010) Improving the brain delivery of gold nanoparticles by conjugation with an amphipathic peptide. Nanomedicine 5(6):897–913

83. Trickler WJ, Lantz SM, Murdock RC et al (2011) Brain microvessel endothelial cells responses to gold nanoparticles: in vitro pro-inflammatory mediators and permeability. Nanotoxicology 5(4):479–492

84. Hutter E, Boridy S, Labrecque S et al (2010) Microglial response to gold nanoparticles. ACS Nano 4(5):2595–2606

85. Velasco-Aguirre C, Morales F, Gallardo-Toledo E et al (2015) Peptides and proteins used to enhance gold nanoparticle delivery to the brain: preclinical approaches. Int J Nanomedicine 10:4919–4936

86. Chen K, Iribarren P, Hu J et al (2006) Activation of toll-like receptor 2 on microglia promotes cell uptake of Alzheimer disease-associated amyloid β peptide. J Biol Chem 281(6):3651–3659

87. Hauck TS, Ghazani AA, Chan WCW (2008) Assessing the effect of surface chemistry on gold nanorod uptake, toxicity, and gene expression in mammalian cells. Small Weinh Bergstr Ger 4(1):153–159

88. Niidome T, Yamagata M, Okamoto Y et al (2006) PEG-modified gold nanorods with a stealth character for in vivo applications. J Control Release 114(3):343–347

89. Jung S, Bang M, Kim BS et al (2014) Intracellular gold nanoparticles increase neuronal excitability and aggravate seizure activity in the mouse brain. PLoS One 9(3):e91360

90. Chin C, Park YS (2016) Identification and localization of gold nanoparticles in potassium ion pores: implications for Kir blockade. Cardiol Ther 5(1):101–108

91. Leifert A, Pan Y, Kinkeldey A et al (2013) Differential hERG ion channel activity of ultrasmall gold nanoparticles. Proc Natl Acad Sci U S A 110(20):8004–8009

92. El-Drieny EAEA, Sarhan NI, Bayomy NA, Elsherbeni SAE, Momtaz R, Mohamed HE-D (2015) Histological and immunohistochemical study of the effect of gold nanoparticles on the brain of adult male albino rat. J Microsc Ultrasruct 3(4):181–190

93. Sofroniew MV (2015) Astrogliosis. Cold Spring Harb Perspect Biol 7(2):a020420. https://www.ncbi.nlm.nih.gov/pmc/articles/PMC4315924/

94. Siddiqi NJ, Abdelhalim MAK, El-Ansary AK, Alhomida AS, Ong WY (2012) Identification of potential biomarkers of gold nanoparticle toxicity in rat brains. J Neuroinflammation 9:123

95. Shrivastava R, Kushwaha P, Bhutia YC, Flora SJS (2014) Oxidative stress following

exposure to silver and gold nanoparticles in mice. Toxicol Ind Health 32(8):1391–1404

96. Arispe N, De Maio A (2018) Memory loss and the onset of Alzheimer's disease could be under the control of extracellular heat shock proteins. J Alzheimers Dis 63 (3):927–934

97. Yi H, Huang G, Zhang K, Liu S, Xu W (2018) HSP70 protects rats and hippocampal neurons from central nervous system oxygen toxicity by suppression of NO production and NF-κB activation. Exp Biol Med Maywood NJ 243(9):770–779

98. Chen Y-S, Hung Y-C, Lin L-W, Liau I, Hong M-Y, Huang GS (2010) Size-dependent impairment of cognition in mice caused by the injection of gold nanoparticles. Nanotechnology 21(48):485102

99. Liang P, Shi H, Zhu W et al (2017) Silver nanoparticles enhance the sensitivity of temozolomide on human glioma cells. Oncotarget 8(5):7533–7539

100. Xu R, Ma J, Sun X et al (2009) Ag nanoparticles sensitize IR-induced killing of cancer cells. Cell Res 19(8):1031–1034

101. Liu P, Huang Z, Chen Z et al (2013) Silver nanoparticles: a novel radiation sensitizer for glioma? Nanoscale 5(23):11829–11836

102. Haase A, Rott S, Mantion A et al (2012) Effects of silver nanoparticles on primary mixed neural cell cultures: uptake, oxidative stress and acute calcium responses. Toxicol Sci 126(2):457–468

103. Xu F, Piett C, Farkas S, Qazzaz M, Syed NI (2013) Silver nanoparticles (AgNPs) cause degeneration of cytoskeleton and disrupt synaptic machinery of cultured cortical neurons. Mol Brain 6(1):29

104. Sharma H, Sharma A (2012) Neurotoxicity of engineered nanoparticles from metals. CNS Neurol Disord Drug Targets 11(1):65–80

105. Huang C-L, Hsiao I-L, Lin H-C, Wang C-F, Huang Y-J, Chuang C-Y (2015) Silver nanoparticles affect on gene expression of inflammatory and neurodegenerative responses in mouse brain neural cells. Environ Res 136:253–263

106. Li L, Cui J, Liu Z et al (2018) Silver nanoparticles induce SH-SY5Y cell apoptosis via endoplasmic reticulum- and mitochondrial pathways that lengthen endoplasmic reticulum-mitochondria contact sites and alter inositol-3-phosphate receptor function. Toxicol Lett 285:156–167

107. Forrester MT, Benhar M, Stamler JS (2006) Nitrosative stress in the ER: a new role for S-nitrosylation in neurodegenerative diseases. ACS Publications, Washington, DC

108. Benarroch EE (2016) Dynamics of microtubules and their associated proteins recent insights and clinical implications. Neurology 86(20):1911–1920

109. Mao B-H, Tsai J-C, Chen C-W, Yan S-J, Wang Y-J (2016) Mechanisms of silver nanoparticle-induced toxicity and important role of autophagy. Nanotoxicology 10 (8):1021–1040

110. Lee Y-H, Cheng F-Y, Chiu H-W et al (2014) Cytotoxicity, oxidative stress, apoptosis and the autophagic effects of silver nanoparticles in mouse embryonic fibroblasts. Biomaterials 35(16):4706–4715

111. Gu C, Shen T (2014) cDNA microarray and bioinformatic analysis for the identification of key genes in Alzheimer's disease. Int J Mol Med 33(2):457–461

112. Walter L, Neumann H (2009) Role of microglia in neuronal degeneration and regeneration. Semin Immunopathol 31:513–525

113. Marr RA, Hafez DM (2014) Amyloid-beta and Alzheimer's disease: the role of neprilysin-2 in amyloid-beta clearance. Front Aging Neurosci 6:187. https://www.ncbi.nlm.nih.gov/pmc/articles/PMC4131500/

114. Shibata M, Yamada S, Kumar SR et al (2000) Clearance of Alzheimer's amyloid-ss(1-40) peptide from brain by LDL receptor-related protein-1 at the blood-brain barrier. J Clin Invest 106(12):1489–1499

115. Fuentes P, Catalan J (2011) A clinical perspective: anti-Tau's treatment in Alzheimer's disease. Curr Alzheimer Res 8(6):686–688

116. Hansen RE, Roth D, Winther JR (2009) Quantifying the global cellular thiol–disulfide status. Proc Natl Acad Sci 106(2):422–427

117. Hong H, Wang F, Zhang Y et al (2015) Red fluorescent zinc oxide nanoparticle: a novel platform for cancer targeting. ACS Appl Mater Interfaces 7(5):3373–3381

118. Kasraei S, Sami L, Hendi S, AliKhani M-Y, Rezaei-Soufi L, Khamverdi Z (2014) Antibacterial properties of composite resins incorporating silver and zinc oxide nanoparticles on Streptococcus mutans and lactobacillus. Restor Dent Endod 39(2):109–114

119. Lai L, Zhao C, Su M et al (2016) In vivo target bio-imaging of Alzheimer's disease by fluorescent zinc oxide nanoclusters. Biomater Sci 4(7):1085–1091

120. Hong J-S, Park M-K, Kim M-S et al (2014) Prenatal development toxicity study of zinc oxide nanoparticles in rats. Int J Nanomedicine 9(Suppl 2):159–171

121. Xiaoli F, Junrong W, Xuan L et al (2017) Prenatal exposure to nanosized zinc oxide in rats: neurotoxicity and postnatal impaired

learning and memory ability. Nanomedicine 12(7):777–795

122. Wang C, Cheng K, Zhou L et al (2017) Evaluation of long-term toxicity of oral zinc oxide nanoparticles and zinc sulfate in mice. Biol Trace Elem Res 178(2):276–282

123. Takeda A, Tamano H, Imano S, Oku N (2010) Increases in extracellular zinc in the amygdala in acquisition and recall of fear experience and their roles in response to fear. Neuroscience 168(3):715–722

124. Mizuno D, Kawahara M (2013) The molecular mechanisms of zinc neurotoxicity and the pathogenesis of vascular type senile dementia. Int J Mol Sci 14(11):22067–22081

125. Mattson MP (2008) Hormesis defined. Ageing Res Rev 7(1):1–7

127. Saddick S, Afifi M, Abu Zinada OA (2017) Effect of zinc nanoparticles on oxidative stress-related genes and antioxidant enzymes activity in the brain of *Oreochromis niloticus* and *Tilapia zillii*. Saudi J Biol Sci 24 (7):1672–1678

128. Cędrowska E, Pruszynski M, Majkowska-Pilip A et al (2018) Functionalized TiO2 nanoparticles labelled with 225Ac for targeted alpha radionuclide therapy. J Nanoparticle Res 20(3):83

129. Wang C, Cao S, Tie X, Qiu B, Wu A, Zheng Z (2011) Induction of cytotoxicity by photoexcitation of TiO2 can prolong survival in glioma-bearing mice. Mol Biol Rep 38 (1):523–530

130. Zhao F, Wang C, Yang Q, Han S, Hu Q, Fu Z (2018) Titanium dioxide nanoparticle stimulating pro-inflammatory responses in vitro and in vivo for inhibited cancer metastasis. Life Sci 202:44–51

131. Wang J, Liu Y, Jiao F et al (2008) Time-dependent translocation and potential impairment on central nervous system by intranasally instilled TiO(2) nanoparticles. Toxicology 254(1–2):82–90

132. Saeed U, Mirza SS, MacIntosh BJ et al (2018) APOE-ε4 associates with hippocampal volume, learning, and memory across the spectrum of Alzheimer's disease and dementia with Lewy bodies. Alzheimers Dement 14:1137

133. Eidi H, Joubert O, Némos C et al (2012) Drug delivery by polymeric nanoparticles induces autophagy in macrophages. Int J Pharm 422(1–2):495–503

134. Hu R, Zheng L, Zhang T et al (2011) Molecular mechanism of hippocampal apoptosis of mice following exposure to titanium dioxide nanoparticles. J Hazard Mater 191 (1–3):32–40

135. Sheng L, Ze Y, Wang L et al (2015) Mechanisms of TiO2 nanoparticle-induced neuronal apoptosis in rat primary cultured hippocampal neurons. J Biomed Mater Res A 103 (3):1141–1149

136. Hong F, Sheng L, Ze Y et al (2015) Suppression of neurite outgrowth of primary cultured hippocampal neurons is involved in impairment of glutamate metabolism and NMDA receptor function caused by nanoparticulate TiO2. Biomaterials 53:76–85

137. Albrecht J, Sidoryk-Węgrzynowicz M, Zielińska M, Aschner M (2010) Roles of glutamine in neurotransmission. Neuron Glia Biol 6(4):263–276

138. Huang D, Liu D, Yin J, Qian T, Shrestha S, Ni H (2017) Glutamate-glutamine and GABA in brain of normal aged and patients with cognitive impairment. Eur Radiol 27 (7):2698–2705

139. Takanashi J-I, Mizuguchi M, Terai M, Barkovich AJ (2015) Disrupted glutamate-glutamine cycle in acute encephalopathy with biphasic seizures and late reduced diffusion. Neuroradiology 57(11):1163–1168

140. Cui Y, Chen X, Zhou Z et al (2014) Prenatal exposure to nanoparticulate titanium dioxide enhances depressive-like behaviors in adult rats. Chemosphere 96:99–104

141. Mohammadipour A, Fazel A, Haghir H et al (2014) Maternal exposure to titanium dioxide nanoparticles during pregnancy; impaired memory and decreased hippocampal cell proliferation in rat offspring. Environ Toxicol Pharmacol 37(2):617–625

142. Mohammadipour A, Hosseini M, Fazel A et al (2016) The effects of exposure to titanium dioxide nanoparticles during lactation period on learning and memory of rat offspring. Toxicol Ind Health 32(2):221–228

143. Lakhina V, Arey RN, Kaletsky R et al (2015) Genome-wide functional analysis of CREB/long-term memory-dependent transcription reveals distinct basal and memory gene expression programs. Neuron 85(2):330–345

144. Teich AF, Nicholls RE, Puzzo D, Fiorito J, Purgatorio R, Arancio O (2015) Synaptic therapy in Alzheimer's disease: a CREB-centric approach. Neurotherapeutics 12 (1):29–41

145. Chen T, Zhu J, Yang L-K, Feng Y, Lin W, Wang Y-H (2017) Glutamate-induced rapid induction of Arc/Arg3.1 requires NMDA receptor-mediated phosphorylation of ERK and CREB. Neurosci Lett 661:23–28

146. Siahposht-Khachaki A, Ezzatpanah S, Razavi Y, Haghparast A (2018) NMDA receptor dependent changes in c-fos and

p-CREB signaling following extinction and reinstatement of morphine place preference. Neurosci Lett 662:147–151

147. Bading H (2017) Therapeutic targeting of the pathological triad of extrasynaptic NMDA receptor signaling in neurodegenerations. J Exp Med 214(3):569–578

148. Pacheco A, Aguayo FI, Aliaga E et al (2017) Chronic stress triggers expression of immediate early genes and differentially affects the expression of AMPA and NMDA subunits in dorsal and ventral hippocampus of rats. Front Mol Neurosci 10:244. https://www.frontiersin.org/articles/10.3389/fnmol.2017.00244/full

149. Singh AK, Kashyap MP, Tripathi VK, Singh S, Garg G, Rizvi SI (2017) Neuroprotection through rapamycin-induced activation of autophagy and PI3K/Akt1/mTOR/CREB signaling against amyloid-β-induced oxidative stress, synaptic/neurotransmission dysfunction, and neurodegeneration in adult rats. Mol Neurobiol 54(8):5815–5828

150. Tang Y-L, Zhang Y-Q (2017) Molecular mechanisms of NMDA receptor-MAPK-CREB pathway underlying the involvement of the anterior cingulate cortex in pain-related aversion. Sheng Li Xue Bao 69(5):637–646

151. Hong F, Ze Y, Zhou Y et al (2017) Nanoparticulate TiO2-mediated inhibition of the Wnt signaling pathway causes dendritic development disorder in cultured rat hippocampal neurons. J Biomed Mater Res A 105 (8):2139–2149

152. Boonen RACM, van Tijn P, Zivkovic D (2009) Wnt signaling in Alzheimer's disease: up or down, that is the question. Ageing Res Rev 8(2):71–82

153. Riise J, Plath N, Pakkenberg B, Parachikova A (2015) Aberrant Wnt signaling pathway in medial temporal lobe structures of Alzheimer's disease. J Neural Transm 122 (9):1303–1318

154. Shimizu T, Smits R, Ikenaka K (2016) Microglia-induced activation of non-canonical Wnt signaling aggravates neurodegeneration in demyelinating disorders. Mol Cell Biol 36:2728

155. Zhou L, Chen D, Huang X-M et al (2017) Wnt5a promotes cortical neuron survival by inhibiting cell-cycle activation. Front Cell Neurosci 11:281. https://www.ncbi.nlm.nih.gov/pmc/articles/PMC5626855/

156. Georgakilas V, Perman JA, Tucek J, Zboril R (2015) Broad family of carbon nanoallotropes: classification, chemistry, and applications of fullerenes, carbon dots, nanotubes, graphene, nanodiamonds, and combined superstructures. Chem Rev 115 (11):4744–4822

157. Ye S, Zhou T, Pan D et al (2016) Fullerene C60 derivatives attenuated microglia-mediated prion peptide neurotoxicity. J Biomed Nanotechnol 12(9):1820–1833

158. Makarova EG, Gordon RY, Podolski IY (2012) Fullerene C60 prevents neurotoxicity induced by intrahippocampal microinjection of amyloid-beta peptide. J Nanosci Nanotechnol 12(1):119–126

159. Vorobyov V, Kaptsov V, Gordon R, Makarova E, Podolski I, Sengpiel F (2015) Neuroprotective effects of hydrated fullerene C60: cortical and hippocampal EEG interplay in an amyloid-infused rat model of Alzheimer's disease. J Alzheimers Dis JAD 45 (1):217–233

160. Martinez ZS, Castro E, Seong C-S, Cerón MR, Echegoyen L, Llano M (2016) Fullerene derivatives strongly inhibit HIV-1 replication by affecting virus maturation without impairing protease activity. Antimicrob Agents Chemother 60(10):5731–5741

161. Strom TA, Durdagi S, Ersoz SS, Salmas RE, Supuran CT, Barron AR (2015) Fullerene-based inhibitors of HIV-1 protease. J Pept Sci 21(12):862–870

162. Kraemer ÂB, Parfitt GM, da Acosta D et al (2018) Fullerene (C60) particle size implications in neurotoxicity following infusion into the hippocampi of Wistar rats. Toxicol Appl Pharmacol 338:197–203

163. Minichiello L (2009) TrkB signalling pathways in LTP and learning. Nat Rev Neurosci 10(12):850–860

164. Fišar Z, Hroudová J (2010) Intracellular signalling pathways and mood disorders. Folia Biol 56(4):135–148

165. Shen D-N, Zhang L-H, Wei E-Q, Yang Y (2015) Autophagy in synaptic development, function, and pathology. Neurosci Bull 31 (4):416–426

166. Duman RS, Aghajanian GK, Sanacora G, Krystal JH (2016) Synaptic plasticity and depression: new insights from stress and rapid-acting antidepressants. Nat Med 22 (3):238–249

167. Cha C, Shin SR, Annabi N, Dokmeci MR, Khademhosseini A (2013) Carbon-based nanomaterials: multifunctional materials for biomedical engineering. ACS Nano 7 (4):2891–2897

168. Singh R, Torti SV (2013) Carbon nanotubes in hyperthermia therapy. Adv Drug Deliv Rev 65(15):2045–2060

169. McKenzie JL, Waid MC, Shi R, Webster TJ (2004) Decreased functions of astrocytes on carbon nanofiber materials. Biomaterials 25 (7–8):1309–1317

170. Roman JA, Niedzielko TL, Haddon RC, Parpura V, Floyd CL (2011) Single-walled carbon nanotubes chemically functionalized with polyethylene glycol promote tissue repair in a rat model of spinal cord injury. J Neurotrauma 28(11):2349–2362

171. Boyles MSP, Young L, Brown DM et al (2015) Multi-walled carbon nanotube induced frustrated phagocytosis, cytotoxicity and pro-inflammatory conditions in macrophages are length dependent and greater than that of asbestos. Toxicol In Vitro 29 (7):1513–1528

172. Rydman EM, Ilves M, Vanhala E et al (2015) A single aspiration of rod-like carbon nanotubes induces Asbestos-like pulmonary inflammation mediated in part by the IL-1 receptor. Toxicol Sci 147(1):140–155

173. Nagai H, Toyokuni S (2012) Differences and similarities between carbon nanotubes and asbestos fibers during mesothelial carcinogenesis: shedding light on fiber entry mechanism. Cancer Sci 103(8):1378–1390

174. Toyokuni S (2013) Genotoxicity and carcinogenicity risk of carbon nanotubes. Adv Drug Deliv Rev 65(15):2098–2110

175. Wang J, Sun P, Bao Y, Liu J, An L (2011) Cytotoxicity of single-walled carbon nanotubes on PC12 cells. Toxicol In Vitro 25 (1):242–250

176. Romano-Feinholz S, Salazar-Ramiro A, Muñoz-Sandoval E et al (2017) Cytotoxicity induced by carbon nanotubes in experimental malignant glioma. Int J Nanomedicine 12:6005–6026

177. Shvedova AA, Kisin ER, Porter D et al (2009) Mechanisms of pulmonary toxicity and medical applications of carbon nanotubes: two faces of Janus? Pharmacol Ther 121(2):192–204

178. Geim AK, Novoselov KS (2007) The rise of graphene. Nat Mater 6(3):183–191

179. Caffo M, Maria C, Merlo L et al (2015) Graphene in neurosurgery: the beginning of a new era. Nanomedicine 10(4):615–625

180. Ren C, Hu X, Li X, Zhou Q (2016) Ultra-trace graphene oxide in a water environment triggers Parkinson's disease-like symptoms and metabolic disturbance in zebrafish larvae. Biomaterials 93:83–94

181. Kang Y, Liu J, Wu J et al (2017) Graphene oxide and reduced graphene oxide induced neural pheochromocytoma-derived PC12 cell lines apoptosis and cell cycle alterations via the ERK signaling pathways. Int J Nanomedicine 12:5501–5510

182. Bourdenx M, Dehay B (2017) Autophagy and brain: the case of neurodegenerative diseases. Med Sci 33(3):268–274

INDEX

Javier O. Morales and Pieter J. Gaillard (eds.), *Nanomedicines for Brain Drug Delivery*, Neuromethods, vol. 157,
https://doi.org/10.1007/978-1-0716-0838-8, © Springer Science+Business Media, LLC, part of Springer Nature 2021

Printed in the United States
by Baker & Taylor Publisher Services